# Quantum Mechanics: A Century Later

In Celebration of Heisenberg's Seminal 1925 Paper

# Quantum Mechanics: A Century Later

## In Celebration of Heisenberg's Seminal 1925 Paper


Edited by

**Tuck Choy**

*University College London, UK*



World Scientific

NEW JERSEY • LONDON • SINGAPORE • BEIJING • SHANGHAI • HONG KONG • TAIPEI • CHENNAI • TOKYO

*Published by*

World Scientific Publishing Co. Pte. Ltd.
5 Toh Tuck Link, Singapore 596224
*USA office:* 27 Warren Street, Suite 401-402, Hackensack, NJ 07601
*UK office:* 57 Shelton Street, Covent Garden, London WC2H 9HE

**British Library Cataloguing-in-Publication Data**
A catalogue record for this book is available from the British Library.

**QUANTUM MECHANICS: A CENTURY LATER**
**In Celebration of Heisenberg's Seminal 1925 Paper**



ISBN 978-981-98-2756-5 (hardcover)
ISBN 978-981-98-2757-2 (ebook for institutions)
ISBN 978-981-98-2758-9 (ebook for individuals)

For any available supplementary material, please visit
https://www.worldscientific.com/worldscibooks/10.1142/14713#t=suppl

Desk Editors: Soundararajan Raghuraman/Rhaimie Wahap

Typeset by Stallion Press
Email: enquiries@stallionpress.com

# Foreword

It is an honour to be invited by the editor on behalf of all participants to write the foreword for this special volume celebrating a hundred years of quantum mechanics, dating from Heisenberg's remarkable discovery of 1925 and de Broglie's discovery of 'matter' waves in 1924. This milestone is of great significance; with passing years, the foundation, history, struggles and conceptual issues faced by the earlier pioneers can be all too easily forgotten or ignored. I note the interesting contents in this volume, mostly based on foundational topics, together with one experimentalist on tunnelling related phenomena_which I have used in my own research. Within the next fifty years, emergent theories, foundations of the Born-Heisenberg-Jordan matrix mechanics, the uncertainty principle, macroreality topics, QED and more, may well become as significant in new quantum technologies as they already are in quantum computing. Therefore, it is my special pleasure and honour to promote and support the tireless effort put in by the editor and all the contributors, most of whom gave their time voluntarily to make this publication possible.

Across the street from where I am writing this in my Oxford study, I can see the house where Sir Rudolf Peierls ended his days. Along with Sir Neville Mott, who was a significant figure and personal influence during my PhD at the Cavendish Laboratory, he was perhaps the last theorist who could contribute to every area of the application of quantum mechanics, from materials science to nuclear fission. There has been astonishing progress since the heyday of those two remarkable scientists, when both men were in their youth and were closely associated with Werner Heisenberg.

I rejoice at the award of the 2025 Nobel Prize in Physics to John Clarke, Michel Devoret, and John Martinis, *"for the discovery of macroscopic quantum mechanical tunnelling and energy quantisation in an electric circuit"* [1]. This showed that unique quantum mechanical properties could be manifested on a macroscopic scale, using Josephson tunnelling (already another earlier remarkable Nobel prize discovery in

1973), and it paved the way for superconducting qubits for quantum computing. Many, perhaps most, of the Nobel Prizes in Physics over the past 100 years have been directly or indirectly related to quantum phenomena. Here are a few further selections from the past fifty years that I have become familiar with through the course of my research work:

1977 "for their fundamental theoretical investigations of the electronic structure of magnetic and disordered systems"

1986 "for his fundamental work in electron optics, and for the design of the first electron microscope" … "for their design of the scanning tunnelling microscope"

1987 "for their important break-through in the discovery of superconductivity in ceramic materials"

1997 "for development of methods to cool and trap atoms with laser light"

2000 "for developing semiconductor heterostructures used in high-speed- and opto-electronics"

2003 "for pioneering contributions to the theory of superconductors and superfluids"

2010 "for groundbreaking experiments regarding the two-dimensional material graphene"

2012 "for ground-breaking experimental methods that enable measuring and manipulation of individual quantum systems"

2013 "for the theoretical discovery of a mechanism that contributes to our understanding of the origin of mass of subatomic particles, and which recently was confirmed through the discovery of the predicted fundamental particle, by the ATLAS and CMS experiments at CERN's Large Hadron Collider"

2022 "for experiments with entangled photons, establishing the violation of Bell inequalities and pioneering quantum information science"

Fifty years ago, at its half century milestone, a volume from a conference was published with the title *Quantum Mechanics, A Half Century Later* [2]. Now we are celebrating the hundredth anniversary of the start date of quantum mechanics if you take that to be Werner Heisenberg's legendary sojourn on Helgoland and the ensuing papers [3,4,5]. What a half century it has been! Here are some of the remarkable achievements that were scarcely more than a twinkle in the eye in 1975.

1. Quantum interference of particles passing one at a time through two slits has now been demonstrated many times, with photons and with electrons, and in modified form with larger particles. Richard Feynman described this experiment as containing the only mystery of quantum mechanics. By this he meant that it depends on superposition, which Paul Dirac had described as the foundational principle of quantum mechanics. In his eponymous lectures published in the early 1960s, Feynman recommended his readers not to try the two-slit experiment with electrons. "This experiment has never been done in just this way … the apparatus would have to be made on an impossibly small scale to show the effects that we are interested in." Well, the experiment has now been done, more than once.
2. Entanglement, once an argument that the description of reality provided by quantum mechanics was incomplete, is now implemented routinely in laboratories all over the world. It would be more than thirty years before entanglement could be experimentally realised. The early demonstrations fell just within the first half century, but the industrialisation of quantum entanglement has occurred in the second fifty years. Somewhat remarkably, Erwin Schrödinger dreamt up his concept of *Quantenverschränkung* while he was living in the same street as Peierls' final abode, having earlier received news of his Nobel Prize in another house in this street.
3. Teleportation was once a legendary dream of *Star Trek*. Quantum teleportation builds directly on quantum entanglement, and it is now an experimental reality. It was theoretically proposed nearly 30 years ago. Within five years, which is a short time in this field, it was experimentally demonstrated by two groups, in Rome and in Innsbruck. Quantum teleportation is now a component in the toolbox for quantum technologies.
4. In the late eighties, Bill Philips and others were making the first attempts to cool atoms by using lasers. The experiments worked, and

they worked significantly better than anyone had expected. Occasionally experimental quantum mechanics comes up with nice surprises, and this was one of them. Laser cooling is now routine for all implementations of quantum computing using neutral atoms and also trapped ions.

5. Quickly following on from laser cooling was the achievement of Bose-Einstein condensation. This was first predicted over 100 years ago by Albert Einstein, building on a paper by Satyendra Nath Bose on the statistics of integer spin particles, now known eponymously as bosons. In this case it was to be many years before the condensation could be found experimentally. By making bosons sufficiently cold you can reach a state in which atoms of the same kind become indistinguishable and are all in essentially the same quantum state. Bose-Einstein condensates are now used for a wide range of cutting-edge research.
6. A large field is opening up of quantum-enhanced sensors. One of the largest scale and most dramatic applications is in detectors for gravitational waves, which use quantum interference enhanced by state squeezing to be able to detect a strain of $10^{-21}$ over a 100 Hz bandwidth. This corresponds to measuring a change in the mirror spacing of 4 km of 1 attometre, which is a thousandth of the diameter of a proton. In 2015 the Laser Interferometer Gravitational-Wave Observatory (LIGO) detected the first evidence of gravitational waves.
7. An emerging quantum technology, or family of technologies, which is attracting massive funding is quantum computing. Richard Feynman advocated using a quantum computer to simulate quantum systems in the early 1980s, "Nature isn't classical, dammit, and if you want to make a simulation of nature, you'd better make it quantum mechanical, and by golly it's a wonderful problem, because it doesn't look so easy." Shortly afterwards, in a paper which served as a manifesto for gate quantum computing, Oxford physicist David Deutsch showed that a universal computing machine must be quantum, and that this opens new possibilities which would not be feasible on a classical Turing machine.
8. Within a decade or so, two quantum algorithms were devised which harnessed Deutsch's ideas. The first used quantum Fourier transform for the factorisation of the product of two prime numbers, which

promises a seismic disruption to conventional public key cryptography. The second offers a quadratic speed-up for unstructured search problems. Further quantum algorithms have been discovered since, but these two remain seminal.

9. These algorithms, alongside other anticipated uses, stimulated a flurry of proposals for building quantum computers, ranging from ion traps to impurity atoms and other defects in semiconductors and insulators. For a while it was thought that pure photons could not be used because they do not interact, but that difficulty was overcome with a series of repeat-until-success protocols, thereby extending the required temperature range from laser cooling through dilution fridges to room temperature.
10. There is now a world-wide endeavour to build scalable quantum computers. Leading candidates include superconducting devices, spins in semiconductors, photons, ion traps, and neutral atoms. It is still too early to say which, if any, of the current implementations will eventually dominate, though each modality has its champions. Much of the engineering has moved from university laboratories to industrial companies, ranging from established data-processing companies to entrepreneurial spinouts founded for the purpose.
11. For good reason, many experts believed that quantum computing could never become a practical reality, because (i) quantum states are too fragile and error prone to support a sustained quantum calculation, and (ii) the no-cloning theorem seemed to preclude all conventional error correction. Two methods, derived independently at about the same time by Andrew Steane and Peter Shor, showed how to circumvent this difficulty by using ancilla qubits and syndrome detection. Implementation of fault-tolerant error correction is now a major endeavour in almost all efforts to build useful quantum computers.
12. Quantum computers will offer the possibility to study entanglement in ever greater scale and complexity. Most scientists are of the opinion that these machines will continue to follow the equations of quantum mechanics however large they become. But physics proceeds by surprises, and it may be that large enough quantum computers will reveal fundamental limits to the number of operational qubits or manifest phenomena that we never would have thought of.

Alongside all these remarkable achievements, where has progress been less conclusive?

We are still far from understanding the relationship between general relativity and quantum theory, despite heroic efforts by many brilliant minds. Some of the attempts make predictions so many orders of magnitude from empirical falsifiability that they cannot benefit from physics' secret weapon of experimentation. It is unappealing to give up in despair and abandon hope of discovering the connection. Is progress to be made through regarding spacetime as a basic reality, and seeking to find a quantum explanation for gravity? Or should we rather seek a gravitational underpinning for quantum mechanics?

The search for a set of postulates from which the whole of quantum mechanics would arise has proved elusive, despite some valiant attempts. The research community is as far as ever from agreement about many aspects of quantum foundations. I participated in a survey in Austria in 2011, which revealed a disparity of views ranging from the nature of superposition, non-local reality, wavefunction collapse, to the measurement problem, and even whether it is a problem. In July 2025 *Nature* conducted a much larger survey, with questions about the nature of the wavefunction, whether there is a boundary between classical and quantum objects, and whether a measurement requires an observer. There seems to be little change in the extent of disagreement, and a persistent range of views about explanations of quantum theory continue to be held. The most favoured is the Copenhagen interpretation, but even so only about 5% of respondents said that they were confident. It may be that in some future poll there could be a question about randomness, and whether that is a basic characteristic of reality.

This should engender a mixture of confidence and curiosity. In the 2011 survey, 24% of the respondents assessed the measurement problem as "A severe difficulty threatening quantum mechanics." Measurements are the currency of experimental quantum mechanics. As Lord Kelvin expressed it long before quantum mechanics was thought of, "I often say that when you can measure what you are speaking about, and express it in numbers, you know something about it; but when you cannot measure it, when you cannot express it in numbers, your knowledge is of a meagre and unsatisfactory kind." [6].

A critic might challenge experimentalists, "How can you continue your research in quantum systems, when there is no consensus about what happens when you make a measurement?" To which the experimentalist might reply, "I long for a deeper understanding of what happens when I measure a quantum system, but meanwhile quantum theory is too robust, and quantum technologies are too important to stop my research."

In science, as in other spheres of human inquiry, an eager curiosity sits alongside a healthy confidence. We are confident, because quantum mechanics has been tested to greater precision than any other creation of the human mind. And yet we remain curious, in the light of the history of surprises which we have time and again encountered in our quest for a satisfactory knowledge of reality [7].

This centenary volume has a special significance. Unlike fifty years ago, the internet has in the last thirty years brought the quantum to many laypersons' consciousness. There seems to be a hunger for an understanding of the marvels of the quantum in almost anyone who is curious about how the world works. Even the expression *quantum leap* has entered the English language in contexts increasingly remote from its original meaning. The contributors to this volume are from a wide range of physics backgrounds, and their papers will be of interest to a large group of both practitioners and non-practitioners of quantum mechanics. I very much appreciate the publication of this special volume and the conference that preceded it, and the work put in by so many participants to mark the centenary of one of humankind's greatest intellectual adventures.

*Andrew Briggs*
University of Oxford

**Acknowledgements:** I am grateful to the editor Tuck Choy, Jean-Marc Lévy-Leblond, Hajo Leschke, Mike Burt and Richard Forbes for their comments on an earlier draft.

## References

[1] See Nobel Prize in Physics 2025 https://www.nobelprize.org/prizes/physics/2025/press-release/

[2] See "Quantum Mechanics, a half century later", papers of a colloquium on fifty years of quantum mechanics, held at the University of Louis Pasteur, Strasbourg, May 2–4, 1974, edited by José Leite Lopes and Michel Paty, D. Riedel Publishing, Dordrecht-Holland/Boston-USA 1977. See extract in Chapter 17 of this volume.

[3] W. Heisenberg, "Über quantentheoretische Umdeutung kinematischer und mechanischer Beziehungen ", *Z. Phys.*, **33,** 879-893 (1925).

[4] Max Born and Pascual Jordan, "Zur Quantenmechanik" *Z. Phys.*, **34,** 858–888 (1925).

[5] M. Born, W. Heisenberg and P. Jordan, "Zur Quantenmechanik II" *Z. Phys.*, **35,** 557–615 (1926).

[6] W. Thompson, lecture to the Institution of Civil Engineers, London, on 3 May 1883, "Electrical Units of Measurement" published in *Popular Lectures and Addresses* **1,** 80–81 (1889).

[7] Andrew Briggs, Hans Halvorson, and Andrew Steane, *It Keeps Me Seeking* (Oxford University Press 2018).

# Foreword

World Scientific Publishing Company was founded on the conviction that the enduring progress of science depends on the careful preservation and dissemination of ideas that shape human understanding. From the outset, physics plays a key role in our mission, and it was through physics that World Scientific first established its scholarly identity. My own lasting affection for the subject is inseparable from the origins of the company and from the friendships and collaborations that sustained it during its formative years.

I remain deeply grateful to the colleagues whose encouragement and trust helped shape World Scientific's early path. Among these memories are exchanges with fellow physicists, including Kung Chao, whose correspondence and support during the 1980s, contributed to the founding of what would become the *International Journal of Modern Physics B*. Such moments reflect the spirit of collegiality and shared intellectual purpose upon which serious scientific publishing has always depended.

It is therefore especially fitting that World Scientific should publish *Quantum Mechanics: A Century Later: In Celebration of Heisenberg's Seminal 1925 Paper*. Werner Heisenberg's work marked a decisive turning point in the history of science, transforming not only physics but the modern worldview itself. Few developments have so profoundly altered our understanding of nature or so deeply shaped the technologies that define contemporary life.

This volume also holds personal significance for me. It was a letter from the editor — whose preface recounts the origins of this project — that set this undertaking in motion. I wish to express my sincere appreciation to Tuck Choy for his vision and dedication, and to all contributors whose scholarship has made this volume both authoritative and enduring.

As World Scientific looks to the future, this centenary volume stands as one of our flagship projects for the 21st century. Together with the

50th anniversary commemorative volume, it forms a pair of legacy works — historical markers that testify to humanity's continuing engagement with the quantum world. World Scientific is proud to be part of this moment in scientific history, and I extend my heartfelt thanks to all who have contributed to bringing this landmark publication to fruition.

K. K. Phua
Editor-in-Chief and Founder
*World Scientific Publishing Company*

# Preface

孔子曰：学而不思则罔，思而不学则殆
Translation: “Learning without thinking is useless, thinking without learning is perilous.” Confucius (551–479 BCE) in Analects of Confucius (论语) Book 2, Chapter 15 (为政).

It is hard for someone in the 21st century to imagine what the world was like 100 years ago in July 1925 when Heisenberg’s landmark paper that launched the new world of quantum mechanics was published. This was clearly a world with electricity, radio telegraphy, AM broadcast, and vacuum tubes but no semiconductors. Yet the recording industry had already been around for over 20 years since Enrico Caruso made his first operatic recording using Edison technology on a cylinder, one of which is the now famously remastered: “una furtiva lagrima” Gaetano Donizetti’s famous aria in the opera “L’elixir d’amour”. It was a very strange world in which there were electric trams and buses, and an environment in which the arts and sciences in particular flourished. July 1925 followed just under one year after the death of the great maestro Giacomo Puccini who left his greatest operatic work Turandot unfinished. On his death bed, he pleaded with the famous conductor his friend Arturo Toscanini to “not let my Turandot die”, having left most of the opera completed but with notes on how he would like to see it finished. So, finally recruiting the help of Franco Alfano, Toscanini had the opera finished and the first performance was in April 1926 in La Scala Milan. Notice the coincidence in the timeline here with the development of quantum mechanics, between 1924–1926. It was Max Born who first put the word ‘quantum mechanics’ in print, in his paper of 1924 [1]. It was also in the same year that Prince Louis de Broglie proposed his hypothesis central to modern quantum theory for which he was awarded the Nobel prize in 1929, later known as wave–particle duality. Two years later in 1926, this inspired Schrödinger’s papers [2] in which he proposed his wave equation, and the mathematical structure of quantum mechanics was finally agreed upon, although there were issues with interpretations that have lasted till the present day. In Puccini's Opera, Princess Turandot challenges any suitor to solve three riddles posed by her. Success wins her as a bride, and the inheritance of the Chinese Empire, while failure will result in execution by beheading. Likewise quantum mechanics poses many troublesome riddles which have metaphorically severed numerous distinguished heads: why the transition from classical deterministic mechanics to quantum mechanics, what causes the quantum universe, what is the origin of the quantum, and Einstein's famous 'spooky action at a distance' etc. Very notable heads too by the way: Bohr who first thought that energy was not conserved in the quantum world, Einstein whose numerous objections to the Uncertainty Principle made him a laughing stock to the younger generation, Heisenberg who thought incorrectly that only observables are allowed in the new theory, Dirac whose quantization rules ran into obstruction difficulties, von Neumann who thought, mistakenly, as shown by Bell [3], that there could be no hidden variables, Schrödinger who for many years hung on to his and de Broglie's untenable "matter waves" concept, and many others. Quantum Mechanics (aka the physicists's Turandot) is there ready to detach any head which dares to challenge her fundamentals. Therefore the physics world seems to be divided into two camps. In one are those who just accept the 'Copenhagen' dogma: shut up and calculate, the theory is complete, a *fait accompli*, end of story. This group dared not challenge Turandot for fear of decapitation. The second group follows Einstein, Schrödinger, de Broglie, Bohm and most notably Bell who continued to call the house of Denmark, rotten [4]. They cannot but be tempted by the charm of Turandot. To be fair, in particular in the case of Einstein and Bell, their challenges have raised far more questions than answers on many foundational aspects of quantum mechanics; in particular, the original so called 'Copenhagen interpretation'. As late as 1992, Karl Popper the famous philosopher of science complained that Bohr "was depressing" and even suggested that the 'Copenhagen' dogma was a crime against humanity [5]. This to me is grossly unfair. A look into the long history of the founding of quantum mechanics and the role of each player will without doubt take us back to the Bohr-Einstein debate. However, few people as far as I am aware pay a lot of attention to Bohr's objections to quantum mechanics [6], while many, particularly non-experts, would have heard about Einstein's objections to quantum mechanics. It is an important twist in history that, had Bohr's objections prevailed, Heisenberg would not have been able to published his famous paper on the uncertainty relations. To quote Jammer [6], "The issue was not about the conclusions, that is, about the validity of the indeterminacy relations, but rather about the conceptual foundations on which they were based". Readers will have to decide for themselves on this issue or non-issue as the case may be, but one cannot blame Bohr for not trying and certainly cannot blame him for taking full responsibility for what we now call the 'Copenhagen' interpretation.

This volume came about following an email in 2023 in which I queried the managing director of World Scientific Publishing Ltd (Singapore), K K Phua, himself a trained particle physicist and an old friend, about what they are doing about the Heisenberg centenary in 2025. The answer was: "Nothing, what a good idea, would you do something, such as prepare a special volume?". Before I could say no and in spite of some attempts with excuses such as "I am not really the guy to do such a job", or "you will need a Nobel laureate" etc, I was already roped in. However the one saving grace was that I was given a free hand, and he said, "you just do it your way and however you feel that it is right". Initially I received encouragement from some older colleagues, in particular the late J.M.J. van Leeuwen of the Leiden Institute, an old friend, who was the first to submit his paper and had a discussion with me on my own work and the project. Sadly he passed away less than a year before our conference and we are grateful to Bernard Nienhuis for attending the conference and acting on his behalf. Michel Paty, a veteran responsible for the 50th anniversary celebrations in Strasbourg in 1974, was most encouraging when I met him in Paris in Nov 2023, but unfortunately due to ill health, he was unable to make a contribution to this volume. The 50th anniversary proceedings edited by the late José Leite Lopes [7] and Michel Paty remain a major achievement and has been an inspiration for me to work on this 100th anniversary volume. We are indeed honoured and fortunate that Jean-Marc Lévy-Leblond, another true veteran who contributed his paper in the 50th anniversary [8] is able once again to contribute another paper at this 100th anniversary celebration. His achievement of a double contribution over half a centenary is a unique record that surely belongs in the Guinness Book of Records. Over the years it saddens me to reflect on the many colleagues I have known and who I know would have liked to be present for the 100th anniversary of Quantum Mechanics but are no longer with us. This work is dedicated to all of them.

The moral lessons from the advancement of quantum mechanics are all too easily forgotten by the younger generations now. We must all be reminded that only 20 years after Heisenberg's paper, mankind detonated its first atomic weapon of mass destruction but it took some forty years before semiconductor microelectronics enabled the first person to step on the moon and drove the development of computers, mobile phones and modern high tech devices that we all now take for granted. Such a historical link to Heisenberg's 1925 paper is unprecedented. I include below a few references that I have accumulated over the last 50 years for the record for this volume and which are relevant to this centenary celebration. Quite apart from Jammer's masterpiece [6] and the work of Lopes-Paty [7], we have Jagdish Mehra's proceedings of Dirac's 70th birthday symposium [9], *Centenary of Heisenberg's birth* edited by Papenfuß et al. [10] and last but not least, Arthur Miller's *62 years of Uncertainty* volume [11]. There is enough material here to cover several lifetimes of intense study.

So here it is. Originally I approached many people, some very eminent, for help. Many were lukewarm, others simply said things like they have retired and are now more interested in just playing the piano. Some said that two years is too long a time frame and they cannot plan for it, yet when the time came in 2025, they then decided to hold their own centenary conferences, such as that held at ICTP in Trieste, 28–30 May 2025 and Helgoland 9–14 June 2025. For a time, I actually thought a conference for our own participants would not be possible. Therefore we owe a deep debt of gratitude to Roberto Benaducci and the QSA committee who offered to absorb our initial proposal for a conference into the IQSA 2025 Tropea program. From the start this project was never about careers, nor any single individual except Heisenberg, and in fact most of the participants are either retired or no longer employed in academia. I was much more concerned to preserve a legacy, that of our founding fathers: Planck, Einstein, Bohr, Sommerfeld, Born, Jordan, Heisenberg, Dirac and others over the last 100 years as well as to provide a guiding light and inspiration for the future, preferably away from practising physics as a living or the pursuit of academic prizes. I have insisted that this shall not be a volume containing papers that are a mere repetition of work that has already been published elsewhere and that it must be work which represents the authors' genuine desire to contribute to the centenary event with preferably some original thinking. The eventual success of this is by no means mine alone and I would like to thank the participants many of whom worked behind the scenes and made it all happen. The list includes CH Cheong, Joseph Sebastian Ang and Rhaimie Wahap of World Scientific, Anthony Harker, Ian Aitchison, Dr Aris Alexopoulos and Dr Andrew John Smith for advice and proof reading chores, my cat Fluffy who was a constant companion into the late hours of the night as well as my wife Debra Ziegeler who had to put up with a lot.

Many thanks also to those who came to the conference in person, notably Mark Davidson from California, Hajo Leschke from Germany, Bernard Nienhuis from the Netherlands, Richard Forbes from the UK and Masanao Ozawa from Japan while others gave their talks online. On-line links were in several cases not without hiccups, as a massive heat wave hit Italy on that day 3rd July 2025 leading to a breakdown of lots of services. Final thanks go to Matin Durrani and Michael Banks of Physics World for granting us permission to reprint the articles of Bell, Peierls and Gottfried, with images omitted in this special volume and other copyright holders mentioned in the text.

I hope the reader will agree that I have now comfortably discharged this responsibility which I had the honour to bear on my shoulders.

Tuck C. Choy 蔡德全
Montagnac, France
September 2025

[1] Max Born "Quantum Mechanics" Zs. f. Phys. **26** 379–395 (1924) reprinted in B. L. van der Waerden (editor), "Sources of Quantum Mechanics", vol V of classics of modern science series 181–198, Dover, New York 1968.

[2] All of Schrödinger's original papers on wave mechanics from 1926–27 have now been translated and are available as a volume in Erwin Schrödinger,

"Collected Papers on Wave Mechanics", Chelsea Publishing Co., New York 1978. His landmark paper establishing the mathematical equivalence of matrix and wave mechanics was submitted on 18 March 1926 which appeared several months later. Although Heisenberg has priority in the discovery of quantum mechanics, we include Schrödinger's profound contributions as part of our centenary celebrations.

[3] John Bell, "On the problem of hidden variables in quantum mechanics", Reviews of Modern Physics **38** 447–52, (1966). Bell's work on the problems with von Neumann's proof has been preceded by that of Jaunch and Piron, Gleason and notably Bohm as cited in his paper. However he appeared to have missed that some issues on philosophical grounds had already been raised back in 1935 by Grete Hermann, see Jammer [6], pg 207–211. I thank Kurt Schönhammer for drawing this to my attention.

[4] John Bell, "Against measurement", Physics World, 33–40, August (1990).

[5] Marie–Christine Combourieu, "Karl R. Popper: About the EPR Controversy", Foundations of Physics, **22** (10) 1303–1323 (1992).

[6] See for example The Philosophy of Quantum Mechanics–The interpretations of Qauntum Mechanics in Historical Perspective, by Max Jammer pg 65, John Wiley and Sons, New York 1974.

[7] See "Quantum Mechanics, a half century later", papers of a colloquim on fifty years of quantum mechanics, held at the University of Louis Pasteur, Strasbourg, May 2–4, 1974, edited by José Leite Lopes and Michel Paty, D. Riedel Publishing, Dordrecht–Holland/Boston–USA 1977.

[8] See "Towards a Proper Quantum Theory" by Jean–Marc Lévy–Leblond in the preceding reference [7] pg 171–206.

[9] Proceedings of the symposium on the development of The Physicist's Conception of Nature in the Twentieth Century, dedicated to the seventieth birthday of Paul Dirac, Trieste, Italy Sep 1972, edited by Jagdish Mehra, D. Reidel Publishing, Dordrecht, Holland 1973, (an unofficial 50th anniversary celebration of quantum mechanics).

[10] 100 years of Werner Heisenberg, works and impact, Volume dedicated to the centenary of his birth, edited by D. Papenfuß, D. Lüst and W. P. Schleich, Wiley-Vch, Verlag GmbH & Co, KGaA, Weinheim, Germany 2002.

[11] Sixty-two years of Uncertainty, Historical, Philosophical and Physical Enquiries into the Foundations of Quantum Mechanics, edited by Arthur I. Miller, Proceedings of a NATO Advanced studies Institute, Erice, Sicily, Italy Aug 1989, Plenum Press, New York, London 1990.

Supplementary material: The reader should note that there is additional supplementary material to that published in this volume, such as Power Point slides, audio and video files taken at the Tropea Conference in July 2025 that is available by open access to the World Scientific Website. These are copyright of the individual authors and permission should be sought from the authors if you intend to use them in your publications.

# Contents

# Chapter 1

# Göttingen and Quantum Mechanics

**K. Schönhammer**
Institut für Theoretische Physik, Universität Göttingen
Friedrich-Hund-Platz 1, D-37077 Göttingen
schoenh@theorie.physik.uni-goettingen.de

## 1 Introduction

In this paper the central role played by Göttingen in the formulation of quantum mechanics is elucidated. It starts with a short historical account of the earlier important steps towards achieving this goal in the first 20 years of the twentieth century [1].
The first step in the formulation of quantum theory took place in the year 1900 with Max Planck's explanation of the spectrum of black-body radiation by treating the walls as a system of harmonic oscillators. He made the assumption that the energy quanta emitted and absorbed are integer multiples of $h\nu$, where $\nu$ is the frequency of the oscillator and $h$ is Planck's constant, which he had introduced a year earlier using thermodynamic arguments.
In 1905 Albert Einstein postulated the existence of light quanta (only much later called photons). This allowed him to give a theoretical explanation of the photoelectric effect.
Rutherford's scattering experiments of alpha particles from gold foils (1911) suggested a planetary model of atoms, with a heavy positively charged nucleus surrounded by electrons. As the electrons perform accelerated motion they radiate according to Maxwell's equations and therefore these atoms cannot be stable within a description by classical physics. This is in contrast to the the plum pudding model proposed by J.J. Thomson in 1904. In this model the electrons are assumed to float freely in a continuous positive charge background, and there also exist stationary configurations in which no radiation is emitted [2]. This model was falsified by Rutherford's experiments.
In order to understand the stability of atoms and present a theoretical description for the spectral lines emitted by hydrogen at high temperatures, Niels Bohr (1885–1962) *postulated* in 1913 that the electron does *not* radiate on *stationary orbits*, for which the angular momentum takes values given by *integer multiples* of $\hbar \equiv h/2\pi$. This assumption leads to discrete energy values $E_n \equiv W(n) = -E_R/n^2$ with integer $n$ and the Rydberg energy $E_R = m_e e^4/2\hbar^2 \approx 13.6057$ eV. Further postulating that in the transition of the electron from orbit with quantum num-

ber $m$ to $n$ a light quantum of frequency $\nu_{mn} = (E_m - E_n)/h$ is emitted, Bohr was able to explain experimental hydrogen spectra like the Balmer series [3]. As in Bohr's later work the *correspondence principle* played an important role: Quantum theory should lead to the same results as classical physics when the *quantum numbers* describing the systems are *large.* Despite the "ad hoc" character of his "rules" his work attracted enormous attention.
Bohr's approach was generalized by Arnold Sommerfeld (1868–1951) with the quantization of the classical action for multiple periodic systems in 1915 and in 1919 he presented the energy of a hydrogen-like atom with one electron and nuclear charge $Ze$ within the framework of *relativistic* mechanics. In his formula the (small) *fine structure constant* $\alpha = e^2/\hbar c \approx 1/137$ provides a measure for the importance of relativistic effects. The status of the Bohr–Sommerfeld theory was presented by Sommerfeld in the same year in the first edition of his book *Atombau und Spektrallinien* (*Atomic Structure and Spectral Lines*)[4]. The experimental spectral lines of the helium atom could not be properly described by the Bohr–Sommerfeld theory. In fact the simplest molecule $H_2^+$ with a single electron presented a serious problem. It was treated by Wolfgang Pauli in his PhD thesis completed in July 1921 with Sommerfeld in Munich. Another student with Sommerfeld at that time was Werner Heisenberg (1901–1976) who played an important role later in Göttingen.

# 2 Göttingen (1921–1924)

The Institute of Theoretical Physics in Göttingen was founded in 1921 after Peter Debye who had his office in an experimental institute had left Göttingen 1920 to become a Professor at the ETH Zurich. The chair in the new Institute was filled by Max Born (1882–1970). James Franck (1882–1964) also came from Frankfurt as a new chair of experimental physics.

Figure 1: Max Born in the 1920s

Max Born, born in Breslau, had studied mathematics in Heidelberg, Zürich and Göttingen, where he soon came in close contact with Felix Klein, David Hilbert and Herrmann Minkowski. For his PhD work (1905) on aspects of the stability of elastic media he received the prize of the Göttingen faculty. After spending time in England and Breslau, Born returned to Göttingen in 1908 and collaborated with Minkowski (1864–1909) on Einstein's special theory of relativity. Unfortunately Minkowski died of appendicitis shortly thereafter. A few years later in 1912, together with Theodore von Karman, Born worked on the dynamical theory of crystal lattices, addressing for example the lattice specific heat (as did Peter Debye at about the same time). This work led to Born's book *Dynamik der Kristallgitter (The Dynamics of Crystal lattices)* in (1915). From 1915 to 1919 Born was Associate professor in Berlin, where he came in contact with Max Planck and Albert Einstein. He had a friendly relationship with the latter until Einstein's death in 1955. From 1919 to 1921, Born was Full professor in Frankfurt (Main).

Born's first two assistants in Göttingen were Wolfgang Pauli (1900–1958) and Friedrich Hund (1896–1997). Pauli came to Göttingen in October 1921 and stayed until April 1922. His position was financed by Henry Goldman (Goldman and Sachs). In March 1922 he published his earlier-mentioned thesis work about the $H_2^+$ molecule which he never cited himself. As Pauli did not like the life in a small provincial town as Göttingen he left for Hamburg to become assistant to Wilhelm Lenz, known for the Laplace–Runge–Lenz vector which he used in 1924 to describe the hydrogen atom in crossed external fields in the framework of the "old quantum theory" [5].

Figure 2: Wolfgang Pauli in the 1920s

Friedrich Hund had studied mathematics, physics and geography and wanted to become a high-school teacher. But he liked scientific work, and after getting his PhD in 1922, he became Born's assistant and participated with him in attempts to generalize Hamilton mechanics to "atom mechanics". He also intensely studied experimental atomic spectra and found the now famous "'Hund's rules" in 1925.

Figure 3: Friedrich Hund in the 1920s

In June 1922 Niels Bohr gave the Wolfskehl Lectures in Göttingen, later called the Bohr Festival because it had started two weeks before the yearly Göttingen Händel Opera Festival. In seven lectures he presented the state of the art of the

Bohr–Sommerfeld theory aiming at an understanding of atoms. In the later lectures he addressed in detail the construction of a theory of the periodic system of elements [6]. One should mention that Bohr emphasized "how incomplete and uncertain everything still is". Sommerfeld had come from Munich with his student Heisenberg and Pauli from Hamburg. Another young, mathematically very gifted, student who participated and played an important role later was Pascual Jordan (1902–1980). Also famous Göttingen mathematicians like David Hilbert, Felix Klein, Carl Runge and Richard Courant were present. Among the many other prominent physicists from outside Göttingen were Paul Ehrenfest and Oskar Klein. After Bohr's third lecture Heisenberg asked a question which prompted a long discussion with him after the lecture on a walk on Göttingen's Hainberg. At the end of the lectures David Hilbert thanked Bohr for allowing insight into the holy grail of his scientific personality [7].
Later in 1922 Bohr received the Nobel Prize in Physics "for his investigations of the structure of atoms and of the radiation emanating from them". Heisenberg spent the winter semester 1922–1923 in Göttingen as Sommerfeld was in the US during this time.

Figure 4: Werner Heisenberg in the 1920s

All physicists working hard to improve the understanding of the structure of atoms, including Bohr, were aware that many open questions remained. In Göttingen Heisenberg joined the efforts to improve this state of affairs in 1923 after receiving his PhD in Munich with Sommerfeld on work related to turbulence. In 1924 Pascual Jordan had finished his PhD work under Born's guidance on the quantum theory of radiation [8]. One can summarize the situation at the end of

1924 with the statement that no real progress was visible.
A situation like this was later described in Thomas Kuhn's book *The structure of scientific revolutions* (1962), a very highly discussed (and also criticised) contribution to the history of science [9]. Kuhn distinguishes periods of "normal science", in which accepted models and theories are applied successfully, from periods of increasing "crisis", in which it becomes obvious that only a "revolution" changing basic assumptions or "paradigms", can bring a solution with a "paradigm shift".
In Kuhn's spirit Max Born's book *Vorlesungen über Atommechanik* is a presentation of the crisis [10]. In the Preface of the book written in November 1924, Born explains that he wants to describe the limits of the present state of atomic theory. The book uses the Hamilton–Jacobi approach to mechanics and devotes a large part to atoms with a single valence electron. This is no surprise as for such systems, the Bohr–Sommerfeld theory is usually successful. But even in this area problems showed up. Born mentions that already the hydrogen atom in crossed magnetic and electric fields poses a problem for the theory and for the simplest two-center molecule, $H_2^+$, the calculated ionization potential deviates considerably from the measured value. Towards the end of the book Born addresses the fact that quantizing the action integrals in multiples of Planck's constant fails in the treatment of the two electrons in the helium atom. As another failure of the usual Bohr-Sommerfeld theory Born mentions the anomalous Zeeman effect.
In order to show the preliminary character of this work he gives the book the subtitle "Volume 1". A later "Volume 2" should then be devoted to the "final atom mechanics". He calls this a daring attempt as rather little is known in this respect and it might take some years until Volume 2 would be written. Born acknowledges that many parts of Volume 1 were written by Hund and only slightly revised by himself and the last chapter on the helium atom was outlined by Heisenberg.
Born was right with the "some years" estimate for the appearance of Volume 2 (1930), but wrong about the time to the breakthrough. The paradigm shift to quantum mechanics, the most successful theory of physics, took place within the next two years. It allows us to obtain results for the properties of matter to an extremely high accuracy.
In the Preface of the book Born points out the fact that for the radioactive decay only a probabilistic description is possible, a concept which should also be used for atomic transitions. This was first proposed by Einstein in his paper "On the Quantum Theory of Radiation" [11], where he introduced probabilities per unit time for the transition between two atomic stationary states.

## 3 1925: Matrix Mechanics and Beyond

Heisenberg spent the winter 1924–1925 in Copenhagen working with Niels Bohr and Hendrik Kramers. In the letter in which he thanked Bohr for his hospitality

Heisenberg wrote: "With respect to science the past half year was for me the most beautiful of my entire life as a student."

## 3.1 Göttingen

After a vacation, Heisenberg was back in Göttingen by the end of April. By the end of May, because of strong attacks of hay fever he retreated to the treeless island Heligoland in the North Sea for two weeks. He had started his attempt to formulate a completely new quantum mechanics. Instead of the hydrogen atom he addressed the anharmonic oscillator. On Heligoland he made progress with his new approach, especially concerning energy conservation. Back in Göttingen he started to write his paper "Über quantentheoretische Umdeutung kinematischer und mechanischer Beziehung (Quantum Theoretical Re-interpretation of Kinematic and Mechanical relations)" [12].
Between June 21 and July 9 Heisenberg sent four letters to Pauli which clearly show the ups and downs of Heisenberg's feelings about his achievements [13]. Unfortunately Pauli's response is not in the collection of his scientific correspondence. Here are excerpts of Heisenberg's letters (translated from the original):
June 21:

*My attempts to formulate a quantum mechanics progress rather slowly...*

June 24:

*I don't want to write about my work because most of it is still unclear to me...*

June 29:

*I made some progress and in my heart I am now convinced that this quantum mechanics is quite correct...*

And finally on July 9 Heisenberg sends his manuscript to Pauli asking for critical remarks:

*...I feel guilty to ask you to send back my draft within 2-3 days, as I want to finish or burn it before leaving. My opinion about my scribbling, about which I am not very happy is as follows: I am convinced about the negative critical part, but the positive one I judge as rather formal and meager, maybe more gifted people are able to make something reasonable out of it.*
*Now I request harsh critisism on your side and a speedy return of my paper.*

As Pauli apparently had no objections Heisenberg gave the paper to Born in mid July and asked him to submit it to *Zeitschrift für Physik* if its content makes sense to him and left for Munich and hiking tours in the Alps. A few days later Born read the paper and as he was impressed (see below) submitted it to *Zeitschrift für Physik.*

Heisenberg begins his paper by stating that he wants to lay the foundations of a quantum theoretical mechanics in which only relations between observable quantities occur. A similar statement can also be found in an earlier paper by Born and Jordan on the quantum theory of aperiodic processes [14]. Heisenberg considers e.g. the position and the period of the electron motion to be unobservable. Treating the radiation emitted by an atom he points out the importance of the associated Einstein–Bohr transition frequencies

$$\nu(n, n-\alpha) = \frac{1}{h}[W(n) - W(n-\alpha)]$$

where $\alpha$ is an integer and $W(m)$ are the energies of Bohr's stationary orbits. Using Newton's equations of motion Heisenberg treated the one-dimensional anharmonic oscillator where the electron undergoes a periodic motion labeled by quantum number $n$. The classical coordinate $x(n,t)$ can then be described by a Fourier series. In order to describe the radiative transitions, Heisenberg proposed to replace the Fourier coefficients by quantities $X(n, n-\alpha)$, depending on the *two* quantum numbers $n$ and $n-\alpha$, like the transition frequencies $\nu(n, n-\alpha)$. Focussing the description on pairs of states and their transition amplitudes was earlier proposed in the above mentioned paper by Born and Jordan [14]. The really bold step in Heisenberg's paper concerned the question: Which quantum object $Y(t)$ corresponds the classical quantity $x(t)^2$? Arguing with the Ritz combination principle Heisenberg comes up with his multiplication rule for transition amplitudes

$$Y(n, n-\beta) = \sum_{\alpha} X(n, n-\alpha) X(n-\alpha, n-\beta) \ .$$

which he considers the "simplest and most natural assumption". A few steps later he points out that while classically $x(t)y(t)$ always equals $y(t)x(t)$ this is not necessarily the case in quantum theory. In order to find a quantitative result for the difference of the two products for the case of position and momentum Heisenberg used the "old quantum theory" (the Thomas–Reiche–Kuhn sum rule).

With the realization that quantum theory has to deal with possibly *noncommuting mathematical objects* quantum mechanics was born. But it took some time until this was formulated in the form students learn it today. No further details of Heisenberg's paper are discussed here. Comments on the readability of the paper are given later.

It took Born about a week to realize that Heisenberg's multiplication rule was nothing but the multiplication rule for *matrices* he had learned as a student in Breslau. The mathematical concept of matrices was not known to Heisenberg when he wrote his paper. In matrix language Heisenberg had only worked with the diagonal element of the commutator of position $q$ and momentum $p$. Born "easily guessed" the off-diagonal elements and was the first to obtain the basic commutation relation [15]

$$pq - qp = \frac{h}{2\pi i} \ .$$

This is the form written on Born's tomb stone in Göttingen.

As Heisenberg was not in Göttingen, Born together with Jordan had a closer look at a proper "derivation" of his guess. Stimulated by Heisenberg's paper, they began to formulate *matrix mechanics* for systems of a single degree of freedom. The fact that the infinite matrices for $q$ and $p$ are mathematically rather subtle objects was not taken very seriously. Their paper "Zur Quantenmechanik"(On Quantum Mechanics) was received by *Zeitschrift für Physik* on September 27, 1925 [14]. In this paper the commutation relation of the matrices for position and momentum appeared in print for the first time and is called "the stronger quantum condition". All further conclusions are based on it.

Figure 5: Pascual Jordan in the 1920s

Heisenberg was in Copenhagen in September and therefore unavailable for discussions. Born had informed Heisenberg about his collaboration with Jordan. Heisenberg was excited about the achievements of his colleagues and began to work on matrix mechanics in Copenhagen after making himself familiar with the mathematical concept. By the end of September he had come up with the commutation relations for coordinates and momenta for systems with several degrees of freedom. Heisenberg raised an objection concerning the definition of the derivative of a product of several matrices with respect to one of its factors used by Born and Jordan and proposed a different definition which is closer to the usual differentiation procedure [17].

In a letter to Pauli Heisenberg pointed out that the most important thing was still missing: the solution for the hydrogen problem within matrix mechanics. It is also not discussed in the "three authors paper" ("Dreimännerarbeit") of Born,

Heisenberg and Jordan " Zur Quantenmechanik II" (On Quantum Mechanics II) which was received by *Zeitschrift für Physik* on November 25, 1925 [18]. It used Heisenberg's differentiation procedure and presented the state of the art of matrix mechanics. In this paper one can find for example a detailed discussion of the quantum mechanical properties of angular momentum.

## 3.2 Hamburg

The missing solution of the hydrogen problem was found by Pauli by the end of October after learning about the progress with the formalism of matrix mechanics from Heisenberg's letters. Proudly he reported to Heisenberg about his successful diagonalization of the Hamiltonian matrix of the hydrogen atom. As it did not seem possible to describe angular variables as matrices Pauli wrote the classical Laplace–Runge–Lenz vector $\vec{A}$ [5] in symmetrized matrix form in order to make it Hermitian

$$\vec{\mathbf{A}} = \frac{1}{Ze^2 m_e} \cdot \frac{1}{2}\left(\vec{\mathbf{l}} \times \vec{\mathbf{p}} - \vec{\mathbf{p}} \times \vec{\mathbf{l}}\right) + \frac{\vec{\mathbf{r}}}{r}$$

and showed that its components are constants of motion in the Coulomb potential $Ze^2/r$, as in the classical case. With the help of the commutation relations of the components of the vector matrix $\vec{\mathbf{A}}$ among each other and the components of the angular momentum matrix $\vec{\mathbf{l}}$, Pauli obtained Bohr's energy values $W(n)$ after an algebraic *tour de force*. This is the reason why it is usually not presented in quantum mechanics textbooks. Pauli submitted this important missing piece for the success of of matrix mechanics only much later (received by *Zeitschrift für Physik* on January 17, 1926) [19], because he wanted to include relativistic corrections. It was almost a bit too late, as discussed in the following section.

## 3.3 Cambridge

Important contributions to the algebraic formulation of quantum mechanics were made by Paul Adrien Maurice Dirac (1902–1984) in Cambridge. On July 28 Heisenberg gave a talk in Cambridge but only mentioned his new work to Ralph Fowler after the talk. Dirac received a copy of the proofsheets of Heisenberg's paper from Ralph Fowler in the middle of August 1925, more than a month before it appeared in print. The paper made no easy reading for Dirac.

Figure 5: Paul Adrien Maurice Dirac in the 1920s

Probably by the end of September he realized that most of the quantum mechanical equations can be written in a form similar to classical Hamilton mechanics using *Poisson brackets.* For differentiable functions of the canonical variables they are defined as

$$\{F, G\} \equiv \sum_{i=1}^{n} \left( \frac{\partial F}{\partial q_i} \frac{\partial G}{\partial p_i} - \frac{\partial G}{\partial q_i} \frac{\partial F}{\partial p_i} \right) .$$

Using Poisson brackets Hamilton's equations can be written as

$$\dot{q}_i = \{q_i, H\} , \qquad \dot{p}_i = \{p_i, H\} ,$$

with $H$ the Hamilton function and

$$\{q_i, p_j\} = \delta_{ij}, \qquad \{q_i, q_j\} = 0 = \{p_i, p_j\}$$

holds. Dirac convinced himself of the correspondence of the quantum mechanical commutator $[A_q, B_q] \equiv A_q B_q - B_q A_q$ of the "$q$-numbers" ($q$ for quantum) $A_q, B_q$ (apart from a factor) to the Poisson bracket of the classical "$c$-numbers" $A_c, B_c$

$$[A_q, B_q] \leftrightarrow i\hbar \{A_c, B_c\} .$$

In his contribution "The Fundamental Equations of Quantum Mechanics", received by *Proceedings of the Royal Society* on November 7 [20], Dirac writes before presenting this correspondence: "We make the fundamental assumption that *the difference between the Heisenberg products of two quantum quantities is equal $ih/2\pi$ times their Poisson bracket expression.*" Later, in the paper, he points out the importance of the fact that "the mathematical operations on the two theories obey in many cases the same laws". Towards the end of the paper Dirac writes down what is usually called *Heisenberg's equation of motion.*

As in Dirac's approach algebraic relations between the quantum objects play the central role, independently of the special matrix form, this was the first step towards the abstract formulation of quantum mechanics students usually learn today.
Before addressing the year 1926, a few remarks about the later perception of Heisenberg's paper are appropriate. Some years ago, two papers appeared in the *American Journal of Physics*, which address details of his 1925 paper [21,22]. They agree that Heisenberg's paper is "notoriously difficult to read". In one paper, a reference is made to the Nobel Prize winner Steven Weinberg. In his book *Dreams of a Final Theory* [23] he writes:
"I have tried several times to read the paper that Heisenberg wrote on returning from Heligoland , and, although I think I understand quantum mechanics, I have never understood Heisenberg's motivation for the mathematical steps in his paper....Heisenberg's paper was pure magic".
This assessment is useful for a better understanding of some of the later events.

# 4 1926: Wave Mechanics and Beyond

## 4.1 Zurich

In Zurich Erwin Schrödinger (1887–1964), born in Vienna, arrived at a new quantum theory in a completely different way. His starting point was the 1924 PhD thesis of Louis de Broglie (1892–1987) extending the wave–particle dualism of light to particles with a nonzero rest mass $m$. He postulated that such a particle, like an electron, also has wave character, with the wave length $\lambda = h/p$ determined by the momentum $p$ of the particle. The issue how Schrödinger arrived at his *wave equation* is not discussed here. In his first paper "Quantisierung als Eigenwertproblem "(Quantization as an Eigenvalue Problem) received by *Annalen oder Physik* on January 27, 1926 [24] he presented the (time-independent) Schrödinger equation for the complex wave function $\psi$ for a particle in an external field. This first in a series of papers focuses on the solution of the hydrogen problem found in every quantum mechanics textbook today.

Figure 5: Erwin Schrödinger in the 1920s

Early in 1926 there seemed to exist two different theoretical approaches to explain atoms: matrix mechanics and wave mechanics. But, rather quickly, Schrödinger was the first to show the equivalence of the two approaches to (nonrelativistic) quantum mechanics [25]. A mathematically rigorous proof was later given by von Neumann [26]. Schrödinger's wave mechanics quickly found more acceptance than matrix mechanics. It was closer to the mathematics known to theoretical physicists from other fields of physics. Now external potentials different from the $1/r$ Coulomb potential could be successfully treated. This was not the case for matrix mechanics which was considered "difficult".

## 4.2 Göttingen

The "three-man" Göttingen group was not happy about this, especially as the physical meaning of the wave function was not generally agreed on. Schrödinger had a "smearing" of the electron in mind. This question brought Max Born back into the game. As he was unable to describe the scattering of particles within matrix mechanics he successfully used the Schrödinger equation [27] for this purpose. He realized that Schrödinger's quantum mechanics can describe scattering events, but not what definitely happens, only how *probable* the effect is. In a footnote of his paper he presented *his* interpretation of the wave function of a particle: $|\psi(\vec{x})|^2\Delta V$ determines the *probability* to detect the particle in the (small) volume $\Delta V$ around the position $\vec{x}$.

Born reported about his switching to wave mechanics in a meeting of the Göttingen Academy of Sciences on January 14, 1927 [28]. In his introduction he states as follows (translated from original):

*While for periodic systems the wave mechanical description of the quantum laws according to Schrödinger provides nothing more and nothing less than the matrix representation of Heisenberg, Jordan and myself, it seems especially well suited for aperiodic processes. But it is necessary to drop completely Schrödinger's ideas which are heading towards a revival of classical continuum theory. One only has to take the formalism and give it a new physical content. One has to assume the existence of a guiding field which determines the probability of discrete elementary acts. As shown recently one can get the laws for the scattering of point particles (electrons, α-particles) from atoms this way.*

In the last sentence Born referred to his own paper [27]. His statement nicely describes his mixed feelings about the events. Despite the fact quantum mechanics had started in Göttingen about half a year before Schrödinger, his approach was more openly accepted by the community.

By the end of 1926, Dirac and Jordan independently submitted papers on "transformation theory", which gave a general formal framework for (nonrelativstic) quantum mechanics [29,30]. Using the Dirac notation one can go from the wave function in position space $\psi(\vec{x}) \equiv \langle\vec{x}|\psi\rangle$ to the corresponding function in momentum space $\tilde{\psi}(\vec{p}) \equiv \langle\vec{p}|\psi\rangle$ to the probability density $|\tilde{\psi}(\vec{p})|^2$ for measurements of the momentum of the particle. This is in complete analogy to Born's probability distribution $|\psi(\vec{x})|^2$ for the position of the particle in the state $|\psi\rangle$ in Hilbert space.

The mathematical formulation of quantum laws was completed. Discussions about the interpretation of the formalism persist till today.

# 5 Interpretation, Applications and Nobel Prizes

## 5.1 Interpretation

While the formalism of nonrelativistic quantum mechanics was ready by the end of 1926 its interpretation was still in its infancy. In March 1927 Heisenberg submitted a paper with his famous uncertainty relation [31]. First leaving aside the experimental implications it is a mathematical result for the square root of the variances $(\Delta x)_{|\psi\rangle}$ and $(\Delta p)_{|\psi\rangle}$ of the probability densities for the position and momentum of a particle in the state $|\psi\rangle$

$$(\Delta x)_{|\psi\rangle}(\Delta p)_{|\psi\rangle} \geq \hbar/2$$

which can be derived using the commutation relation for position and momentum [32]. If one can repeatedly prepare (at least approximately) the same state $|\psi\rangle$ and the repeated position measurement shows that the wave function $\psi(\vec{x})$ is strongly localized in space, measurement of the momentum, when one again repeatedly prepares the state $|\psi\rangle$, shows a large variance and vice versa. Discussions of Heisenberg with Bohr and others about this inequality, Born's probability interpretation of the state and the proposal of the "collapse of the wave function" by a measurement with a macroscopic experimental device led to what is usually called the *Copenhagen interpretation* of quantum mechanics, despite the fact that there is no general agreement about its concise meaning. The discussion of the many alternative interpretations would be a topic for a separate paper.

## 5.2 Applications

Soon after the formulation of quantum mechanics it was successfully applied to various problems in atomic, molecular, nuclear and solid state physics, which can only be touched on here. Again the focus is on the Göttingen players. Friedrich Hund discovered the phenomenon of "tunneling" in quantum mechanics and ealier than Robert Mulliken he showed the importance of what was later called "molecular orbitals" [33]. Also, Born together with his PhD-student Robert J. Oppenheimer addressed problems in molecular physics. The much larger nuclear compared to electron mass led to the formulation of the "Born–Oppenheimer approximation" which lies at the heart of chemistry [34]. Post-docs with Born at that time were Walter Heitler, Victor Weisskopf, Eugene Wigner, Vladimir Fock and Edward Teller. Born himself worked on the nuclear decay, while Heisenberg meanwhile as Professor in Leipzig addressed e.g. the helium atom which had resisted an explanation in Bohr–Sommerfeld theory. He also formulated a theory of ferromagnetism.

## 5.3 Nobel Prizes

A very large number of Nobel Prizes were awarded for work related to quantum mechanics, especially for various applications (see below). Three of the physicists mentioned in sections 3 and 4 were awarded the Nobel Prize for physics in 1933. Heisenberg received the 1932 prize "for the creation of quantum mechanics, the application of which has, inter alia, led to the discovery of the allotropic forms of hydrogen". At that time the price was rarely awarded for purely theoretical work. Therefore the experimental discovery of the "allotropic forms of hydrogen" is mentioned in the announcement. In the year 1926 Heisenberg had addressed the electronic structure of ortho- and parahelium [35], which involve the configurations of the spins of the two electrons. For orthohelium the total spin is $S = 1$, while for parahelium it is $S = 0$. In part two of his papers on "Many-Body Problems and Resonance in Quantum Mechanics" [36] Heisenberg considered the case that the nuclei in a diatomic molecule each have an "eigenrotation" $s = 1/2$, which leads to an analogous scenario as for the electrons in the helium atom, with an ortho- and a parasystem. He mentions that this possibly could happen for $H_2$ molecules. This was experimentally confirmed two years later [37].

Schrödinger and Dirac shared the 1933 prize "for the discovery of new productive forms of atomic theory". In this paper we discussed Dirac's contributions to the

formulation of nonrelativistic quantum mechanics. He had made an even more important contribution with his relativistic *Dirac equation* for spin 1/2 particles [38]. The other two of the five fathers of quantum mechanics, especially Max Born, were unhappy not to receive the prize. Born had to wait more than twenty years and was awarded the 1954 Nobel Prize for physics, "for his fundamental research in quantum mechanics, especially for his statistical interpretation of the wave function". Jordan's contribution to the formulation of quantum mechanics was not honored by a Nobel Prize. This might be due to his involvement during the time of national socialism.
Of the many Nobel Prizes for the application of quantum mechanics I mention only one, the 1966 chemistry prize, awarded to Robert Mulliken,"for his fundamental work concerning chemical bonds and the electronic structure of molecules by the molecular orbital method". In his Nobel lecture, he wrote as follows *Molecular orbital theory has long been known as Hund-Mulliken theory in recognition of the major contribution of Professor Hund in its early development'.* After receiving the prize Mullikan mentioned that he would have wished to share it with Hund.

# 6 Summary

Many of the modern quantum mechanics textbooks avoid a detailed discussion of its historical development. The first excellent textbook proceeding that way is Dirac's *The Principles of Quantum Mechanics* [39]. The first edition was published in 1930 and it treats the "Schrödinger representation" as well as the "Heisenberg representation". Its fourth edition from 1958 has been reprinted many times and despite the fact that the concept of *entanglement* which is very important in the context of quantum information is not discussed, students should have a look at it even today.
Dirac's book is in sharp contrast to Born's "Volume 2" promised in 1924 to appear several years later. The book *Elementare Quantenmechanik, Zweiter Band der Vorlesung über Atommechanik* with Jordan as coauthor was also published in 1930 [40]. It promises the treatment of wave mechanics in "Volume 3", which was never written. Therefore "Volume 2" contains no wave mechanics but only treats the algebraic methods of matrix mechanics. In a review of the book Pauli pointed out very clearly that he considered this restriction a bad idea [41]. Due to this one-sided approach the book could be of use only for a very restricted readership. The only positive sentence in his review is the last one. Pauli, well known for his sarcasm, writes (translated from the original):

*The making of the book with respect to print and paper is excellent.*

In hindsight one has to admit that this book definitely did not help to popularize the important first steps towards nonrelativistic quantum mechanics made in Göttingen.

## Acknowledgements

The author would like to thank H. Leschke and K. Scharnberg for the critical reading of of the manuscript and useful suggestions.

## References

[1] J. Mehra and H. Rechenberg, *The Historical Development of Quantum Theory*, Volume 1, Springer Verlag (1982)

[2] M. Born, "Über das Thomsonsche Atommodell", *Physikalische Zeitschrift* **10**, 1031 (1909)

[3] N. Bohr, "On the Constitution of Atoms and Molecules", *Philosophical Magazine*, **26**, 1 (1913)

[4] Arnold Sommerfeld, *Atombau und Spektrallinien*, Friedr. Vieweg and Sohn, Braunschweig, 1919 (1. Auflage)

[5] W. Lenz, "Über den Bewegungsverlauf und die Quantenzustände der gestörten Keplerbewegung", *Zeitschrift für Physik* **24**, 197 (1924)

[6] N. Bohr, Wolfskehl Lectures; in Niels Bohr Collected Works, Volume 4, pp. 341–419, J. Rud Nielsen (editor), North Holland Publishing Company, 1977

[7] D. Hilbert, unpublished works, Cod-MS-537, Niedersächsische Staats- und Universitätsbibliothek, Göttingen

[8] P. Jordan, " Über das thermische Gleichgewicht zwischen Quantenatomen und Hohlraumstrahlung", *Zeitschrift für Physik* **33**, 649 (1925)

[9] T. Kuhn, *The Structure of Scientific Revolutions*, The University of Chicago Press, Chicago and London, 1962 (first edition).

[10] M. Born, *Vorlesungen über Atommechanik*, Julius Springer publishers, Berlin, 1924 (first edition)

[11] A. Einstein, " Zur Quantentheorie der Strahlung", *Mitteilungen der Physikalischen Gesellschaft Zürich* **18**, 47 (1916) and *Physikalische Zeitschrift* **18**, 121

(1917)

[12] W. Heisenberg, "Über quantentheoretische Umdeutung kinematischer und mechanischer Beziehungen", *Zeitschrift für Physik* **33**, 879 (1925); received July 29.

[13], W. Pauli, Scientific Correspondence with Bohr, Einstein, Heisenberg and others, K. von Meyenn (editor), Springer Verlag 1985.

[14] M. Born and P. Jordan, "Zur Quantentheorie aperiodischer Vorgänge", *Zeitschrift für Physik* **34**, 479 (1925), received June 11.

[15] M. Born, "Die statistische Deutung der Quantenmechanik" Nobel lecture, December 11, 1954

[16] M. Born and P. Jordan, "Zur Quantenmechanik", *Zeitschrift für Physik* **34**, 858 (1925); received September 27.

[17] J. Mehra and H. Rechenberg, *The Historical Development of Quantum Theory*, Volume 3, Springer Verlag (1982)

[18] M. Born, W. Heisenberg and P. Jordan, "Zur Quantenmechanik II", *Zeitschrift für Physik* **35**, 557 (1926); received November 16.

[19] W. Pauli, "Über das Wasserstoffspektrum vom Standpunkt der neuen Quantenmechanik", *Zeitschrift für Physik* **36**, 336 (1926); received January 17.

[20] P.A.M. Dirac, "The fundamental equations of quantum mechanics", *Proceedings of the Royal Society A* **109**, 642 (1925); received November 7.

[21] I.J.R. Aitchison, D.A. MacManus, and T.M. Snyder, "Understanding Heisenberg's magical paper of July 1925: A new look at the calculational details", *American Journal of Physics* **72**, 1370 (2004)

[22] K. Gottfried, "P.A.M. Dirac and the discovery of quantum mechanics", *American Journal of Physics* **79**, 261 (2011)

[23] S. Weinberg, '*Dreams of a Final Theory*, Pantheon, New York, 1992, pp. 67–68

[24] E. Schrödinger, "Quantisierung als Eigenwertproblem", *Annalen der Physik* **79**, 361 (1926); received January 27

[25] E. Schrödinger, "Über das Verhältnis der Heisenberg-Born-Jordanschen Quantenmechanik zu der meinen", *Annalen der Physik* **79**, 734 (1926); received March 18.

[26] J. von Neumann, *Mathematische Grundlagen der Quantenmechanik*, Julius Springer, Berlin, 1932; english tranlation: *Mathematical foundations of quantum mechanics*, Princeton University Press, Princeton, 1955

[27] M. Born, "Zur Quantenmechanik der Stoßvorgänge", *Zeitschrift f. Physik* **37**, 863 (1926); received July 21,

[28] Max Born, "Zur Wellenmechanik der Stoßvorgänge", *Nachrichten von der Gesellschaft der Wissenschaften zu Göttingen, aus dem Jahre 1927*, 77 (Mathematisch-Physikalische Klasse)

[29] P.A.M. Dirac, "The physical interpretation of the quantum dynamics", *Proceedings of the Royal Society A* **113**, 621 (1927); received December 2, 1926.

[30] P. Jordan, "Über eine neue Begründung der Quantenmechanik", *Zeitschrift für Physik* **40**, 809 (1927); received December 8, 1926.

[31] W. Heisenberg, "Über den anschaulichen Inhalt der quantentheoretischen Kinematik und Mechanik", *Zeitschrift für Physik* **43**, 172 (1927); received March 23, 1927.

[32] H.P. Robertson, "The uncertainty principle", *Physical Review* **34** 163 (1929)

[33] F. Hund, " Zur Deutung der Molekülspektren", *Zeitschrift für Physik* **40**, 742 (1927); received November 19, 1926.

[34] M. Born und R. Oppenheimer, "Zur Quantentheorie der Molekeln", *Annalen der Physik* **84** , 310, 1927; received August 25, 1927.

[35] W. Heisenberg, "Mehrkörperproblem und Resonanz in der Quantenmechanik" *Zeitschrift für Physik* **38**, 411 (1926)

[36] W. Heisenberg, "Mehrkörperproblem und Resonanz in der Quantenmechanik II" *Zeitschrift für Physik* **41**, 239 (1927)

[37] K.F. Bonhoeffer an P. Harteck, "Experimente über Para- und Orthowasserstoff", *Naturwissenschaften* **17**, 182, (1929)

[38] P.A.M. Dirac, "The Quantum Theory of the Electron", *Proceedings of the*

*Royal Society A* **117**, 610 (1928)

[39] P.A.M. Dirac, *The Principles of Quantum Mechanics*, Clarendon Press, Oxford, 1930 (first edition)

[40] M. Born and P. Jordan, *Elementare Quantenmechanik, Zweiter Band der Vorlesungen über Atommechanik*, Springer Verlag, Berlin 1930

[41] W. Pauli, Review of M. Born and P. Jordan, "Elementare Quantenmechanik, Zweiter Band der Vorlesungen über Atommechanik" *Naturwissenschaften*, **18**, 602 (1930)

# Chapter 2

# The Quantum Formalism Revisited

Differences from the classical theory with some of them quantified 100 years after HEISENBERG's discovery

Hajo LESCHKE

Friedrich–Alexander–Universität Erlangen–Nürnberg
Institut für Theoretische Physik
Erlangen, Germany

To Wolfgang KUNDT, who made me understand what he had learned from Pascual JORDAN

Intermediate IQSA 2025 Conference
30 June – 4 July, 2025
Tropea, Calabria (Italy)

Second chapter of the book
*Quantum Mechanics: A Century Later*, ed. by Tuck CHOY
World Scientific, Singapore 2026

*"Heisenberg hat ein großes Quantenei gelegt. In Göttingen glauben sie daran (ich nicht)."*[1]

Albert EINSTEIN to Paul EHRENFEST
(in a letter dated 20 November 1925)

---

*"If a subject is robust it should be insensitive to its foundations."*

Mark KAC
(p. 111 in his autobiography [Ka85])

[1] In English: "Heisenberg has laid a big quantum egg. In Göttingen they believe it (I don't)." [See also: Tom SIEGFRIED in the magazine *Science News*, March 30, 2020.]

**Abstract** For the simple system of a point-like particle confined to a straight line, I compile, initially in a concise table, the structural elements of quantum mechanics and contrast them with those of classical (statistical) mechanics. Despite many similarities, there are the well-known fundamental differences, resulting from the algebraic non-commutativity in the quantum structure. The latter was discovered by Werner HEISENBERG (1901–1976) in June 1925 on the small island of Helgoland in the North Sea, as a consequence of understanding atomic spectral data within a matrix scheme consistent with energy conservation. I discuss the differences and exemplify their quantifications by the variance and entropic indeterminacy inequalities, by (pseudo-)classical bounds on quantum canonical partition functions, and by the correlation inequalities of John BELL (1928–1990) and others.

## Contents



## Acknowledgments

This text grew out of lectures given at the Friedrich–Alexander–Universität Erlangen–Nürnberg (FAU). It also owes its existence to a series of lectures for probability theorists (Sept. 2013, Villa Volpi, Ghiffa, Piedmont, Italy) suggested by Peter EICHELSBACHER (Ruhr-Universität Bochum (RUB), Germany). I am especially grateful to my former students Helmut FINK (Heisenberg-Gesellschaft, München), Sebastian ROTHLAUF, and Rainer RUDER for valuable discussions and their help with the LaTeX document-preparation system. I also appreciate many clarifying comments made by Stefan WEIGERT (University of York, UK) and by my colleague Daniel BURGARTH (FAU). Last but not least, my thanks go to the editor Tuck C. CHOY for his kind invitation, assistance, and patience. A travel grant of the Hans-Hermann TOEDTER Demokritos-Stiftung (München) is gratefully acknowledged.

## 1 Introduction

*Quantum mechanics* is the most successful theory of physics for model building from subatomic to cosmic length scales. As its name suggests, probably first coined by BORN [Bo24], the theory was needed because the time-honored and well-established *classical mechanics* of NEWTON, LAGRANGE, and HAMILTON had turned out to be unable to explain the *quantized* or portioned energy values observed in atomic physics. Very likely, most physicists had expected a more or less refined version to emerge from classical mechanics, but after HEISENBERG's landmark paper [Hei25] and the subsequent development of the formalism by him and other brilliant physicists in the mid-1920s, it became clear that quantum mechanics is not just an evolution, but a scientific revolution. Its startling implications are not only of fundamental importance for most of physics, but also for our understanding of Nature and the philosophy of science. They are also responsible for the theoretical and technological achievements obtained for the more recent topics of quantum information, communication, and computer science with its now omnipresent effects on our everyday lives.

In this contribution I review some aspects of the HILBERT-space formalism for quantum systems in terms of its structural elements, its main notions, and its most important mathematical consequences. For this purpose I start out in the next section – making up roughly one half of the text – with comparing *canonical* classical (statistical) mechanics to the corresponding quantum mechanics, with an emphasis on the latter. This seems to be reasonable, because (i) it reflects the historical development and (ii) most of the mentioned quantum notions and consequences retain their significance for non-mechanical quantum systems such as pure spin systems, which only appear in the subsequent sections.

Two remarks are in order. First, in his development of the HILBERT-space formalism VON NEUMANN did not (want to) rely on the analogy to classical mechanics, but aimed for a quantum formalism which could stand on its own, see [vN18, DJ13]. For this reason, he also may be considered to be one of the founding fathers of the more recent developments [Au07, SS08, Pet08, NC10, Hay17] mentioned above.[2] Second, for these topics it is typically sufficient to deal with a HILBERT space of *finite* dimension, that is, with linear algebra and its associated matrix theory [Hal87]. In contrast, HEISENBERG and the other early giants of quantum mechanics were from the very beginning confronted with unbounded operators acting on an infinite-dimensional HILBERT space with its topological intricacies and allowing for continuous spectra. Only amazing intuition enabled them to overcome these complications on their way to creating a fundamentally new physical theory and inspiring HILBERT, WEYL, STONE, and other great mathematicians to develop *functional analysis* further. In fact, the first rigorous proof of the spectral theorem for general self-adjoint operators was given only about 1930 by VON NEUMANN, who was a visionary polymath [Bha21].

The following sections are mainly meant for readers who want to grasp, or remember, the essential concepts and consequences without getting lost in a morass of technicalities[3], when having to accept a HILBERT space of infinite dimension, like for the spectral theory of the *hydrogen atom* [Pa26, Sc26a, GP90, Boh93] or for general *scattering theory* [Bo26, GP91, Boh93, Ya00]. Accordingly, I often do not adhere to the standards of mathematical rigor. In particular, domain questions of unbounded operators are almost always ignored. Readers interested in mathematical details may consult [BEH08, T14, Mor17] or [Stro08, GS20]. For readers wanting to understand, how the quantum formalism was (hypothetically) deduced from experiments and intuition I recommend [LévB90]. In the supporting Biblio-

[2] But not discussed in this text, for lack of space and because I am only faintly familiar with them.

[3] Although – quoting Kai Lai CHUNG – one person's technicalities are another person's profession.

graphy at the end, some of the selected books are not cited in the main text. But every single one of them, cited or not, has something special to earn the attention of the reader.

## 2 Canonical Classical Mechanics and Quantum Mechanics in Comparison

Classical *statistical* mechanics (CM) and quantum mechanics (QM) have many structural similarities. Nevertheless, there are fundamental differences due the *non-commutativity* in the quantum structure discovered in 1925 by HEISENBERG [Hei25, Hei71, BJLR17]. Their main consequences have then been worked out jointly with BORN and JORDAN [BJ25, BHJ26]. Other highly important contributions to the quantum formalism are due to DIRAC [D25], SCHRÖDINGER [Sc26a, Sc26b], and VON NEUMANN in his classic [vN32, vN18].

### 2.1 Structural elements of classical and quantum mechanics – contrasted in tabular form

It is sufficient and convenient to consider the simple model system "point mass in the EUCLIDean line $\mathbb{R}$" for a particle with mass $m > 0$, but "practically" without spatial extent (and spin), moving rectilinearly at (variable) speed much smaller than that of light in vaccum.

| | Classical Mechanics (CM) | Quantum Mechanics (QM) |
|---|---|---|
| • Arena (or Stage) | phase space $\Gamma$ [here: $\Gamma = \mathbb{R} \times \mathbb{R}$] | separable HILBERT space $\mathscr{H}$ over the complex numbers $\mathbb{C}$ [here: $\mathscr{H} = L^2(\mathbb{R})$] |
| • Canonical Elements (momentum and position) | real coordinates of $\Gamma$ <br> $p, q$ | basic self-adjoint operators on $\mathscr{H}$ <br> $P, Q$ |
| • Elements of the LIOUVILLE Space | complex functions $a : \Gamma \to \mathbb{C}$ <br> $a(p,q)$ | operators $A$ on $\mathscr{H}$ <br> $A(P,Q)$ |
| • Scalar Product | $\langle a, b \rangle := \int_\Gamma \frac{\mathrm{d}p\,\mathrm{d}q}{2\pi\hbar} a^*(p,q)\, b(p,q)$ | $\langle A, B \rangle := \mathrm{Tr}(A^* B)$ |
| • LIE Product | POISSON bracket <br> $[a,b] := \frac{\partial a}{\partial p}\frac{\partial b}{\partial q} - \frac{\partial b}{\partial p}\frac{\partial a}{\partial q}$ | (standardized) commutator <br> $[A,B] := \frac{\mathrm{i}}{\hbar}(AB - BA)$ |
| • Canonical Relation | $[p,q] = 1$ | $[P,Q] = \mathbb{1}$ |
| • Observables | real (BOREL) functions <br> $a^* = a$ | self-adjoint operators <br> $A^* = A$ |
| Important example: Standard HAMILTONian | $h(p,q) = \frac{1}{2m} p^2 + \upsilon(q)$ | $H = \frac{1}{2m} P^2 + \upsilon(Q)$ |
| | "kinetic plus potential energy" | |
| • Events | indicator functions of (BOREL) subsets $\mathscr{E} \subseteq \Gamma$ <br> $\chi_{\mathscr{E}}(p,q) := \begin{cases} 1 & \text{if } (p,q) \in \mathscr{E} \\ 0 & \text{else} \end{cases}$ | projections on (closed) subspaces $E\mathscr{H} \subseteq \mathscr{H}$ <br> $E^* = E = E^2$ |

| | | |
|---|---|---|
| • States | probability densities on $\Gamma$ (possibly in the distributional sense)<br>$w^* = w \geq 0, \quad \langle w, 1\rangle = 1$ | statistical operators on $\mathscr{H}$ (still often: "density matrices")<br>$W^* = W \geq 0, \quad \langle W, \mathbb{1}\rangle = 1$ |
| • Pure States | DIRAC measures at some $(p_0, q_0) \in \Gamma$<br>$w(p,q) = 2\pi\hbar\,\delta(p-p_0)\,\delta(q-q_0)$ | $W = W^2$ |
| • Entropy of a State | $-\langle w, \ln w\rangle$ | $-\langle W, \ln W\rangle$ |
| • Eigenstate of an Observable | $w(p,q) = \rho(p,q)\,\delta\big(a(p,q)-\alpha\big)$ with some $\rho \geq 0, \quad \alpha \in \mathbb{R}$ | $AW = \alpha W$<br>$\alpha \in \mathbb{R}$ |
| • Expectation Value of an Observable in a State | $\langle a\rangle := \langle w, a\rangle$ | $\langle A\rangle := \langle W, A\rangle$ |
| • Probability of an Event Occurring in a State | $\langle \chi_{\mathscr{E}}\rangle$ | $\langle E\rangle$ |
| • Time-Evolution of a State: | LIOUVILLE equation | VON NEUMANN equation |
| instantaneous | $\frac{\mathrm{d}}{\mathrm{d}t} w_t = [w_t, h]$ | $\frac{\mathrm{d}}{\mathrm{d}t} W_t = [W_t, H]$ |
| prolonged[4] | $w_t = \mathrm{e}^{-t[h,\cdot]}\, w_0$ | $W_t = \mathrm{e}^{-\mathrm{i}tH/\hbar}\, W_0\, \mathrm{e}^{\mathrm{i}tH/\hbar}$ |
| [in the SCHRÖDINGER picture] | | |
| • Invariance Group | canonical transformations | (anti-)unitary transformations |

[4] For short times, that is, for small $|t|$, the evolution is (informally) given by the LIE expansion formula:

$$w_t = w_0 - \frac{t}{1!}[h, w_0] + \frac{t^2}{2!}\big[h, [h, w_0]\big] + \ldots \quad \text{resp.} \quad W_t = W_0 - \frac{t}{1!}[H, W_0] + \frac{t^2}{2!}\big[H, [H, W_0]\big] + \ldots$$

## 2.2 Explanations and comments

1) Arena
   - The arena of the CM system is [here] just the EUCLIDean plane $\mathbb{R} \times \mathbb{R} \cong \mathbb{R}^2$ with real coordinates $(p, q)$ in the sense of CARTESIUS/DESCARTES, for momentum and position, respectively.
   - The arena of the corresponding QM system is a HILBERT space $\mathscr{H}$ over the (algebraic field of) complex numbers $\mathbb{C}$ of [here] countably *infinite* dimension. This means that $\mathscr{H}$ is a *linear space* of vectors $\psi, \varphi, \eta, \ldots$ with a scalar product $\langle\psi|\varphi\rangle$ and norm $\|\psi\| := \langle\psi|\psi\rangle^{1/2}$, where $\langle\psi|\varphi\rangle$ is assumed to be anti-linear in the left argument. Moreover, there is a countable set of pairwise orthogonal unit vectors (orthonormal basis =: ONB) such that every $\psi \in \mathscr{H}$ can be written as a linear combination of these unit vectors which converges (in norm). Finally, every CAUCHY sequence in $\mathscr{H}$ is convergent. To summarize, $\mathscr{H}$ is the "natural" extension to infinite dimension of the $d$-dimensional *unitary* space $\mathbb{C}^d$ with *finite* $d \in \{1, 2, 3, \ldots\}$, the complex analog of the EUCLIDean space $\mathbb{R}^d$. Although all HILBERT spaces of the same dimension are isometrically isomorphic, it is convenient throughout this Section 2 to follow SCHRÖDINGER [Sc26a] and to think of $\mathscr{H} = \mathrm{L}^2(\mathbb{R})$. This is the linear space of all functions $\psi : \mathbb{R} \to \mathbb{C}, q \mapsto \psi(q)$ being square-integrable with respect to the LEBESGUE measure, that is, $\int_{\mathbb{R}} \mathrm{d}q \left|\psi(q)\right|^2 < \infty$. Here two functions $\psi, \varphi \in \mathrm{L}^2(\mathbb{R})$ are identified with each other if their values coincide *almost everywhere* (a.e.), that is, differ at most on a set of LEBESGUE measure zero. The scalar product of $\mathrm{L}^2(\mathbb{R})$ is given by $\langle\psi|\varphi\rangle = \int_{\mathbb{R}} \mathrm{d}q\, \psi^*(q)\varphi(q)$, where $\psi^* \in \mathrm{L}^2(\mathbb{R})$ is defined by pointwise complex conjugation: $\psi^*(q) := \psi(q)^*$. Whether $\mathrm{L}^2(\mathbb{R})$ or another realization of $\mathscr{H}$ is used, all elements and notions in the right column of the table that do not refer to $P$ and $Q$ of a *mechanical* system in the strict sense, retain their significance for other quantum systems with infinite-dimensional or finite-dimensional $\mathscr{H}$. For example, for pure spin systems one may think of $\mathscr{H} = \mathbb{C}^d$.

2) Canonical Elements
   - For the CM see Explanation 1).
   - For the QM (in the SCHRÖDINGER realization) the *position operator* $Q$ is defined as the basic *multiplication* operator by $(Q\psi)(q) := q\psi(q)$ for all $\psi \in \mathrm{dom}(Q) := \{\eta \in \mathrm{L}^2(\mathbb{R}) : \|Q\eta\| < \infty\}$, the *domain* of $Q$. It's obvious that $\langle\varphi|Q\psi\rangle = \langle Q\varphi|\psi\rangle$ for all $\varphi, \psi \in \mathrm{dom}(Q)$ (with the finiteness following from the CAUCHY–SCHWARZ inequality $|\langle\varphi|Q\psi\rangle| \leq \|\varphi\|\|Q\psi\| < \infty$). Since $\mathrm{dom}(Q)$ is dense in $\mathrm{L}^2(\mathbb{R})$, it is not hard to show that the (unique) adjoint operator $Q^*$ has the same domain. Hence $Q$ is *self-adjoint* on $\mathrm{dom}(Q)$ which is written as $Q^* = Q$, see [BEH08, T14].

     The definition of the *momentum operator* $P$ is more involved. Informally, it is the basic *differential* operator "$-\mathrm{i}\hbar\frac{\mathrm{d}}{\mathrm{d}q}$", where $2\pi\hbar \approx 6.63 \times 10^{-34}$ JOULE-seconds is the fundamental PLANCK constant. Its precise domain turns out to be

$$\mathrm{dom}(P) := \left\{\eta \in \mathrm{L}^2(\mathbb{R}) : \eta \text{ is absolutely continuous with derivative } \eta' \in \mathrm{L}^2(\mathbb{R})\right\}.$$

     A function $\eta : \mathbb{R} \to \mathbb{C}, q \mapsto \eta(q)$ is *absolutely continuous* if its (a.e. existing) derivative $\eta'$ satisfies the fundamental theorem of calculus in the LEBESGUE sense

$$\int_{q_1}^{q_2} \mathrm{d}q\, \eta'(q) = \eta(q_2) - \eta(q_1) \quad \text{for all } q_1, q_2 \in \mathbb{R} \text{ with } q_1 \leq q_2 .$$

     For $\psi \in \mathrm{dom}(P)$ it can be shown that $\psi(q) \to 0$ as $q \to \pm\infty$.

3) Scalar Product (for the LIOUVILLE space)
   - In the CM case, $\langle a,b\rangle$ is a (positive definite) scalar product in the strict sense only for $a,b \in L^2(\mathbb{R}\times\mathbb{R})$. For other functions on $\mathbb{R}\times\mathbb{R}$ it is merely a convenient notation for the underlying integral if it exists. Simple example: $\langle |a|,1\rangle < \infty$ and $|b(p,q)| \leq \gamma$ for LEBESGUE-almost all $(p,q)\in\mathbb{R}\times\mathbb{R}$ with some constant $\gamma<\infty$.
   - Similar remarks apply to the QM case. Here $\mathrm{Tr}(A^*B)$ makes sense as a scalar product if $A$ and $B$ are HILBERT–SCHMIDT operators. But the *trace* also exists, for example, if $A$ is even of the *trace class*, $\mathrm{Tr}\sqrt{A^*A}<\infty$, and $B$ is only *bounded* in the sense that its uniform norm $\|B\| := \sup\{\|B\eta\| : \eta\in\mathscr{H}, \|\eta\|=1\}$ is finite.

4) LIE Product
   - is, by definition, bilinear, anticommutative, and obeys the JACOBI identity

$$\big[A,[B,C]\big]+\big[B,[C,A]\big]+\big[C,[A,B]\big]=\mathbb{0}\,.$$

   - is, in addition, compatible with the "ordinary" product by the LEIBNIZ rule

$$[AB,C]=A[B,C]+[A,C]B\,.$$

[similar in the CM for differentiable $a$, $b$ and $c$]

5) Canonical Relation
   - is obvious in the CM.
   - is in the QM known as the HEISENBERG–BORN–JORDAN *commutation relation*. It follows *informally* by ignoring domain questions and using the product rule of calculus. One rigorous approach is through the "unitary version" of the relation [Wey27].

6) Observables
   - represent the *quantities* that are *measurable* in the sense of experimental physics. Basic examples are the canonical elements themselves. But is there an operational instruction to measure, for example, $PQ^2P = QP^2Q$ in the laboratory? I don't know.
   - can be interpreted in the CM directly as real-valued *random variables* in the sense of *classical probability theory* (CPT) due to KOLMOGOROV 1933, see [Bau96].
   - manifest themselves in the QM *physically* in that only the (real) *spectral values* of the self-adjoint operator are postulated as possible outcomes of measurements in the laboratory. Thereby they become *quantum* random variables, but in a more fundamental and intrinsic sense than in the CM. I will try to explain this in the following.

7) Events
   - are $\{1,0\}$-(spectral)valued (or BERNOULLI) observables. They correspond to simple yes/no alternatives and form a *lattice* or algebra endowed with a complementation operation, classically even a BOOLEan one, see Section 4.
   - are often induced by an observable and a BOREL set $I\subseteq\mathbb{R}$ in the following sense:

$$\begin{array}{lll} & \mathscr{E}=\chi_{a^{-1}(I)} & = \text{ indicator function of the pre-image of } I \text{ under } a \\ \text{resp.} & E=\chi_I(A) & = \textit{ spectral projection} \text{ of } A \text{ associated with } I\,. \end{array}$$

   They are interpreted as follows:
   i) Mathematically: "$a$ resp. $A$ has or has not a (spectral)value in $I$."
   ii) Physically: "The act of a single (ideal) measurement of $a$ resp. $A$ in the laboratory does or does not give a result in $I$."

- may be combined to form a *joint* event, if there are two of them, by the well-known formula $\chi_{\mathscr{E}\cap\mathscr{F}} = \chi_{\mathscr{E}}\chi_{\mathscr{F}}$ in the CM and CPT. It generalizes in the QM to $E \wedge F := \lim_{n\to\infty} E(FE)^n$, because in general $EF \neq FE$, see Section 4. Example: For the spectral projections $E = \chi_I(P)$ and $F = \chi_J(Q)$ on $\mathscr{H} = \mathrm{L}^2(\mathbb{R})$ associated with non-empty and bounded BOREL sets $I, J \subset \mathbb{R}$ one finds the *impossible* event $E \wedge F = \mathbb{0}$ (equivalently $E\mathscr{H} \cap F\mathscr{H} = \{0\}$), as a consequence of the "delocalization property" of the FOURIER–PLANCHEREL transformation, see Sections 2.3 and 4 as well as pp.104–109 in [BGL95]. In other words, these two events are *disjoint* (or *mutually exclusive*).
- have a natural generalization to observables taking (spectral) values in the unit interval $[0,1]$. They are mainly of interest in the QM, where they obey the operator inequalities $\mathbb{0} \leq E \leq \mathbb{1}$ and form a kind of fuzzy or unsharp events, often called *effects* [Lu85/87]. More importantly, they are basic for the effect-valued or, less precise, *positive-operator-valued* (POV) measures on $\mathbb{R}$ and the related *unsharp* observables [BGL95, BZ06, Au07, NC10, BLPY16, Hay17]. These are important for modern measurement theory and quantum information theory, but they are not considered in this text.

8) States

- A *state* steers the randomness of the observables of the system. The notion includes those needed for *statistical* mechanics and for a *composite* system, when restricting/reducing its (total) state to one of its subsystems, see Section 3. A quantum state is a positive operator of trace class and therefore only has a discrete spectrum. Example: BOLTZMANN–GIBBS state

$$W = \frac{\mathrm{e}^{-\beta H}}{\mathrm{Tr}\,\mathrm{e}^{-\beta H}}$$

of a quantum system with HAMILTONian $H$ (= energy observable) in thermodynamic equlibrium at absolute temperature $1/(\beta k_B) > 0$. Here the BOLTZMANN constant $k_B > 0$ fixes the KELVIN scale and the discrete spectrum of $H$ is such that the trace in the denominator, the canonical *partition function*, is finite. Example: $\mathscr{H} = \mathrm{L}^2(\mathbb{R})$ and $H = P^2/(2m) + \gamma Q^2$ with $\gamma > 0$, defining the quantum *harmonic oscillator*.

- A *pure* quantum state is a one-dimensional projection:

$$W = |\psi_0\rangle\langle\psi_0| := \text{projection on the span of } \psi_0 \in \mathscr{H} \text{ with } \langle\psi_0|\psi_0\rangle = 1$$

$$\psi_0 = \textit{state vector} = \text{"normalized wave function"}.$$

A *pure* classical state is fully determined by only two real numbers $(p_0, q_0)$. In contrast, a pure quantum state needs an infinite sequence of complex numbers for its characterization [given an ONB in $\mathrm{L}^2(\mathbb{R})$], many of them possibly equal to 0.

- *Mixed* (or *impure*) quantum states are non-trivial (possibly infinite) convex combinations of pure ones according to their *spectral resolution* (or *decomposition*):

$$W = \sum_n \rho_n |\psi_n\rangle\langle\psi_n|$$

$$\rho_n \geq 0, \quad \sum_n \rho_n = 1, \quad \psi_n \in \mathscr{H}, \quad \langle\psi_n|\psi_m\rangle = \delta_{n,m}, \quad \sum_n |\psi_n\rangle\langle\psi_n| = \mathbb{1}.$$

9) Entropy of a State (following BOLTZMANN, GIBBS, VON NEUMANN, and SHANNON)
   - It may be understood purely information-theoretically in the sense of C.E. SHANNON and quantifies how much *information* the state lacks compared to the maximum possible, that is, how impure it is. Only for pure states is the entropy minimal. The larger it is, the less (or more coarse-grained) is the information encoded in the state.
   - In the CM there exists, in principle, only subjective ignorance (or "uncertainty"), whereas in the QM there may be additional objective *indeterminacy*, even for pure states.[5] See Explanation 13).
   - The entropy of the BOLTZMANN–GIBBS state of Explanation 8) equals the *thermal* entropy (in the case of a single particle), when multiplied by $k_B$.
   - Range of the entropy:

$$\text{CM:} \qquad -\infty \leq -\langle w, \ln w\rangle \leq \ln\left(\int_\Gamma \frac{\mathrm{d}p\,\mathrm{d}q}{2\pi\hbar}\right) \quad \left[=\infty \quad \text{for } \Gamma = \mathbb{R}\times\mathbb{R}\right]$$

$$\begin{aligned} \text{QM:} \qquad & 0 \leq -\ln\langle W, W\rangle \leq -\langle W, \ln W\rangle \\ & \leq \ln\left(\dim\mathscr{H}\right) \quad \left[=\infty \quad \text{for } \mathscr{H} = \mathrm{L}^2(\mathbb{R})\right] \end{aligned}$$

   [The lower bound in the QM case is the second-order quantum RÉNYI entropy.]

10) Eigenstate of an Observable
   i) For the CM one has:
      - An eigenstate is concentrated on the level line corresponding to one of the possible values the function $a$ can take. Therefore one has $aw = \alpha w$ similarly to the QM.
      - Common eigenstates of two observables only exist if these have level lines crossing each other.
      - Every pure state is a common eigenstate of *all* observables (if defined on the whole phase space).
   ii) For the QM one has:
      - The operator $A$ has eigenstates only if $A$ has *eigenvectors* $\psi \in \mathscr{H}$. Upon normalization they determine the pure eigenstates of $A$ according to: $W = |\psi\rangle\langle\psi|$, $\langle\psi|\psi\rangle = 1$, $A\psi = \alpha\psi$.
      - Mixed eigenstates of $A$ only exist if the multiplicity of the corresponding eigenvalue is at least 2; namely as an arbitrary convex combination of pure eigenstates of this eigenvalue.
      - Common eigenstates of two observables exist in general only if these *commute*.

11) Expectation Value (mean or average value) of an Observable in a State
   - is a linear, real, positive, and normalized functional on the star-algebra of (bounded) operators [resp. functions in the CM and CPT] in the sense that

$$\langle\alpha A + \beta B\rangle = \alpha\langle A\rangle + \beta\langle B\rangle \quad (\alpha, \beta \in \mathbb{C}),$$
$$\langle A^*\rangle = \langle A\rangle^*, \quad \langle A^* A\rangle \geq 0, \quad \langle \mathbb{1}\rangle = 1.$$

[5] For a quantum state its spectral resolution yields for the entropy

$$-\langle W, \ln W\rangle = -\operatorname{Tr} W \ln W = -\sum_n \rho_n \ln \rho_n \qquad (\text{with the definition} \quad 0\ln 0 := 0).$$

These properties are sufficient [Far78] to imply the CAUCHY–SCHWARZ inequality $|\langle A^*B\rangle|^2 \le \langle A^*A\rangle\langle B^*B\rangle$.

- is in the CM case, $\langle a\rangle = \langle w,a\rangle$, independent of the PLANCK constant $2\pi\hbar$ by the normalization $\langle w,1\rangle = 1$ (unless $a$ depends on it).
- is a "weighted mean" of the (spectral) values an observable can take (in repeated measurements). The probability that a particular outcome will occur equals the expectation value of the associated event, see the next item as well as the Explanations 6), 7), and 12).
- may be written in two simple alternative forms

$$\boxed{\langle A\rangle = \langle W,A\rangle = \mathrm{Tr}(WA) = \sum_n \rho_n \langle \psi_n|A\,\psi_n\rangle = \sum_j \alpha_j \,\mathrm{Tr}(WE_j)\,,}$$

in case the QM observable $A$, just as $W$, has a purely *discrete* spectral resolution according to $A = \sum_j \alpha_j E_j$ with $\alpha_j \in \mathbb{R}$, $\alpha_i \neq \alpha_j$ if $i \neq j$, $E_i^* E_j = \delta_{ij} E_j$, and $\sum_j E_j = \mathbb{1}$, where the trace $\mathrm{Tr}\,E_j$ is the dimension of the $j$-th eigenspace, that is, the multiplicity of the eigenvalue $\alpha_j$. Each of the two forms has an obvious probabilistic interpretation. In case the spectrum of $A$, as a subset of the real line $\mathbb{R}$, has also or even only, a *continuous* part [such as the operators $P$ and $Q$], then the last sum must be replaced by the integral $\int_{\mathbb{R}} \mu_A(\mathrm{d}\alpha)\,\alpha$, see the next Explanation.

12) Probability of an Event Occurring in a State

- Each quantum state $W$ induces a *probability measure* $\mu$ on the lattice/algebra of events $E$ defined by $\mu(E) := \langle E\rangle = \mathrm{Tr}\,(WE)$, see Explanation 7) and Section 4. The corresponding definition in the CM leads to a probability space in the sense of CPT. This is not true for the QM, because the corresponding lattice is not of BOOLEan type which, in turn, is the reason for the quantum particularities such as interference, violations of BELL-type inequalities, and the like, see [Ac10], Sections 5 and 6.
- A given state $W$ steers the randomness of *all* observables through their spectral projections considered as events. More precisely, the probability measure $\mu_A$ on the BOREL sets $I$ of the real line $\mathbb{R}$, defined by

$$\mu_A(I) := \mu\big(\chi_I(A)\big) = \langle \chi_I(A)\rangle\,, \quad I \subseteq \mathbb{R}\,,$$

is the probability *distribution* of the observable $A$ in the state $W$. In close analogy to the CM and CPT, it contains the complete probabilistic information about $A$. BORN and VON NEUMANN have postulated $\mu_A(I)$ as the probability that, in the state $W$, the outcome of a measurement of $A$ lies in $I$, confer Explanations 6) and 7). Informally and by not allowing for a *singular* continuous spectrum of $A$, one can associate to $\mu_A$ the (possibly distributional) probability *density* $\rho_A$ given by

$$\rho_A(\alpha) := \frac{\mathrm{d}}{\mathrm{d}\alpha}\mu_A\big(]-\infty,\alpha]\big) = \frac{\mathrm{d}}{\mathrm{d}\alpha}\big\langle \chi_{]-\infty,\alpha]}(A)\big\rangle = \langle \delta(\alpha\mathbb{1} - A)\rangle \ge 0\,, \quad \alpha \in \mathbb{R}$$

[with respect to the LEBESGUE measure] in terms of the DIRAC delta. Then one may write

$$\langle f(A)\rangle = \int_{\mathbb{R}} \mu_A(\mathrm{d}\alpha)\, f(\alpha) = \int_{\mathbb{R}} \mathrm{d}\alpha\, \rho_A(\alpha) f(\alpha)$$

for any $\mu_A$-integrable function $f : \mathbb{R} \to \mathbb{R}$, symbolically: "$\mu_A(\mathrm{d}\alpha) = \mathrm{d}\alpha\, \rho_A(\alpha)$".

Example: $\langle \chi_I(Q)\rangle = \mathrm{Tr}\,(W\chi_I(Q)) = \int_I \mathrm{d}q\,\langle q|W|q\rangle$ is the probability of finding the particle in $I \subseteq \mathbb{R}$. Accordingly, its density $\rho_Q(q) = \langle q|W|q\rangle$ is the "diagonal" of the *integral kernel* $\langle q|W|q'\rangle$ or "position representation" of $W$ in the sense that $(W\eta)(q) = \int_{\mathbb{R}} \mathrm{d}q'\langle q|W|q'\rangle\eta(q')$, $\eta \in \mathrm{L}^2(\mathbb{R})$. In particular, for a pure $W = |\psi\rangle\langle\psi|$ with $\psi \in \mathrm{L}^2(\mathbb{R})$ and $\langle\psi|\psi\rangle = 1$ one obtains the "usual" absolute square of SCHRÖDINGER's wave function: $\rho_Q(q) = |\psi(q)|^2$.

- All probabilities are interpretated as *predictions of relative frequencies* in often repeated measurements at a single (always equally prepared) system or in a simultaneous measurement of a large number of dynamically independent equal copies of the system (= *statistical ensemble*). However, according to an important theorem of OZAWA, an observable that admits a repeatable measurement is a discrete observable [Oz84], see also [Ho01, BLPY16].

13) *Variance* of an observable in a state
is the mean-square *deviation* from, or the quadratic *fluctuation* around, the mean of an observable (in repeated measurements). It is most commonly used to quantify the "width" of the (distributional) density $\rho_A$ of its probability distribution $\mu_A$:

$$\boxed{\sigma_A^2 := \left\langle \left(A - \langle A\rangle \mathbb{1}\right)^2\right\rangle} = \int_{\mathbb{R}} \mu_A(\mathrm{d}\alpha)\left(\alpha - \langle A\rangle\right)^2 = \langle A^2\rangle - \langle A\rangle^2 .$$

*Facts:*

i) $\sigma_{\lambda A}^2 = \lambda^2\sigma_A^2, \quad \sigma_{A+\lambda\mathbb{1}}^2 = \sigma_A^2; \quad \lambda \in \mathbb{R}.$

ii) $\boxed{\sigma_A^2 \geq 0}$

iii) $\boxed{\sigma_A^2 = 0 \iff \mu_A \text{ is a DIRAC measure} \iff W \text{ is an eigenstate of } A}$

[similar in the CM and CPT; CM observables as random variables]

***Proof*** i) obvious by definition.

ii) by the operator inequality $\left(A - \langle A\rangle\mathbb{1}\right)^2 \geq \mathbb{0}$.

iii) The first equivalence is standard by the CHEBYSHEV inequality [Bau96].
The equivalence
"$\sigma_A^2 = 0 \iff W$ is an eigenstate of $A$" can be shown as follows:

"$\Leftarrow$": The assumption $AW = \alpha W$ with $\alpha \in \mathbb{R}$ implies $A^2W = \alpha^2 W$ . The claim follows by taking traces.

"$\Rightarrow$": With $C := \left(A - \langle A\rangle\mathbb{1}\right)W^{1/2}$ one has $C^*C = W^{1/2}\left(A - \langle A\rangle\mathbb{1}\right)^2 W^{1/2}$ and therefore $\mathrm{Tr}\,C^*C = \sigma_A^2$. The assumption $\sigma_A^2 = 0$ hence implies
$C = \mathbb{0} \quad\Rightarrow\quad \left(A - \langle A\rangle\mathbb{1}\right)W = \mathbb{0} \quad\Rightarrow\quad AW = \alpha W \quad \text{with} \quad \alpha := \langle A\rangle.$ □

Consequences of iii) with additions:

- Only if the system is an eigenstate of an observable, with eigenvalue $\alpha$ say, then its (renewed) measurement will (again) give $\alpha$ with probability 1 (*non-randomness* or *absence of fluctuations* in the sense of $\sigma_a^2 = 0$ resp. $\sigma_A^2 = 0$).
- For a QM system with $\dim\mathscr{H} \geq 2$ there exists no state, not even a pure one, that is a common eigenstate of *all* observables.[6] Simple example: If $\psi, \varphi \in \mathscr{H}$ is an

[6] Since for $\dim\mathscr{H} = \infty$ there exist also observables $A$ with continuous spectra, this statement reads more generally: There exists no *sequence* of states $(W_n)$ such that $\lim_{n\to\infty}\sigma_A^2(W_n) = 0$ for *all* $A$. Famous example: For *P and Q* no such sequence exists by the indeterminacy inequality, see Explanation 15).

orthonormal pair, then $W := |\psi\rangle\langle\psi|$ is an eigenstate of $A := |\varphi\rangle\langle\varphi|$, but not of $B := |\eta\rangle\langle\eta|$ defined by the "superposition" $\eta := (\varphi + \psi)/\sqrt{2}$, since $\sigma_B^2 \equiv \sigma_B^2(W) = 1/4$ [Non-existence of *universally fluctuation-free* or *dispersion-less* states].

- For an individual QM system (with $\dim\mathscr{H} \geq 2$) it is at most acceptable in eigenstates of $A$ to imagine that the system "possesses" an eigenvalue of $A$ as an intrinsic *property* pre-existing before a measurement. In all other states the "value of $A$" is *objectively indeterminate*, because the outcomes of repeated measurements on the same, always equally prepared, QM system objectively fluctuate with (a statistics obeying) $\sigma_A^2 > 0$.
- If the QM system is in a general state $W$, then the outcome value of an ideal measurement of a discrete[7] observable $A$ only *arises* in the moment of measurement, namely *by chance* with the probability $\langle W, E\rangle$. Here $E$ is the eigenprojection of $A$ corresponding to the outcome. According to an "update rule" of LÜDERS [Lü51], $W$ then changes to the state $W_E := EWE/\langle W, E\rangle$. Such a non-linear and non-invertible change is called a *state collapse*. It is usually undisputed that it occurs in experiments, at least approximately and, in fact, regardless of the knowledge of the persons involved. However, whether it can be explained in a purely quantum-theoretical way is still controversial today, despite the ideas of ZEH [Zeh70] and subsequent authors to understand it as a consequence of the *entanglement*[8] with the noisy environment (from the measuring apparatus etc.) and the resulting irreversible "loss of coherence" in the macroscopic limit (In short: decoherence). See also Section 5. Apart from this, the state $W_E$ is (for $\dim\mathscr{H} \geq 3$) the only (!) one that provides for the occurrence of any event $F$ with $EF = F$ ($\Leftrightarrow F\mathscr{H} \subseteq E\mathscr{H}$) in a subsequent measurement the (conditional) probability $\langle W_E, F\rangle$, in particular[9] $\langle W_E, E\rangle = 1$. This claim is shown in [BC81, Hug89, V07] to be a consequence of GLEASON's theorem[10].

14) *Covariance* of two observables in a state
quantifies the correlation of their (symmetrized) fluctuations

$$\boxed{\tau_{A,B} := \frac{1}{2}\Big\langle \big(A - \langle A\rangle\mathbb{1}\big)\big(B - \langle B\rangle\mathbb{1}\big) + \big(B - \langle B\rangle\mathbb{1}\big)\big(A - \langle A\rangle\mathbb{1}\big)\Big\rangle}$$
$$= \frac{1}{2}\langle AB + BA\rangle - \langle A\rangle\langle B\rangle\,.$$

It should be noted that the definition of the covariance neither needs nor implies the existence of a joint probability distribution of $A$ and $B$. Regardless of this, correlations play a major role in all of science and technology. But in quantum physics they even have a fundamental meaning. See, for example, Section 6.

*Facts:*

i) $\tau_{A,A} = \sigma_A^2\,, \quad \tau_{A,B} = \tau_{B,A}\,, \quad \tau_{\lambda A,B} = \lambda\,\tau_{A,B}\,, \quad \tau_{A+\lambda\mathbb{1},B} = \tau_{A,B}\,; \quad \lambda \in \mathbb{R}.$

ii) $\boxed{\sigma_A^2 + \sigma_B^2 = \sigma_{A+B}^2 - 2\tau_{A,B}}$

---

7 for a continuous observable see [Oz84].

8 see Section 3.

9 confer the above *absence of fluctuations*.

10 see Section 4.

iii) $\boxed{\sigma_A^2\sigma_B^2 \geq \tau_{A,B}^2 + \frac{\hbar^2}{4}\langle [A,B]\rangle^2}$ *Indeterminacy inequality* (IN-IN)

[similar in the CM and CPT, but without the commutator term]

***Proof*** i) obvious by definition. For ii) and iii) consider at first the case $\langle A\rangle = 0$ and $\langle B\rangle = 0$. The expectation value of $A^2 + B^2 = (A+B)^2 - AB - BA$ then gives ii). The CAUCHY–SCHWARZ inequality from Explanation 11) yields iii) according to

$$\langle A^2\rangle\langle B^2\rangle \geq |\langle AB\rangle|^2 = \big(\mathrm{Re}\langle AB\rangle\big)^2 + \big(\mathrm{Im}\langle AB\rangle\big)^2$$
$$= \Big\langle \frac{1}{2}(AB+BA)\Big\rangle^2 + \Big\langle \frac{1}{2\mathrm{i}}(AB-BA)\Big\rangle^2 = \tau_{A,B}^2 + \frac{\hbar^2}{4}\langle [A,B]\rangle^2 .$$

If $\langle A\rangle \neq 0$ and/or $\langle B\rangle \neq 0$, then replace $A$ with $A - \langle A\rangle \mathbb{1}$ and/or $B$ with $B - \langle B\rangle \mathbb{1}$, which gives the then correct variances and covariance, but does not change the commutator.

□

15) Explanation and comments on the indeterminacy inequality (IN-IN)
   - The IN-IN (for the product of variances) is often still listed under the imprecise or incorrect designations "uncertainty relation" or "uncertainty principle".
   - The IN-IN refers to the statistics of outcomes of measurements and is, in this general form, due to ROBERTSON [Rob29, Rob30, Rob34] and SCHRÖDINGER [Sc30, Sc99/08]. The IN-IN has nothing to do with possible perturbations of the QM system or with imperfections of single measurements. Instead, it tells the physicists that the HILBERT-space formalism of the QM does not allow for states violating it. This is in accord with EINSTEIN's "Erst die Theorie entscheidet darüber, was man beobachten kann."[11]
   - For $\langle [A,B]\rangle = 0$ the IN-IN reduces to a standard inequality in the CPT and CM.
   - The IN-IN is often weakened by omitting $\tau_{A,B}^2$, by which one (carelessly) also loses the classical inequality.
   - Its best known example is the "canonical" one that goes back to HEISENBERG [Hei27, WF19] as an approximate measurement-uncertainty equality (for special states) and to KENNARD [Ke27], barely four months later, as a precise inequality for a variance product:

$$\sigma_P^2\sigma_Q^2 \geq \tau_{P,Q}^2 + \frac{\hbar^2}{4} \geq \frac{\hbar^2}{4} \quad (\textit{independent} \text{ of the state } W).$$

   Therefore, there is *no* state with $\sigma_P^2\sigma_Q^2 < \hbar^2/4$, although the momentum variance $\sigma_P^2 > 0$ and the position variance $\sigma_Q^2 > 0$ can be arbitrarily small *individually* (for a state with a small enough width of its probability density $\rho_P$ resp. $\rho_Q$ on momentum resp. position space). For rigorous HEISENBERG-type error-disturbance relations see [BLW14] and [Oz03, Oz19]. The authors do not seem to agree on all aspects, but [Er-etal12] reports experimental spin measurements in support of the latter author.

[11] In a conversation with HEISENBERG in April 1926, quoted after the latter's autobiography [Hei69]. In English: "It is the theory which decides what we can observe", see p. 63 in [Hei71] and also [Hol00, Kl19]. One should note that EINSTEIN used the term "theory" in the hypothetico-deductive sense [Ei19]. Thereby he anticipated K. POPPER's main idea on the progress of natural science (In short: "trial and error").

- In the IN-IN for $P$ and $Q$ *equality* holds, for example, for the pure state $|\psi_0\rangle\langle\psi_0|$ defined by the complex-valued square-integrable function

$$\psi_0(q) := (2\pi\alpha^2)^{-1/4} e^{-q^2/(4\alpha^2)} e^{i\gamma q^2/\hbar} \qquad (q, \gamma \in \mathbb{R},\ \alpha^2 > 0)\,.$$

  This state has the covariance $\tau_{P,Q} = 2\gamma\alpha^2 = 2\gamma\sigma_Q^2$ and a positive WIGNER density, see Remark iv) in Section 2.5 .
- The analytical power of the (HEISENBERG–)KENNARD inequality should not be overestimated. Lower bounds on the ground-state energy of a standard HAMILTONian are in general better derived by SOBOLEV inequalities and related ones [Far78]. For example, while it is true that the stability of the hydrogen atom is due to quantum fluctuations, it does *not* follow from $\sigma_P^2\sigma_Q^2 \geq \hbar^2/4$, when generalized to three spatial dimensions.

16) *Correlation coefficient* (as "de-dimensionalized" covariance)

$$\boxed{\kappa_{A,B} := \frac{\tau_{A,B}}{\sigma_A\sigma_B}} \qquad (\text{if} \quad \sigma_A\sigma_B > 0)\,.$$

with the " standard deviations": $\sigma_A := \sqrt{\sigma_A^2} \geq 0$ and $\sigma_B := \sqrt{\sigma_B^2} \geq 0$.

*Facts:*

i) Symmetry: $\kappa_{A,B} = \kappa_{B,A}$

ii) Invariance under scaling and translation:
$A \mapsto \lambda A,\quad \lambda > 0;\quad A \mapsto A + \xi\mathbb{1},\quad \xi \in \mathbb{R}$; similar for $B$

iii) Indeterminacy inequality: $\kappa_{A,B}^2 \leq 1 - \frac{\hbar^2}{4}\langle[A/\sigma_A, B/\sigma_B]\rangle^2 \leq 1$

iv) States for the two extreme values:

| | | |
|---|---|---|
| $\kappa_{A,B} = \pm 1$ | $\iff$ | There exists $\lambda > 0$ such that $W$ is an eigenstate of $A \mp \lambda B$ |

Interpretation:

In the state $W$ the observables $A$ and $B$ are *perfectly* $\begin{cases}\text{correlated.}\\ \text{anticorrelated.}\end{cases}$

Moreover, in this state one obviously has by iii): $\langle[A,B]\rangle = 0$.

[similar in the CM and CPT , but without the commutator term]

***Proof*** i) and ii) are obvious by definition.
iii) is only a reformulation of the IN-IN from Explanation 14).
It remains to show the equivalence iv). This can be done "quite classically" as follows:
"$\Leftarrow$": Fact iii) for the variance, Facts i) and ii) for the covariance, and the inequality of the arithmetic and geometric means yield

$$0 = \sigma_{A\mp\lambda B}^2 = \sigma_A^2 + \lambda^2\sigma_B^2 \mp 2\lambda\,\tau_{A,B} \geq 2\sqrt{\sigma_A^2\lambda^2\sigma_B^2} \mp 2\lambda\,\tau_{A,B} = 2\lambda\,(\sigma_A\sigma_B \mp \tau_{A,B})\,,$$

hence $\sigma_A\sigma_B \leq \pm\tau_{A,B}$. On the other hand, iii) obviously implies $|\tau_{A,B}| \leq \sigma_A\sigma_B$. Thus $\tau_{A,B} = \pm\sigma_A\sigma_B$, hence $\kappa_{A,B} = \pm 1$.

"⇒": With $\lambda := \sigma_A/\sigma_B > 0$ one has

$$0 = (\sigma_A - \lambda\sigma_B)^2 = \sigma_A^2 + \lambda^2\sigma_B^2 - 2\lambda\sigma_A\sigma_B = \sigma_{A\mp\lambda B}^2 \,.$$

Here the last equality is a consequence of $\sigma_A\sigma_B = \pm\tau_{A,B}$ (due to the assumption $\kappa_{A,B} = \pm 1$) in combination with the Facts i) and ii) for the covariance. Fact iii) for the variance now shows that the underlying state $W$ is an eigenstate of $A \mp \lambda B$ (with eigenvalue $\langle A \mp \lambda B\rangle$). □

17) Time-Evolution [of a State]

- is purely deterministic (for a *closed* or isolated system as considered here) and is generated by the time-independent energy observable called HAMILTONian (= HAMILTON function $h$ resp. HAMILTON operator $H$) according to a canonical resp. unitary transformation. It enables the physicists, in principle, to predict a future *expectation value* for times $t > 0$, given the present one at $t = 0$, for example, of the position of a particle in *motion*.
- does not change the entropy, in particular pure resp. mixed states remain pure resp. mixed.
- is equivalent to the well-known differential equations of HAMILTON resp. SCHRÖDINGER (for pure states).
- can be "transferred" from the state to the observables according to the transformation

$$a_t := \mathrm{e}^{t[h,\cdot]}\, a \qquad \text{resp.} \qquad A_t := \mathrm{e}^{\mathrm{i}tH/\hbar}\, A\, \mathrm{e}^{-\mathrm{i}tH/\hbar}\,.$$

The time-dependence of the physically more important expectation values therefore takes two alternative forms

$$\langle a\rangle_t := \langle w_t, a\rangle = \langle w_0, a_t\rangle \qquad \text{resp.} \qquad \langle A\rangle_t := \langle W_t, A\rangle = \langle W_0, A_t\rangle\,,$$

known as the "transition between the SCHRÖDINGER and the HEISENBERG picture". However, only in the latter picture can one define and study multi-time correlation functions such as $\tau_{a_t,b_{t'}}$ resp. $\tau_{A_t,B_{t'}}$ for $t \neq t'$ .

- is for a *free* particle, classical or quantum, particularly simple in the HEISENBERG picture, because $H = P^2/(2m)$ generates the equations of motion

$$P_t = P \qquad \text{and} \qquad Q_t = Q + \frac{t}{m}P\,.$$

For example, they give for the time-dependence of the position variance and the momentum-position covariance, for an arbitrary initial state $W_0$ with $\langle P^2\rangle < \infty$ and $\langle Q^2\rangle < \infty$, the equations:

$$\sigma_{Q_t}^2 - \sigma_Q^2 = \frac{2t}{m}\tau_{P,Q} + \frac{t^2}{m^2}\sigma_P^2 \qquad \text{and} \qquad \tau_{P_t,Q_t} = \tau_{P,Q} + \frac{t}{m}\sigma_P^2\,.$$

They are contained already in [Sc30, Sc99/08], although SCHRÖDINGER did not write down the second one, but see also [Lév86]. In the case $\tau_{P,Q} < 0$, that is, for anticorrelated $P$ and $Q$ in the initial state, as in the example of Explanation 15) for $\gamma < 0$, their right-hand sides remain negative only for short enough (positive) times; more precisely, for all $t \in\, ]0, 2t_0[$ resp. $t \in [0, t_0[$ with $t_0 := m|\tau_{P,Q}|/\sigma_P^2 \leq m\sigma_Q/\sigma_P$. These time-dependencies also hold classically for any initial state $w_0$ with $\sigma_p\sigma_q \in ]0,\infty[$. They can illustrate the quantum time-dependencies by a swarm of dynamically independent classical particles.

*Time-independent* or *stationary* states (in the SCHRÖDINGER picture)
- are characterized by a vanishing LIE product with the HAMILTONian:

$$[w_0, h] = 0 \qquad \text{resp.} \qquad [W_0, H] = \mathbb{0}\,.$$

Example: BOLTZMANN–GIBBS state from Explanation 8).
- include in the QM all energy eigenstates, but not so in the CM.
- are mixed in CM cases of physical interest.

18) Invariance Group
Leaves expectation values *numerically invariant*, but LIE products and equations of motion (that is, differential equations for the time-evolution of states, observables, and expectation values) are in general only *form-invariant*. Since the two invariance groups are not isomorphic, there is no unique prescription to "quantize" a given classical canonical system. Related problems arise from the multiplicity of ordering the factors in operator products and from finding domains of (essential) self-adjointness for the HAMILTONian and other operators that capture the underlying physical reality. For some discussions see Ch.8.3 in [BEH08].

## 2.3 Entropic indeterminacy inequality for momentum and position

This subsection is a kind of addendum to Explanation 15) in Section 2.2. The variance of a probability density provides just one way to quantify its width. Another quantifier is its entropy. So it might not come as a surprise that the famous (HEISENBERG–)KENNARD indeterminacy inequality $\hbar/2 \leq \sigma_P \sigma_Q$ has an entropic analog. It was conjectured by HIRSCHMAN in 1957 and proved by BECKNER and BIAŁYNICKI-BIRULA–MYCIELSKI in 1975, see also [FoSi97].

*Assumption and notations:*
- Given a state vector corresponding to a pure state (position representation)

$$\psi \in \mathrm{L}^2(\mathbb{R})\,,\ \langle\psi|\psi\rangle = \int_{\mathbb{R}} \mathrm{d}q\,|\psi(q)|^2 = 1 \quad \left(|\psi(q)|^2 = \langle q|W|q\rangle\,, \quad W := |\psi\rangle\langle\psi|\right).$$

- Transformed state vector by FOURIER–PLANCHEREL (momentum representation)

$$\widehat{\psi}(p) := \int_{\mathbb{R}} \frac{\mathrm{d}q}{\sqrt{2\pi\hbar}}\,\psi(q)\,\mathrm{e}^{-\mathrm{i}pq/\hbar}\,, \qquad \int_{\mathbb{R}} \mathrm{d}p\,|\widehat{\psi}(p)|^2 = 1\,.$$

- *Pseudo-classical* state assigned to $W$

$$w(p,q) := 2\pi\hbar\,|\widehat{\psi}(p)|^2\,|\psi(q)|^2$$

$$w \geq 0\,, \quad \langle w, 1\rangle = \int_{\mathbb{R}\times\mathbb{R}} \frac{\mathrm{d}p\,\mathrm{d}q}{2\pi\hbar}\,w(p,q) = 1\,.$$

*Assertion* (*entropic* IN-IN for momentum and position):

The *classical* entropy of $w$ has the state-independent *lower* bound $\ln\frac{\mathrm{e}}{2}$, that is

$$\ln\frac{\mathrm{e}}{2} \leq -\langle w, \ln w\rangle\,.$$

*Interpretation:* Although the entropies of the (suitably de-dimensionalized) quantum probability densities on momentum and position space, $\rho_P$ resp. $\rho_Q$, may be arbitrarily small *individually* and even negative, their *sum* never falls below $\ln(\mathrm{e}/2) = 1 - \ln 2 > 0.3068$.

***Proof*** (following [Hir57, Bec75, BB–M75])
The function $f : [2,\infty[ \to \mathbb{R}$, defined by the difference

$$f(\mathsf{v}) := \left(\frac{\mathsf{u}}{2\pi\hbar}\right)^{1/(2\mathsf{u})} ||\psi||_{\mathsf{u}} - \left(\frac{\mathsf{v}}{2\pi\hbar}\right)^{1/(2\mathsf{v})} ||\widehat{\psi}||_{\mathsf{v}} \qquad (\mathsf{v} \geq 2, \quad \mathsf{u} := \mathsf{v}/(\mathsf{v}-1) \in ]1,2]\,)$$

in terms of the usual LEBESGUE norms $||\varphi||_r := (\int_{\mathbb{R}} \mathrm{d}x\,|\varphi(x)|^r)^{1/r}$ with $r \in [1,\infty[$, obeys $f(2) = 0$ by the well-known PARSEVAL–PLANCHEREL equality, as used already above for $\langle w, 1\rangle = 1$. Less well known is the sharp HAUSDORFF–YOUNG inequality $f(\mathsf{v}) \geq 0$ for all $\mathsf{v} \geq 2$, see [LL01]. The combination of both implies $f'(2+0) \geq 0$ for the (right-sided) derivate of $f$ at 2. This last inequality turns out to be equivalent to the assertion when using the formula

$$r^2 \frac{\mathrm{d}}{\mathrm{d}r} ||\varphi||_r = ||\varphi||_r^{1-r} \int_{\mathbb{R}} \mathrm{d}x |\varphi(x)|^r \ln\left(|\varphi(x)|^r / ||\varphi||_r^r\right)$$

with $r = 2$ for $\varphi = \psi$ as well as for $\varphi = \widehat{\psi}$ and noting again that $||\psi||_2 = ||\widehat{\psi}||_2 = 1$. □

*Remarks:*

- The assertion is extended from pure to mixed states $W\,(\geq W^2)$ by using the more general definition

$$\boxed{w(p,q) := 2\pi\hbar \langle p|W|p\rangle\langle q|W|q\rangle \geq 0, \quad \langle w, 1\rangle = 1}$$

  for the pseudo-classical state $w$ assigned to $W$. This claim simply follows from the spectral resolution[12] of $W$, the convexity of the function $x \mapsto x\ln x$ (for $x \geq 0$), the JENSEN inequality [Bau96], and the above result for pure states.

- The *entropic* IN-IN implies KENNARD's *variance* IN-IN, see Explanation 15) in Section 2.2, because the classical entropy of $w$ can simply be bounded from *above*:

$$\boxed{\ln\frac{\mathrm{e}}{2} \leq -\langle w, \ln w\rangle \leq \ln\left(\frac{\mathrm{e}}{\hbar}\sigma_P\sigma_Q\right).}$$

  ***Proof*** I first recall the GIBBS-type inequality[13] $-\langle w, \ln w\rangle \leq -\langle w, \ln\widetilde{w}\rangle$ for any probability density (or classical state) $\widetilde{w}$ on the phase space $\mathbb{R}\times\mathbb{R}$. The claimed upper bound follows from computing $-\langle w, \ln\widetilde{w}\rangle$ for the bivariate GAUSSian density

$$\widetilde{w}(p,q) = \frac{\hbar}{\sigma_P\sigma_Q}\exp\left\{-\frac{(p-\langle P\rangle)^2}{2\sigma_P^2} - \frac{(q-\langle Q\rangle)^2}{2\sigma_Q^2}\right\}$$

  characterized by a vanishing covariance and the means and variances of $P$ and $Q$ in the given state $W$. □

[12] see Explanation 8) in Section 2.2.

[13] Obviously, it is due to the elementary inequality $d(x,y) := x\ln x - x\ln y - x + y \geq 0$ for $x, y \geq 0$. The latter results from $d(0,y) = y \geq 0$ and $d(x,y) = x[(y/x) - 1 - \ln(y/x)] \geq 0$ for $x > 0$ because $z - 1 \geq \ln z$ for $z \geq 0$. Historical references for the GIBBS inequality and its quantum version can be found in [Fal70, Hub70].

*Conclusion:*

> If the *entropic* indeterminacy inequality (IN-IN) is strict, then the assigned pseudo-classical state provides a sharpening of the most famous *variance* IN-IN. Equality in this variance IN-IN implies that in the entropic IN-IN.

*Remark:* The entropic IN-IN is complemented by the inequality $-\langle W, \ln W\rangle \leq -\langle w, \ln w\rangle$, also valid for any state $W$. It is simpler to prove [FL12], but for *mixed* $W$ not less interesting. For example, weakening it by the last inequality from above establishes $\ln(\mathrm{e}\,\sigma_P\sigma_Q/\hbar)$ as a simple upper bound[14]on the entropy of the underlying (mixed) state $W$. For a lower bound on that entropy in terms of a phase-space integral see Remark iii) in Section 2.5 .

## 2.4 FEYNMAN–KAC formula and classical bounds on quantum partition functions

The non-commutativity of $P$ and $Q$ in a HAMILTONian such as $H = P^2/(2m) + \upsilon(Q)$ is also reflected by the thermal properties of the underlying physical system, which are determined by the BOLTZMANN–GIBBS state $W = \mathrm{e}^{-\beta H}/\mathrm{Tr}\,\mathrm{e}^{-\beta H}$, see Explanation 8) in Section 2.2. The derivation is complicated by the fact that the functional equation of the exponential for complex numbers does not extend to (functions of) non-commuting operators. However, there is a useful method of writing $\mathrm{e}^{-\beta H}$ as a (classically) probabilistic average of a product of two factors, one depending only on $P$ and the other only on $Q$. It was invented by R.P. FEYNMAN in his 1942 Princeton Ph.D. dissertation and now goes under the name "the path-integral approach to quantum physics" [Roe94, FH10]. After having heard about it, M. KAC found in 1948 a rigorous proof for the case of interest here.[15] This result became known, see pp.115–116 in [Ka85] and [Roe94], as the FEYNMAN–KAC formula (F–K formula), here concisely written as an operator identity:

$$\boxed{\mathrm{e}^{-\beta H} = \int \rho(\mathrm{d}w)\,\mathrm{e}^{-\mathrm{i}w(\beta)P/\hbar}\,\exp\Big(-\int_0^\beta \mathrm{d}\tau\,\upsilon\big(Q + w(\tau)\mathbb{1}\big)\Big).}$$

The first integration is over all continuous functions $w : [0,\infty[\to \mathbb{R}, \tau \mapsto w(\tau)$ on the positive half-line obeying $w(0) = 0$, with respect to the (dimensionless) GAUSSian probability measure $\rho$ that is uniquely defined by the

$$\textit{mean} \quad \int \rho(\mathrm{d}w)w(\tau) = 0 \quad \text{and } \textit{covariance} \quad \int \rho(\mathrm{d}w)w(\tau)w(\tau') = (\hbar^2/m)\min\{\tau, \tau'\}.$$

In other words, $\{w(\tau)\}_{\tau\geq 0}$ is the *stochastic process* introduced by N. WIENER [Mas90] in 1923 for modelling the [seemingly] random paths followed by a single particle in the phenomenon of free BROWNian motion [here only considered in one spatial dimension and with diffusion constant $\hbar^2/(2m)$]. The technical paper [BLM04] provides rather mild conditions on the potential $\upsilon$ for the validity of the F–K formula. Its analytical efficiency is well

[14] It may be viewed as a rigorous version of a similar one claimed in [Zac07]. A sharper rigorous bound, $\ln(\mathrm{e}/\gamma)$ with $\gamma := 2\sinh(\hbar/(2\sigma_P\sigma_Q))$, results directly from the [nowadays] well-known *quantum* GIBBS inequality, $-\langle W, \ln W\rangle \leq -\langle W, \ln \widetilde{W}\rangle$, by the choice $\widetilde{W} = \gamma\exp\{-(P - \langle P\rangle\mathbb{1})^2/2\sigma_P^2 - (Q - \langle Q\rangle\mathbb{1})^2/2\sigma_Q^2\}$. The inequality itself basically follows from the spectral resolutions of $W$ and $\widetilde{W}$, combined with the above elementary inequality which already yielded the classical GIBBS inequality. The quantum version goes back to [DM36], as I have learned from Albrecht HUBER (Universität Kiel, Germany); see his review [Hub70].

[15] FEYNMAN was then mainly interested in quantum dynamics, that is, in the informally similar, but mathematically less robust and less powerful case with the imaginary $\mathrm{i}t/\hbar$ replacing $\beta > 0$.

illustrated by deriving the following (pseudo-)classical bounds on the quantum canonical partition function, or the equivalent ones on the quantum *free energy* $-\beta^{-1}\ln(\mathrm{Tr}\,\mathrm{e}^{-\beta H})$.

**Theorem** (FEYNMAN–HIBBS 1965, SYMANZIK 1965)

The partition function of the HAMILTONian $H := \frac{1}{2m}P^2 + \upsilon(Q)$ satisfies the inequalities

$$z(\beta,\beta) \le \mathrm{Tr}\,\mathrm{e}^{-\beta H} \le z(\beta,0)$$

with the (pseudo-)classical partition function

$$z(\beta,\tau) := \int_{\mathbb{R}\times\mathbb{R}} \frac{\mathrm{d}p\mathrm{d}q}{2\pi\hbar}\mathrm{e}^{-\beta\left(p^2/(2m)+\upsilon_\tau(q)\right)} = \frac{1}{\lambda}\int_{\mathbb{R}} \mathrm{d}q\,\mathrm{e}^{-\beta\upsilon_\tau(q)}\,,$$

the *thermal* DE BROGLIE wavelength $\lambda := \sqrt{2\pi\beta\hbar^2/m}$, and a potential being a GAUSS transform of the given potential according to:

$$\upsilon_\tau(q) := \int_{\mathbb{R}} \frac{\mathrm{d}\xi}{\sqrt{2\pi}}\mathrm{e}^{-\xi^2/2}\,\upsilon\Big(q+\xi\sqrt{\tau\hbar^2/(12m)}\Big) = \exp\Big(\frac{\tau\hbar^2}{24m}\frac{\mathrm{d}^2}{\mathrm{d}q^2}\Big)\upsilon(q)\,, \quad \tau\in[0,\beta]\,.$$

***Proof*** The F–K formula in the position representation gives for the trace

$$\mathrm{Tr}\,\mathrm{e}^{-\beta H} = \int_{\mathbb{R}} \mathrm{d}q\,\langle q|\mathrm{e}^{-\beta H}|q\rangle = \int_{\mathbb{R}} \mathrm{d}q \int \rho(\mathrm{d}w)\,\delta\big(w(\beta)\big)\,\mathrm{e}^{-\int_0^\beta \mathrm{d}\tau\,\upsilon\left(q+w(\tau)\right)}.$$

Applying the JENSEN inequality to the (convex) exponential with respect to the uniform average $\beta^{-1}\int_0^\beta \mathrm{d}\tau(\cdot)$ and interchanging the $\tau$-integration with the two other ones results in

$$\mathrm{Tr}\,\mathrm{e}^{-\beta H} \le \beta^{-1}\int_0^\beta \mathrm{d}\tau \int_{\mathbb{R}} \mathrm{d}q \int \rho(\mathrm{d}w)\delta\big(w(\beta)\big)\mathrm{e}^{-\beta\upsilon\left(q+w(\tau)\right)} = z(\beta,0)\,.$$

The equality follows from interchanging the $q$-integration with the $w$-integration and noting the fact that $\lambda\int\rho(\mathrm{d}w)\,\delta\big(w(\beta)\big) = 1$ as well as $\int_0^\beta \mathrm{d}\tau = \beta$. This completes the proof of the upper bound. For the proof of the lower bound let me associate to each path $w$ its "time" average $\overline{w} := \beta^{-1}\int_0^\beta \mathrm{d}\tau\,w(\tau)$ and define the function $x\mapsto I(x)$ by inserting an additional DIRAC delta into the above trace formula in order to restrict the path integration to a fixed value of $\overline{w}$:

$$\begin{aligned} I(x) &:= \int_{\mathbb{R}} \mathrm{d}q \int \rho(\mathrm{d}w)\,\delta\big(w(\beta)\big)\delta\big(q+\overline{w}-x\big)\,\mathrm{e}^{-\int_0^\beta \mathrm{d}\tau\,\upsilon\left(q+w(\tau)\right)} \\ &= \int \rho(\mathrm{d}w)\,\delta\big(w(\beta)\big)\,\mathrm{e}^{-\int_0^\beta \mathrm{d}\tau\,\upsilon\left(x+w(\tau)-\overline{w}\right)}\,. \end{aligned}$$

Clearly, integrating over $x\in\mathbb{R}$ gives the desired trace. However, to simplify the restricted $w$-integration, let me again apply the JENSEN inequality to the exponential, but this time with respect to the average $\lambda\int\rho(\mathrm{d}w)\,\delta\big(w(\beta)\big)(\,\cdot\,)$ (over all paths "returning" to the origin at "time" $\tau=\beta$). Its normalization has already been recognized above. This gives

$$I(x) \ge \frac{1}{\lambda}\exp\Big(-\int_0^\beta \mathrm{d}\tau \int_{\mathbb{R}} \mathrm{d}y\,\upsilon(y)\,J(x-y)\Big)$$

with the $\tau$-independent GAUSSian probability density on the real line

$$J(z) := \lambda\int\rho(\mathrm{d}w)\,\delta\big(w(\beta)\big)\delta\big(z+w(\tau)-\overline{w}\big) = \frac{\sqrt{12}}{\lambda}\exp\big(-12\pi z^2/\lambda^2\big)\,, \quad z\in\mathbb{R}\,.$$

The last equality is due to the fact that $w(\beta)$ and $w(\tau)-\overline{w}$ are two correlated GAUSSian random variables with vanishing means and easily calculable variances and covariance. The claimed lower bound $z(\beta,\beta)$ now emerges from integrating over $x$ (and renaming $x$ to $q$).

□

*Remarks:*

- The above proof of the upper bound $z(\beta,0)$ is due to [Sy65] confirming an unpublished conjecture of FEYNMAN. Another proof follows from the more general GOLDEN-THOMPSON inequality $\mathrm{Tr}\,\mathrm{e}^{A+B} \leq \mathrm{Tr}(\mathrm{e}^{A}\mathrm{e}^{B})$ of 1965. For this and other trace inequalities see [Si05]. In any case, the tacit assumption $z(\beta,0)<\infty$ is seen to be a simple sufficient condition for $H$ to only have a discrete spectrum.
- The proof of the lower bound $z(\beta,\beta)$ is a streamlined version of the one given in Ch.11-2 of [FH10]. The GAUSS transform $\upsilon_\beta$ of the given potential $\upsilon=\upsilon_0$ in $z(\beta,\beta)$ is an *effective* potential in the sense that it takes into account some quantum effects in the classical arena. The less $\upsilon(q)$ changes with $q$ on the scale given by $\lambda$, the sharper the bound is. In any case, it yields the correct high-temperature asymptotics ($\beta\downarrow 0$).
- The classical partition function $z(\beta,\tau)$ with the transformed potential $\upsilon_\tau$ for general $\tau\geq 0$ is decreasing in $\tau$. This follows from the semigroup property of the GAUSS transformation and the JENSEN inequality, see [LeW89]. For a sufficiently smooth potential $\upsilon$ one even has strict monotonicity,

$$\frac{\partial}{\partial\tau}z(\beta,\tau) = -\frac{\beta\lambda}{48\pi}\int_{\mathbb{R}} \mathrm{d}q\,\mathrm{e}^{-\beta\upsilon_\tau(q)}\left(\frac{\partial}{\partial q}\upsilon_\tau(q)\right)^2 < 0,$$

  because $\upsilon$, and hence $\upsilon_\tau$, is not constant everywhere by $z(\beta,0)<\infty$. Combining this result with the above bounds and referring to the intermediate-value theorem yields $\mathrm{Tr}\,\mathrm{e}^{-\beta H} = z(\beta,\tau^*(\beta))$ with a unique $\tau^*(\beta)\in\,]0,\beta]$ associated to the given $\beta>0$. Hence, for any temperature and any potential, the quantum partition function is equal to the classical partition function with a well-defined *temperature-dependent* effective potential. However, $\tau^*(\beta)$ itself is not explicitly known for a general $\upsilon$.
- The lower bound $z(\beta,\beta)$ can be sharpened by applying the JENSEN inequality to fewer paths than in the above proof. For an impressive example along these lines see [GT85, FK86]. Reviews on such effective-potential methods are [LeW89, CGTVV95].

## 2.5 WEYL–WIGNER mapping

This mapping provides a *linear* and injective (or one-to-one) assignment between operators $A$ acting on the HILBERT space $\mathrm{L}^2(\mathbb{R})$ and complex-valued functions $a$ defined on the phase space $\mathbb{R}\times\mathbb{R}$. It thereby establishes a *quantitative correspondence* between the QM and CM in the same arena rather than a mere confrontation of respective structural elements [Wey27, Wig32, Moy49]. As a general reference I recommend [CFZ19] and for more mathematical aspects [dG17].

*Definition:* The WEYL–WIGNER *mapping* $A\mapsto a$ is defined by

$$\boxed{a(p,q) := 2\int_{\mathbb{R}} \mathrm{d}r\,\mathrm{e}^{2\mathrm{i}pr/\hbar}\langle q-r|\,A\,|q+r\rangle\,,} \qquad (p,q)\in\mathbb{R}\times\mathbb{R},$$

with $\langle q|A|q'\rangle$ denoting the (possibly distributional) position representation of $A$. The phase-space function $a$ is called the WEYL–WIGNER *symbol* or, more briefly, the *symbol* of the operator $A$. It may depend on $\hbar$.

*Facts:*

i) Linearity: $\alpha A+\beta B \mapsto \alpha a+\beta b$ for all $\alpha,\beta\in\mathbb{C}$, in particular $\mathbb{0}\mapsto 0$

ii) Compatibility with conjugation: $A\mapsto a \iff A^*\mapsto a^*$

iii) $f(Q)\mapsto f(q)$, $g(P)\mapsto g(p)$ for (BOREL) functions $f,g:\mathbb{R}\to\mathbb{R}$, in particular $\mathbb{1}\mapsto 1$

iv) Isometry: $\langle a,b\rangle = \langle A,B\rangle$

v) Product: $AB\mapsto a\star b := a\exp\left(-\frac{\mathrm{i}\hbar}{2}\mathscr{D}\right)b$ with[16] $\mathscr{D} := \frac{\overleftarrow{\partial}}{\partial p}\frac{\overrightarrow{\partial}}{\partial q} - \frac{\overleftarrow{\partial}}{\partial q}\frac{\overrightarrow{\partial}}{\partial p}$

vi) The (standardized) commutator is mapped to the MOYAL *bracket* :

$$[A,B]\mapsto \{a,b\} := \frac{\mathrm{i}}{\hbar}(a\star b - b\star a) = a\,\frac{2}{\hbar}\sin\left(\frac{\hbar}{2}\mathscr{D}\right)b$$

vii) WEYL *quantization* as the inverse mapping $a\mapsto A$:

$$A = \int_{\mathbb{R}\times\mathbb{R}} \frac{\mathrm{d}p\,\mathrm{d}q}{2\pi\hbar}\,\mathrm{e}^{\mathrm{i}(p\mathbb{1}-P)q/\hbar}\,a\left(p, Q+\frac{q}{2}\mathbb{1}\right) = a\left(\frac{\hbar}{\mathrm{i}}\frac{\partial}{\partial q}, \frac{\hbar}{\mathrm{i}}\frac{\partial}{\partial p}\right)\mathrm{e}^{\mathrm{i}(qP+pQ)/\hbar}\bigg|_{p=0,\,q=0}$$

$$\langle q|A|q'\rangle = \int_{\mathbb{R}} \frac{\mathrm{d}p}{2\pi\hbar}\,\mathrm{e}^{\mathrm{i}p(q-q')/\hbar}\,a\left(p, \frac{q+q'}{2}\right), \quad (q,q')\in\mathbb{R}\times\mathbb{R}.$$

*Remarks:*

i) $A\mapsto a$ Phase-space *representation* of the QM (WIGNER 1932, MOYAL 1949)

$a\mapsto A$ HILBERT-space *representation* of the CM or (canonical) *quantization* prescription of classical phase-space functions (WEYL 1927) with totally symmetrized factor ordering of $P$ and $Q$.

ii) Observables are mapped to observables.
Example:
For a twice continuously differentiable function $f:\mathbb{R}\to\mathbb{R}$ one has the assignment

$$p^2 f(q) \longleftrightarrow \frac{1}{4}\left(P^2 f(Q) + 2Pf(Q)P + f(Q)P^2\right) = Pf(Q)P - \frac{\hbar^2}{4}f''(Q).$$

iii) Positive operators are in general *not* mapped to positive functions and vice versa.

In particular, the WIGNER *density* $w$, that is, the symbol of a QM state $W$ may take also *negative* values. It is therefore, in general, not a probability density on the phase space. Rather, it is similar to a bounded electric-charge density with positive total charge. This is because $|w(p,q)|\le 2$ for any $W$ and, by the isometry, also $0\le\langle w,w\rangle = \langle W,W\rangle \le \langle W,\mathbb{1}\rangle = \langle w,1\rangle = 1$ and $\langle w,a\rangle = \langle W,A\rangle$ for expectation values. This implies that the lower bound on the entropy of a (mixed) state $W$, mentioned in Explanation 9) in Section 2.2, can be written as $-\ln\langle w,w\rangle$, see also the final Remark in Section 2.3. Moreover, the

[16] The phase-space function $a\star b$ is known as the *star product* of $a$ and $b$. The arrows in $\mathscr{D}$ indicate, whether (in the informal TAYLOR series of the exponential) the differentiation refers to $a$ or $b$. Example: $a\mathscr{D}b = [a,b]$.

*marginal* densities of any WIGNER density $w$ for momentum and position are the correct quantum probability densities, used in the entropic IN-IN of Section 2.3:

$$\int_{\mathbb{R}} \frac{\mathrm{d}q}{2\pi\hbar}\, w(p,q) = \langle p|W|p\rangle = \rho_P(p) \quad \text{and} \quad \int_{\mathbb{R}} \frac{\mathrm{d}p}{2\pi\hbar}\, w(p,q) = \langle q|W|q\rangle = \rho_Q(q)\,.$$

iv) Back to $w$ itself, HUDSON has shown [Hud74] that a pure state $W = |\psi\rangle\langle\psi|$ has a positive WIGNER density if, and only if, $\psi \in \mathrm{L}^2(\mathbb{R})$ is the exponential of a quadratic polynomial. For an example see Explanation 15) on the (variance) IN-IN in Section 2.2. CARTWRIGHT has shown [Car76] that any WIGNER density becomes positive everywhere upon convolution with a GAUSSian probability density of product form on the phase space, confer the special $\widetilde{w}$ in Section 2.3, provided its momentum variance times its position variance is not smaller than $\hbar^2/4$. Her original proof is for pure states, but it easily extends to mixed states $W$ by their spectral resolutions. Moreover, if the variance product is equal to $\hbar^2/4$, then its factors can be chosen in such a way that the "smoothed" WIGNER density becomes the HUSIMI density of $W$. Its values may be viewed as the expectation of $W$ with respect to (pure) canonical coherent states [Bal98, BZ06, CFZ19].

v) MOYAL bracket
   - is a LIE–Product (without LEIBNIZ rule).
   - satisfies for "small" $\hbar$ informally: $\{a,b\} = [a,b] - \frac{\hbar^2}{24} a\,\mathscr{D}^3\, b + \mathscr{O}(\hbar^4)$ and hence $\{a,b\} = [a,b]$, if $a$ or $b$ is at most quadratic in $p$ and $q$.

vi) Instantaneous time-evolution (in the SCHRÖDINGER picture)

$$\frac{\mathrm{d}}{\mathrm{d}t} W_t = [W_t, H] \longleftrightarrow \frac{\mathrm{d}}{\mathrm{d}t} w_t = \{w_t, h\}$$

vii) Classical or quasi-classical limit (informally)

$$A \mapsto a \overset{\hbar\downarrow 0}{\longrightarrow} a_{\text{cl.}}, \quad \{\ ,\ \} \overset{\hbar\downarrow 0}{\longrightarrow} [\ ,\ ]$$

viii) Since about 1960, WEYL–WIGNER symbols and some of their "relatives" have found their way into the theory of *pseudo-differential operators* and have thereby gained great importance for the rigoros *asymptotics of quasi-classical approximations* ($\hbar \downarrow 0$) (by R. KOHN, NIRENBERG, HÖRMANDER, SHUBIN, EGOROV, H. WIDOM, and others). For the "relatives" see the concise overview [FiLe93] in terms of bi-orthogonal systems in the quantum LIOUVILLE space and references therein.

ix) In Section 2.3 we have seen that the entropic indeterminacy inequality for momentum and position and, hence, also its weaker variance version, can be derived by FOURIER analysis [FoSi97] without referring to HILBERT-space operators. Since this analysis also plays a major role in classical *optics* and *communications engineering*, there are analogs of the quantum indeterminacy inequalities in these fields for optical or other signals as functions of frequency or time (with frequency $\widehat{=} p$, time $\widehat{=} q$, and $2\pi\hbar \equiv 1$). Similarly, in these fields one also makes use of analogs (and generalizations) of WIGNER densities.

## 3 Composite Systems, Reduced States, and Entanglement in Quantum Mechanics

Up to now, I have considered for simplicity only the system of a single particle moving along the EUCLIDean line $\mathbb{R}$. The generalization to the $d$-dimensional EUCLIDean space $\mathbb{R}^d$ with $d \in \{2,3,\dots\}$ is more or less straightforward and mainly a matter of notation. The position and the momentum both have $d$ components. The corresponding single-particle system has therefore the set-theoretic product $\Gamma_1 := \mathbb{R}^d \times \mathbb{R}^d \cong \mathbb{R}^{2d}$ as its phase space for the classical description and $\mathscr{H}_1 := \mathrm{L}^2(\mathbb{R}^d)$ as its HILBERT space for the quantum description (in the SCHRÖDINGER realization). We may also consider a second particle in $\mathbb{R}^d$ with the same mass or not and being dynamically coupled to the first one or not. In any case, $\Gamma := \mathbb{R}^{2d} \times \mathbb{R}^{2d} = \Gamma_1 \times \Gamma_2$ is the total (or common) phase space and $\mathscr{H} := \mathrm{L}^2(\mathbb{R}^d \times \mathbb{R}^d) \cong \mathrm{L}^2(\mathbb{R}^d) \otimes \mathrm{L}^2(\mathbb{R}^d) = \mathscr{H}_1 \otimes \mathscr{H}_2$ is the total HILBERT space of the (total) system composed of the two particles, which is a special case of a general bipartite system. We see that the role of the set-theoretic multiplication in the classical case, is taken over by the tensorial multiplication in the quantum case.

The extension from two to more particles should be obvious. Also the structural elements and notions compiled in the table of Section 2.1 have natural extensions for multi-particle systems in multi-dimensional space. However, two remarks are appropriate. First, the canonical relations hold as they stand (in the table), for each given particle with the *same* component of its momentum and position. All other LIE products of two such components vanish. Second, if the particles are *indistinguishable*, one should in the classical case divide all phase-space integrals by the factorial of the total number of particles. In the quantum case one should restrict the multi-particle HILBERT space either to its totally symmetric or anti-symmetric subspace, according to the BOSE–EINSTEIN resp. FERMI–DIRAC statistics. The latter is the algebraic formulation of the PAULI exclusion principle [Pa25].

In the following I only consider the interesting enough quantum case of a general *bipartite* system with two distinguishable subsystems ① and ②, possibly *dynamically coupled*, that is, interacting with each other. A prominent example is the hydrogen atom composed of a proton and an electron. Another example is the electron (or proton) itself, but viewed as composed of a particle in $\mathbb{R}^3$ and an internal rotational degree of freedom, its *spin*.[17]

Of basic interest are:
- The arena, observables, and states of the composite or *total system* "①+②" composed of the subsystems.
- The states of the subsystems, given the state of the total system.

| | | |
|---|---|---|
| Arena of the total system: | $\mathscr{H} = \mathscr{H}_1 \otimes \mathscr{H}_2$ | (tensor product of HILBERT spaces) |
| General observable: | $A$ | (self-adjoint operator on $\mathscr{H}$) |
| ①-observable as an observable of the total system: | $A_1 \otimes \mathbb{1}_2$ | (tensor product of operators) |
| ②-observable as an observable of the total system: | $\mathbb{1}_1 \otimes A_2$ | (tensor product of operators) |

General observables $A$ of the total system are real (possibly infinite) linear combinations of *product observables* $A_1 \otimes A_2 = (A_1 \otimes \mathbb{1}_2)(\mathbb{1}_1 \otimes A_2) = (\mathbb{1}_1 \otimes A_2)(A_1 \otimes \mathbb{1}_2)$.

### 3.1 Partial traces of an operator acting on a product HILBERT space

*Useful fact:* If $(\varphi_n)_n$ resp. $(\psi_m)_m$ is an *orthonormal basis* (ONB) in $\mathscr{H}_1$ resp. $\mathscr{H}_2$, then $(\varphi_n \otimes \psi_m)_{n,m}$ is an ONB in $\mathscr{H} = \mathscr{H}_1 \otimes \mathscr{H}_2$. It is known as a *product*-ONB and to be conve-

[17] for this spin see, for example, Sections 4 and 5.1.

nient for defining and calculating.

In particular, it can be used to write the *trace* of an operator $A$ on $\mathscr{H}$ as

$$\operatorname{Tr} A = \sum_{n,m} \left\langle \varphi_n \otimes \psi_m \middle| A(\varphi_n \otimes \psi_m) \right\rangle .$$

*Partial traces* of $A$ with respect to $\mathscr{H}_2$ or $\mathscr{H}_1$

- are operators on $\mathscr{H}_1$ resp. $\mathscr{H}_2$
- are denoted by $\operatorname{Tr}_2 A$ resp. $\operatorname{Tr}_1 A$
- can be defined through their (possibly infinite) matrix representations with respect to $(\varphi_n)_n$ resp. to $(\psi_m)_m$ as follows

$$\boxed{\begin{aligned} \left\langle \varphi_n \middle| \operatorname{Tr}_2 A\, \varphi_{n'} \right\rangle &:= \sum_m \left\langle \varphi_n \otimes \psi_m \middle| A(\varphi_{n'} \otimes \psi_m) \right\rangle \\ \left\langle \psi_m \middle| \operatorname{Tr}_1 A\, \psi_{m'} \right\rangle &:= \sum_n \left\langle \varphi_n \otimes \psi_m \middle| A(\varphi_n \otimes \psi_{m'}) \right\rangle \end{aligned}}$$

*Facts:*

- Partial "tracing" is the analog of the "classical" integration with respect to a part of the phase-space variables.
- The definitions of $\operatorname{Tr}_2 A$ and $\operatorname{Tr}_1 A$ are independent of the employed bases.
- The partial trace of $A$ is self-adjoint (and positive) if $A$ is self-adjoint (and positive).
- $\operatorname{Tr}_2 \left[A(B_1 \otimes \mathbb{1}_2)\right] = (\operatorname{Tr}_2 A)B_1, \quad \operatorname{Tr}_1 \left[A(\mathbb{1}_1 \otimes B_2)\right] = (\operatorname{Tr}_1 A)B_2.$
- $\operatorname{Tr}\left(\operatorname{Tr}_2 A\right) = \operatorname{Tr}\left(\operatorname{Tr}_1 A\right) = \operatorname{Tr} A$ (in analogy to the FUBINI–TONELLI theorem).

*Reduced states* (or marginal states)

of a state $W$ of the (bi-partite) total system are the two states

$$\boxed{W_1 := \operatorname{Tr}_2 W, \quad W_2 := \operatorname{Tr}_1 W}$$

on $\mathscr{H}_1$ resp. $\mathscr{H}_2$ of subsystem ① resp. ②.

Although they are operators, they are still often called "reduced density matrices". They correspond to the marginal densities in the CM and CPT.

*Non-entangled and entangled states* (according to WERNER [Wer89]):

- A state $W$ on $\mathscr{H} = \mathscr{H}_1 \otimes \mathscr{H}_2$ is called *non-entangled* (*separable* or *classically correlated*) if it is for some $N \in \{1, 2, 3, \dots\}$ a convex combination of *product states*[18]:

$$\boxed{W = \sum_{n=1}^{N} p_n\, W_1^{(n)} \otimes W_2^{(n)} .}$$

  Here $W_1^{(n)}$ and $W_2^{(n)}$ are states on $\mathscr{H}_1$ resp. $\mathscr{H}_2$ and the numbers $p_n \geq 0$ sum up to 1.
- A state $W$ on $\mathscr{H}_1 \otimes \mathscr{H}_2$ is called *entangled* if it is not non-entangled.

[18] If $\dim \mathscr{H} = \infty$, then $W$ must be at least the (trace-norm) limit of the right-hand side as $N \to \infty$.

*Remarks:*

- A non-entangled state with $N = 1$ is the product of its two reduced states

$$W = W_1 \otimes W_2 .$$

  It corresponds to stochastic independence in the CM and CPT and implies that the factors of all product observables $A_1 \otimes A_2$ are uncorrelated:

$$\langle A_1 \otimes A_2 \rangle = \mathrm{Tr}(W_1 \otimes W_2)(A_1 \otimes A_2) = \mathrm{Tr}(W_1 A_1 \otimes W_2 A_2) = \langle A_1 \rangle \langle A_2 \rangle .$$

- A pure state is non-entangled if, and only if, it is, as above, the product of its two reduced states (and hence determined by a product vector).

3.2 Two particularities of the reduction of quantum states

in comparison to the reduction of classical states:

1) The reduction of a pure QM state is in general mixed, because many pure states are not a product and thus *entangled pure states.*
2) Each QM state may be viewed as the reduction of a pure state on a "larger" HILBERT space, which goes under the name *state purification.*

***Proof*** 1) It is enough to consider a simple QM system with product HILBERT space $\mathscr{H}_1 \otimes \mathscr{H}_2$ from two-dimensional factors $\mathscr{H}_1 = \mathscr{H}_2 \cong \mathbb{C}^2$. This product is, for example, the arena of a system composed of *two* spins with the same (intrinsic) "main quantum number" 1/2, but without the "translational degree of freedom" considered in Section 2. For details of a single spin see Sections 4 and 5.1. For a given ONB $\{\varphi_1, \varphi_2\}$ in $\mathscr{H}_1$ and an ONB $\{\psi_1, \psi_2\}$ in $\mathscr{H}_2$, the two normalized linear combinations ("superpositions")

$$\Phi^{\pm} := \frac{1}{\sqrt{2}}(\varphi_1 \otimes \psi_1 \pm \varphi_2 \otimes \psi_2) \qquad (\Phi^{\pm} \in \mathscr{H}_1 \otimes \mathscr{H}_2)$$

define on $\mathscr{H}_1 \otimes \mathscr{H}_2$ the two different states (confer the singlet state in Section 6)

$$W^{\pm} := |\Phi^{\pm}\rangle\langle\Phi^{\pm}| \qquad (W^+ W^- = \mathbb{0}) .$$

Both of them are pure and have analogous *completely* (or "chaotically") mixed reduced states in the sense that

$$W_1^{\pm} = \frac{1}{2}\mathbb{1}_1 \quad \text{and} \quad W_2^{\pm} = \frac{1}{2}\mathbb{1}_2 .$$

Since $\frac{1}{2}\mathbb{1}_1 \otimes \frac{1}{2}\mathbb{1}_2 \neq W^{\pm}$, $W^+$ as well as $W^-$ is entangled.

2) If $W = \sum_n \rho_n |\psi_n\rangle\langle\psi_n|$ is the spectral resolution of the given state on some HILBERT space $\mathscr{H}$, then $\Phi := \sum_n \sqrt{\rho_n}\,\psi_n \otimes \psi_n$ is a unit vector in $\mathscr{H} \otimes \mathscr{H}$. The pure state $|\Phi\rangle\langle\Phi|$ on $\mathscr{H} \otimes \mathscr{H}$ has the partial trace(s) $\mathrm{Tr}_{\mathscr{H}} |\Phi\rangle\langle\Phi| = W$. □

I conjecture that both particularities are closely related to the quantum *contextuality* explained in Section 5.

3.3 Entropies, expectation values, and time-evolutions of subsystems

1) Entropies

a) *total* entropy and *reduced* entropies in the QM case,

$$s := -\operatorname{Tr} W \ln W\,, \quad s_1 := -\operatorname{Tr} W_1 \ln W_1\,, \quad s_2 := -\operatorname{Tr} W_2 \ln W_2\,,$$

satisfy a triangle inequality in the sense of EUCLIDean geometry

$$\boxed{|s_1 - s_2| \leq s \leq s_1 + s_2\,.}$$

*Visualization:*

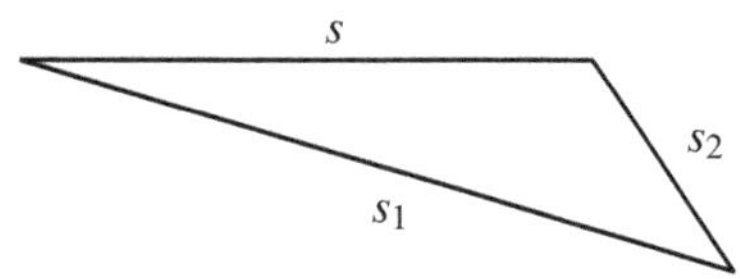


- plane of the triangle = drawing level
- $s$, $s_1$, and $s_2$ correspond to the three side lengths of the (here: obtused-angled) triangle.

*Remarks:*

i) The upper bound (known as *subadditivity* and following from the quantum GIBBS inequality with $\widetilde{W} = W_1 \otimes W_2$, see the final Remark in Section 2.3 .)
   - holds also in the CM and CPT,
   - is interpretated as: "the information in $W$ is not smaller than the sum of those in $W_1$ and $W_2$",
   - turns into an equality if, and only if, $W = W_1 \otimes W_2$.

ii) The lower bound (ARAKI–LIEB [AL70])
   - has no classical analog,
   - implies $s_1 = s_2$ if $W$ is pure (see the geometric visualization for $s = 0$),
   - yields, when combined with subadditivity, the implication "$W_1$ or $W_2$ pure $\Longrightarrow$ $W = W_1 \otimes W_2$".

b) Entropy of *entanglement* (or: relative entropy, mutual information)

$$\boxed{\delta s := s_1 + s_2 - s}$$

is a simple quantification of the entanglement between the two subsystems encoded in $W$. For mixed states $W$ this definition is in general too simple, in particular, because it may be "spoiled" by classical correlations.

*Facts:*

i) $\delta s \geq 0$
ii) $\delta s = 0 \quad \Longleftrightarrow \quad W = W_1 \otimes W_2$
iii) $W^2 = W \quad \Longrightarrow \quad \delta s = 2s_1 = 2s_2$

c) The two pure states $W^{\pm}$ of the previous subsection provide extreme examples with regard to entropy due to the following four facts:

$$s = -\operatorname{Tr} W^{\pm} \ln W^{\pm} = -1 \ln 1 + 3(-0 \ln 0) = 0$$

(minimal, *since* the information about the total system is maximum),

$$s_1 = -\operatorname{Tr} W_1^{\pm} \ln W_1^{\pm} = -\frac{1}{2} \ln \frac{1}{2} - \frac{1}{2} \ln \frac{1}{2} = \ln 2$$
$$s_2 = -\operatorname{Tr} W_2^{\pm} \ln W_2^{\pm} = \ln 2$$

(maximum, *although* the information about the total system is maximum!),

$$\delta s = s_1 + s_2 - s = 2s_1 = 2s_2 = 2 \ln 2$$

(maximum).

2) Expectation values of subsystem observables

$$\boxed{\begin{aligned} \langle A_1 \otimes \mathbb{1}_2 \rangle &= \operatorname{Tr} W (A_1 \otimes \mathbb{1}_2) = \operatorname{Tr}(W_1 A_1) \\ \langle \mathbb{1}_1 \otimes A_2 \rangle &= \operatorname{Tr} W (\mathbb{1}_1 \otimes A_2) = \operatorname{Tr}(W_2 A_2) \end{aligned}}$$

The information encoded in $W_j$ suffices to answer all questions regarding system ⓙ for $j \in \{1,2\}$. [analogous in the CM and CPT]

3) Time-evolutions of subsytems
   - They are simple in the case of *dynamically independent* subsystems, that is, if the HAMILTONian of the total system "①+②" has the form

$$H = H_1 \otimes \mathbb{1}_2 + \mathbb{1}_1 \otimes H_2$$

   without a coupling term. Then the subsystems evolve autonomously. More precisely, the time-evolutions of the two reduced states of an arbitrary initial state of the total system are independent of each other and unitarily generated by $H_1$ resp. $H_2$. As a consequence, all three entropies $s$, $s_1$, and $s_2$ do not change with time [analogously in the CM].
   - They are more interesting (and complicated) in the general case of dynamically dependent (or coupled) subsystems, provided they cannot be decoupled by a suitable (canonical resp. unitary) transformation. Then the evolutions of the subsystems, inherited from the unitary evolution of the total system, are no longer unitary, intertwined with each other, and allow for time-dependent $s_1$ and $s_2$. This is nicely illustrated by the explicit example in Ch.7.5 of [BC81] for a QM system with two spins, as mentioned in Section 3.2 and considered in Sections 5.2 and 6. In this example $s_1$ oscillates between its extreme values 0 and $\ln 2$, corresponding to a pure resp. completely mixed (reduced) state of subsystem ①.

## 4 Lattice of Projections and the Theorem of GLEASON

- The set $\mathscr{P}(\mathscr{H})$ of *all* projections $E, F, G, \ldots$ on a separable HILBERT space $\mathscr{H}$ forms [Ja77, Hug89, V07] an *orthomodular* lattice or *partial* BOOLEan algebra with respect to the three operations:

| | | | | |
|---|---|---|---|---|
| i) | $E^{\perp} := \mathbb{1} - E$ | projects on | $(E\mathscr{H})^{\perp}$ | (orthog. complement) |
| ii) | $E \wedge F := \lim_{n\to\infty} E(FE)^n$ | projects on | $E\mathscr{H} \cap F\mathscr{H}$ | (intersection)[19] |
| iii) | $E \vee F := \left(E^{\perp} \wedge F^{\perp}\right)^{\perp}$ | projects on | $\overline{E\mathscr{H} + F\mathscr{H}}$ | (closed linear sum). |

If $\dim \mathscr{H} \geq 3$, then $\mathscr{P}(\mathscr{H})$ is not a BOOLEan algebra.

Geometric visualization of ii):

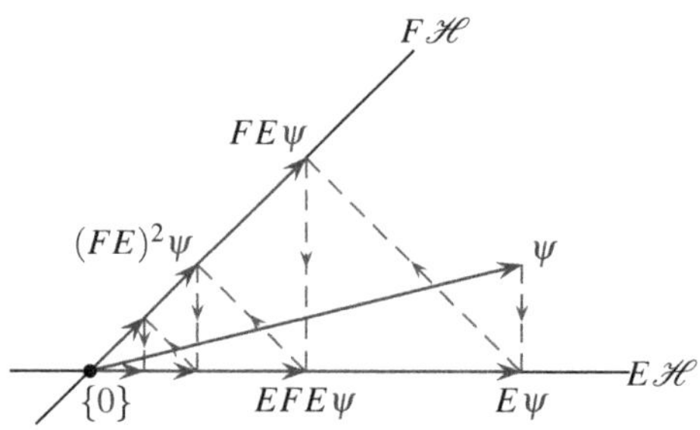


Alternating action of two projections $E$ and $F$ (on $\mathscr{H} = \mathbb{R}^2 =$ drawing level) with $EF \neq FE$ and $E \wedge F = \mathbb{0}$.
[For $\dim \mathscr{H} = 2$ one has: "$EF = FE \Longleftrightarrow EF = \mathbb{0}$".]

- Simplifications for *compatible* events, that is, for pairwise *commuting* projections ($EF = FE$ etc.):

$$E \wedge F = EF, \quad E \vee F = E + F - EF\,.$$

These imply the BOOLEan distributive laws:

$$E \wedge (F \vee G) = (E \wedge F) \vee (E \wedge G)$$
$$E \vee (F \wedge G) = (E \vee F) \wedge (E \vee G)\,.$$

- Each subset of pairwise commuting projections in $\mathscr{P}(\mathscr{H})$ with $\mathbb{1}$ and $\mathbb{0}$ is isomorphic to the BOOLEan algebra of subsets of some basic set with the set-theoretic operations complement, intersection, and union.

**Theorem** If $W$ is a state on $\mathscr{H}$, then the mapping $\mu : \mathscr{P}(\mathscr{H}) \to \mathbb{R}$, defined by $E \mapsto \mu(E) := \mathrm{Tr}(WE)$, is a *probability measure* on $\mathscr{P}(\mathscr{H})$ in the sense that it has the following three properties:

a) $0 \leq \mu(E) \leq 1$

b) $\mu(\mathbb{0}) = 0,\ \mu(\mathbb{1}) = 1$

c) $\mu\left(\bigvee_n E_n\right) = \sum_n \mu(E_n)$
for any sequence $(E_n)$ of pairwise *disjoint* and pairwise *commuting* projections.

[19] If $\dim \mathscr{H} = \infty$, the convergence is understood in the strong operator-topology, see p.71 in [BEH08].

*Explanation:* If $EF = FE$, then the following equivalences hold (see above):

$$\text{“}E \text{ and } F \text{ are } \textit{disjoint} :\Longleftrightarrow E \wedge F = \mathbb{0} \Longleftrightarrow EF = \mathbb{0} \Longleftrightarrow E \vee F = E + F\text{”.}$$

Property c) in the theorem can therefore be recognized as $\mu$'s *countable additivity*: $\mu(\sum_n E_n) = \sum_n \mu(E_n)$ for any sequence $(E_n)$ of pairwise *orthogonal* projections, that is, $E_n E_m = \delta_{nm} E_n$. The proof of the theorem is hence obvious if $\dim \mathscr{H} < \infty$. Even the proof of countable additivity in the case $\dim \mathscr{H} = \infty$ is not hard, see Theorem A11 in [BC81]. In contrast, the following *reverse* statement of the theorem is profound and its proof therefore arduous [Gl57, V07]. A simplified proof is given in [CKM85], see also [Hug89].

**Theorem** (GLEASON 1957)

If $\dim \mathscr{H} \geq 3$, then there exists for each probability measure $\mu$ on $\mathscr{P}(\mathscr{H})$ a unique state $W$ on $\mathscr{H}$ such that $\mu(E) = \mathrm{Tr}(WE)$ for all $E \in \mathscr{P}(\mathscr{H})$.

*Remarks:*

- Each ONB in $\mathscr{H}$ provides an example of a sequence of pairwise orthogonal (one-dimensional) projections. The depth (and surprise?) of the result is related to the fact that additivity is only required for commuting projections.
- Given the notion of a *quantum* event, the result is fundamental because it justifies the notion of a *quantum* state (for systems with finitely many degrees of freedom).
- The *quantum* (or *non-commutative*) probability theory (QPT) with the lattice/algebra of projections (or associated subspaces) is more general than the classical one (CPT), but reduces to CPT for any (BOOLEan) *sublattice* of $\mathscr{P}(\mathscr{H})$ containing only pairwise *commuting* projections including $\mathbb{0}$ and $\mathbb{1}$, see FARIS in [W95] and [Ja77, Heg85, Stre00, Ac10].

  *Analogy:* QPT extends CPT in a similar way to how non-EUCLIDean geometry extends the EUCLIDean one [Ja77, Stre07, Ac10]. The outcomes of experiments decide which theory is better suited to model aspects of Nature.

For $\dim \mathscr{H} = 2$ there exist probability measures on $\mathscr{P}(\mathscr{H})$ that differ from those offered by the GLEASON theorem for $\dim \mathscr{H} \geq 3$. This fact can be seen as follows.

a) Explicit description of all elements of $\mathscr{P}(\mathbb{C}^2)$:

   A QM system with a two-dimensional HILBERT space $\mathscr{H} \cong \mathbb{C}^2$ is the simplest non-trivial one. It is the arena for a single spin with (intrinsic) "main quantum number" $1/2$. The corresponding vector operator $\vec{S} := (S^x, S^y, S^z)$ is represented in the eigenbase of $S^z$, apart from the factor $\hbar/2$, by the triple of $2 \times 2$ (DIRAC–)PAULI matrices according to:

$$S^x \mathrel{\widehat{=}} \begin{pmatrix} 0 & 1 \\ 1 & 0 \end{pmatrix}, \quad S^y \mathrel{\widehat{=}} \begin{pmatrix} 0 & -\mathrm{i} \\ \mathrm{i} & 0 \end{pmatrix}, \quad S^z \mathrel{\widehat{=}} \begin{pmatrix} 1 & 0 \\ 0 & -1 \end{pmatrix}.$$

   They are self-adjoint, their squares are equal to the $2 \times 2$ unit matrix representing the unit operator $\mathbb{1}$, and they imply the identities $S^x S^y = \mathrm{i} S^z$, $S^y S^z = \mathrm{i} S^x$, and $S^z S^x = \mathrm{i} S^y$.

The lattice/algebra $\mathscr{P}(\mathbb{C}^2)$ consists of the following projections:

| | |
|---|---|
| one 0-dimensional | $\mathbb{0}$ |
| one 2-dimensional | $\mathbb{1}$ |
| a two-parameter family of 1-dimensional | $E := \frac{1}{2}(\mathbb{1} + \vec{e} \cdot \vec{S})$, $\vec{e} \in \mathbb{R}^3$ with $\lvert\vec{e}\rvert = 1$ |

Implication: $(\vec{e} \cdot \vec{S})E = E$

Interpretation: $E$ is a pure eigenstate of the spin component $\vec{e} \cdot \vec{S} := e^x S^x + e^y S^y + e^z S^z$ along the $\vec{e}$ direction with eigenvalue 1.

b) Explicit description of all probability measures on $\mathscr{P}(\mathbb{C}^2)$:

$$\mu(0) := 0,\ \mu(\mathbb{1}) := 1,\ \mu(E) := \frac{1}{2}(1 + m(\vec{e}))$$

Here $m$ is an arbitrary mapping of the 2-sphere $\mathbb{S}^2$ to the interval $[-1, 1]$ with $m(-\vec{e}) = -m(\vec{e})$ (to guarantee $\mu(E) + \mu(E^\perp) = 1$).
Simplest example: $m(\vec{e}) = 0$ for all unit vectors $\vec{e}$.

A comparison with the three formulas $\mathrm{Tr}(W\mathbb{0}) = 0$, $\mathrm{Tr}(W\mathbb{1}) = 1$, and $\mathrm{Tr}(WE) = 1 + \frac{1}{2}\vec{e} \cdot \mathrm{Tr}(W\vec{S})$ now yields the

**Equivalence**:

| | | |
|---|---|---|
| $\mu$ is induced by a state $W$ on $\mathbb{C}^2$ | $\Longleftrightarrow$ | There is $\vec{p} \in \mathbb{R}^3$ with $\lvert\vec{p}\,\rvert \leq 1$ such that $m(\vec{e}) = \vec{p} \cdot \vec{e}$ for all $\vec{e} \in \mathbb{R}^3$ with $\lvert\vec{e}\rvert = 1$ |

***Proof*** Put $\vec{p} = \mathrm{Tr}(W\vec{S})$ and conversely $W = \frac{1}{2}(\mathbb{1} + \vec{p} \cdot \vec{S})$. □

The above claimed existence of probability measures *not* arising from a state $W$ on $\mathbb{C}^2$, is now easy to see. Here are two examples (with $\vec{\lambda} \in \mathbb{R}^3$, $\vec{\lambda} \neq \vec{0}$):

i) $m(\vec{e}) = \sin(\vec{\lambda} \cdot \vec{e})$, $\quad \sin(\xi) := \frac{1}{2\mathrm{i}}(\mathrm{e}^{\mathrm{i}\xi} - \mathrm{e}^{-\mathrm{i}\xi}) \quad$ for $\quad \xi \in \mathbb{R}$

ii) $m(\vec{e}) = \mathrm{sgn}(\vec{\lambda} \cdot \vec{e})$, $\quad \mathrm{sgn}(\xi) := \begin{cases} 1 & \text{for} \quad \xi \geq 0 \\ -1 & \text{für} \quad \xi < 0 \end{cases}$

The resulting measures $\mu$ do not seem to have any physical relevance, but some of their "relatives" play a role in classical models with hidden variables for QM systems with $\dim \mathscr{H} = 2$, see Section 5.1 .

## 5 On the Realistic Interpretability of Quantum Mechanics

Edward NELSON (1932–2014) (in [Far06, p. 230]):

*"The problem of finding a realistic interpretation of quantum mechanics is, in my view, as unresolved as it was in the 1920s."*

*Findings*

The HILBERT-space formalism of quantum mechanics (QM) in the sense of VON NEUMANN together with its pragmatic and minimal(ist) *statistical interpretation* – as summarized above – has developed over the decades into the most successful theory of physics.

*Discomfort of classical realists* (like PLANCK, EINSTEIN, DEBROGLIE, SCHRÖDINGER, D. BOHM, BELL, 'T HOOFT[20],...)

Even in pure states, "most" observables do not have a specific value, but are *objectively indeterminate*. Values only arise as factual outcomes of measurements with an *intrinsic* probability $\langle W,E\rangle$, as briefly described in Explanation 13) in Section 2.2 about the "state collapse". This contrasts with a sentence by EINSTEIN to BORN in December 1926, often shortened but pithily quoted as "Gott würfelt nicht."[21] Even to this day, no proposal at understanding the quantum indeterminacy or accepting it as a feature of (a veiled?) reality [d'Esp03] has met with general approval. This is, of course, related to the infamous *measurement problem.* For some discussions of the latter see [Zeh70, Mit98, Ho01, Zeh07, BLPY16, Hay17].

*Hope for a way out by cryptodeterminism* (Models with hidden variables (HV))

The indeterminacy, reflected by a strictly positive variance, $\sigma_A^2 > 0$, in non-eigenstates of an observable $A$ can be explained by "embedding" the QM into a more fundamental probabilistic theory, which is *realistically* (or "classically") interpretable in the following sense:

i) Each observable $A$ of an *individual* QM system has in each (pure) state $W$ an *objectively definite* (eigen)value, pre-existing before any measurement.
ii) The QM probabilities only express that the distribution of these values is in general not concentrated at a single point, unknown, and empirically in-accessable, that is, *hidden.*

In short: *Subjective ignorance* instead of *objective indeterminacy*

[somewhat similar to *statistical* CM].

### 5.1 Basic idea behind most models with hidden variables

- Starting point: QM system with HILBERT space $\mathscr{H}$, observables $A : \mathscr{H} \to \mathscr{H}$, and a state $W : \mathscr{H} \to \mathscr{H}$ mostly assumed to be pure.
- Classical *probability space* [Bau96]: $\big(\Omega, \mathscr{A}(\Omega), \mathbb{P}\big)$, that is, a triple consisting of the set $\Omega$ of the *hidden variables* $\omega$, a sigma-algebra $\mathscr{A}(\Omega)$ of subsets of $\Omega$, and a probability measure $\mathbb{P}$ on $\Omega$, more precisely, on $\mathscr{A}(\Omega)$.

[20] see ['t Hoo24]

[21] In English: "God does not play dice."

- Desired identity:

$$\boxed{\int_{\Omega} \mathbb{P}(\mathrm{d}\omega)\, a_W(\omega) = \mathrm{Tr}(WA)}$$

  for a suitable *value-function* $a_W : \Omega \to \mathbb{R}$ assigned to the pair $(A,W)$. In CPT such a function is usually called a random variable.

*Remarks:*

- The *hidden variables* $\omega \in \Omega$ discriminate between the individual systems (realizations).
- The probability measure $\mathbb{P}$ also may depend on $W$, or $W$ may only be "transferred" from $a_W$ to $\mathbb{P}$. The WEYL–WIGNER mapping of Section 2.5 provides something similar to the latter case with the phase space $\Gamma$ playing the role of $\Omega$, the WIGNER density $w$ that of $\mathbb{P}_W$, and the symbol $a$ of $A$ that of $a_W$. However, one serious problem with this "could-be" HV-model for the canonical QM is that $w$ is in general not positive everywhere. Another one is that the symbol of $A^2$ is in general not $a^2$ etc.

*Interpretation:*

While the description by a (pure) QM state $W(=W^2)$ may not provide a "complete picture" of the physical reality on its own, the pair $(W,\omega)$, the so-called *microstate*, does so in the sense of the *assignment*:

$$\big(A,(W,\omega)\big) \mapsto a_W(\omega) = \text{some (eigen)value of the observable } A \text{ in the microstate } (W,\omega)\,.$$

*Model with hidden variables in the case of a two-dimensional* HILBERT *space:*

- HILBERT space $\mathscr{H} = \mathbb{C}^2$ associated with a single spin 1/2 as in Section 4.
- Useful relation for the product of two general spin-operator components:

$$\boxed{\big(\vec{a}\cdot\vec{S}\big)\big(\vec{b}\cdot\vec{S}\big) = \vec{a}\cdot\vec{b}\,\mathbb{1} + \mathrm{i}\big(\vec{a}\times\vec{b}\big)\cdot\vec{S}}$$

  for *any* vectors $\vec{a}, \vec{b} \in \mathbb{R}^3$ in terms of their dot product and cross product.
- Each QM observable $A$ on $\mathbb{C}^2$ is a linear combination

$$\boxed{A = a_0\mathbb{1} + \vec{a}\cdot\vec{S}}$$

  with $a_0 := \frac{1}{2}\mathrm{Tr}A \in \mathbb{R}$ and $\vec{a} := \frac{1}{2}\mathrm{Tr}(A\vec{S}) \in \mathbb{R}^3$.
- Equivalence (if $\vec{a} \neq \vec{0} \neq \vec{b}$):

$$\boxed{AB = BA \;\Leftrightarrow\; \vec{a} \text{ and } \vec{b} \text{ are parallel or anti-parallel (that is, } \textit{collinear}\text{)},}$$

  by the commutation relation $AB - BA = 2\mathrm{i}\big(\vec{a}\times\vec{b}\big)\cdot\vec{S}$ from the above "useful relation".

- Each QM state $W$ on $\mathbb{C}^2$ can be written as (see also Section 4):

$$\boxed{W = \frac{1}{2}(\mathbb{1} + \vec{p} \cdot \vec{S}).}$$

The vector $\vec{p} := \mathrm{Tr}(W\vec{S}) \in \mathbb{R}^3$ is known as the *polarization* of $W$ and obeys $|\vec{p}\,| \leq 1$. The pure states on $\mathbb{C}^2$ (or *quantum bits* of quantum information theory) are exactly the *totally* polarized ones:

$$\boxed{W = W^2 \Leftrightarrow |\vec{p}\,| = 1.}$$

Conclusion:

> Each state $W$ corresponds exactly to one point of the closed unit ball in $\mathbb{R}^3$ (called BLOCH ball in this setting). The pure states correspond to the points on its surface $\mathbb{S}^2$.

- Expectation values are given by:

$$\boxed{\mathrm{Tr}(WA) = a_0 + \vec{p} \cdot \vec{a}\,.}$$

With the choices (and the definition $\mathrm{sgn}(0) := 1$)

$\Omega = \mathbb{S}^2 :=$ unit sphere in $\mathbb{R}^3$

$\mathscr{A}(\Omega) = \mathscr{B}(\mathbb{S}^2) :=$ sigma-algebra of all BOREL subsets of $\mathbb{S}^2$

$\mathbb{P} =$ uniform distribution on $\Omega$

$a_W(\vec{\omega}) := a_0 + |\vec{a}\,| \,\mathrm{sgn}\left((\vec{p} + \vec{\omega}) \cdot \vec{a}\right), \qquad \vec{\omega} \in \Omega$

one now actually gets, by explicit integration in spherical coordinates, the desired identity

$$\int_\Omega \mathbb{P}(\mathrm{d}\vec{\omega})\, a_W(\vec{\omega}) = \frac{1}{4\pi} \int_{\mathbb{R}^3} \mathrm{d}^3\omega\, \delta\left(|\vec{\omega}| - 1\right) a_W(\vec{\omega}) = a_0 + \vec{p} \cdot \vec{a} = \mathrm{Tr}(WA)\,.$$

**Facts**
on the above non-linear assignment $A \mapsto a_W$ (for fixed $W$, equivalently, fixed $\vec{p}$):

- the value $a_W(\vec{\omega})$ of $a_W$ coincides with one of the two eigenvalues $a_0 \pm |\vec{a}\,|$ of the observable $A \leftrightarrow (a_0, \vec{a})$. This is called *spectral consistency*. In particular,

$$e_W(\vec{\omega}) = \frac{1}{2}\Big(1 + \mathrm{sgn}\left((\vec{p} + \vec{\omega}) \cdot \vec{e}\right)\Big)$$

is (for $\vec{p} = \vec{0}$) equal to one of the (eigen)values of a 1-dimensional projection $E$ of the example ii) in Section 4 for a probability measure on $\mathscr{P}(\mathbb{C}^2)$, which does not arise from a QM state.
- the (eigen)value $c_W(\vec{\omega})$ of the sum $C := A + B$ is *additively consistent* in the sense that $c_W(\vec{\omega}) = a_W(\vec{\omega}) + b_W(\vec{\omega})$ if $\boxed{AB = BA\,.}$
- the (eigen)value $d_W(\vec{\omega})$ of the product $D := AB$ is *multiplicatively consistent* in the sense that $d_W(\vec{\omega}) = a_W(\vec{\omega}) b_W(\vec{\omega})$ if $\boxed{AB = BA\,.}$ In particular, the microstate $(W, \vec{\omega})$ is universally fluctuation-free because $d_W(\vec{\omega}) = \left(a_W(\vec{\omega})\right)^2$ for $A = B$.
- the assignment is *injective* (or one-to-one) in the sense that "$a_W(\vec{\omega}) = b_W(\vec{\omega})$ for all $\vec{\omega}$" implies $A = B$ .

Remarks:

- Such HV-models for $\dim \mathscr{H} = 2$ go back to a publication of J.S. BELL in 1966 (submitted in 1964), see the first paper re-printed in [Bel04].
- For $\dim \mathscr{H} \geq 3$ such HV-models do not exist (for *all* observables), because then there may emerge a conflict between the above consistencies and the injectivity. This will be explained in the next subsection.

### 5.2 Impossibility theorem of BELL and KOCHEN–SPECKER

The following theorem refers to a *single* individual QM system and does not provide a statement on the statistics of quantum measurement outcomes. Accordingly, it does not explicitly refer to a state (before measurement). Instead it analyzes assignments $A \mapsto a$ to (pre-existing) real (eigen)values for a putative *realistic interpretation* from a more fundamental point of view, regardless whether they depend on a fixed state $W$ or not. No averaging takes place. For the original works see the first paper re-printed in [Bel04] and [KS67].

**Theorem** (BELL 1966, KOCHEN–SPECKER 1967)

Let $\dim \mathscr{H} \geq 3$ and $\mathscr{O}(\mathscr{H})$ be the set of self-adjoint (and bounded) *operators* $A : \mathscr{H} \to \mathscr{H}$. Moreover, let $\big(\Omega, \mathscr{A}(\Omega)\big)$ be a measurable space [= "phase space" = set of hidden variables] and $\mathscr{F}(\Omega)$ be the set of $\big(\mathscr{A}(\Omega), \mathscr{B}(\mathbb{R})\big)$-measurable *functions* $a : \Omega \to \mathbb{R}$, $\omega \mapsto a(\omega)$ [= value-functions and possible random variables].[a]

Then there exists *no* mapping [= "classical representation"]

$$\Theta : \mathscr{O}(\mathscr{H}) \to \mathscr{F}(\Omega), \qquad A \mapsto a := \Theta(A)$$

with the following two properties:

1) $\Theta$ is *injective* (or one-to-one) in the usual sense of the implication

$$\text{“}\, a(\omega) = b(\omega) \quad \text{for all } \omega \in \Omega \implies A = B\,\text{”},$$

2) $\Theta$ is *functionally consistent* in the sense of the equality

$$\Theta\big(f(A)\big)(\omega) = f\big(\Theta(A)\big)(\omega) \quad \text{for all } \omega \in \Omega$$

and any BOREL function $f : \mathbb{R} \to \mathbb{R}$.

[a] The notion "measurable" is meant here in the sense of mathematical measure- and integration theory. The function $a : \Omega \to \mathbb{R}$ is defined to be $\big(\mathscr{A}(\Omega), \mathscr{B}(\mathbb{R})\big)$-*measurable* if the pre-image $a^{-1}(I) := \{\omega \in \Omega : a(\omega) \in I\}$ of any BOREL set $I \subseteq \mathbb{R}$ belongs to the sigma-algebra $\mathscr{A}(\Omega)$ of suitable subsets of $\Omega$.

*Comments*

- The operator $f(A)$ as the image of the self-adjoint operator $A$ under the function $f$ is defined, as usual, by the spectral resolution of $A$ according to the *functional calculus*, which is in agreement with what one expects for a polynomial $f$, see [BEH08].
- *Functional consistency* in the above sense is the physical one among the two properties required. It simply ensures that the value $f(a(\omega))$ of the composed function $f(a) := f \circ a$ at $\omega \in \Omega$ agrees with $f(\alpha)$ for some eigenvalue $\alpha$ of $A$. Example: If a measurement

reveals an eigenvalue of $A^2$, then it should be the square of some eigenvalue of $A$. Functional consistency and injectivity are illustrated by the following diagram:

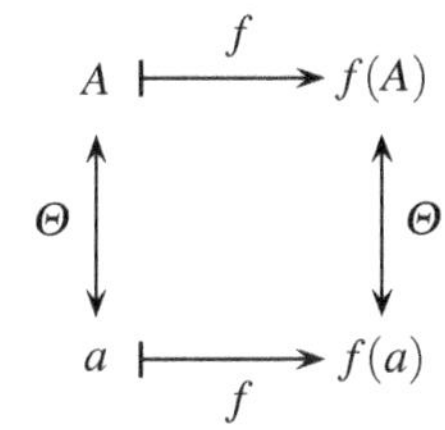


- The formulation of the theorem, given above, may appear slightly complicated and tend to veil its central message. If so, in a first reading one may ignore the "sigma-algebras" $\mathscr{A}(\Omega)$ and $\mathscr{B}(\mathbb{R})$ and understand "measurable function" and "BOREL function" simply as "function". The remarkable result is that any possible mapping $\Theta$ assigning to each self-adjoint operator $A$ a "corresponding" real-valued function $a$ must violate either the injectivity or the functional consistency. Most physicists in search of a realistic interpretation will not renounce the latter for good physical reasons, see above. As it turns out, see below, they therefore have to accept that one and the same quantum observable may be assigned to different classical representations depending on the specific physical *context* in which other commuting (!) observables are considered for (simultaneous) measurements. This is, for example, the case in the well-confirmed BELL-test experiments, see Section 6. Injective representations, that is,"universal" or *non-contextual* ones are mostly rejected because they lack full functional consistency and therefore cannot fully describe the statistics of quantum measurement outcomes. This is often summarized as: The B–KS theorem rules out non-contextual HV-models. For further informations and discussions see [Hug89, Mer93, I95, Per02, Stre07, Cab21, BCGKL22].

*Some consequences of injectivity and functional consistency*

i) Finitely additive consistency: $\Theta\Big(\sum_{j=1}^{n} A_j\Big) = \sum_{j=1}^{n} \Theta(A_j)$,
for pairwise *commuting* $A_1, \ldots, A_n \in \mathscr{O}(\mathscr{H})$

ii) Finitely multiplicative consistency: $\Theta\Big(\prod_{j=1}^{n} A_j\Big) = \prod_{j=1}^{n} \Theta(A_j)$,
for pairwise *commuting* $A_1, \ldots, A_n \in \mathscr{O}(\mathscr{H})$

iii) Homogeneous consistency: $\Theta(\lambda A) = \lambda \Theta(A), \ \lambda \in \mathbb{R}$

iv) Spectral consistency: $\Theta(A)(\Omega) \equiv a(\Omega) \subseteq \operatorname{spec} A$ (:= spectrum of $A$)

All of them follow from the

**Lemma** (VON NEUMANN 1931)
For any collection of pairwise *commuting* $A_1, \ldots, A_n \in \mathscr{O}(\mathscr{H})$ there exist a bounded $X \in \mathscr{O}(\mathscr{H})$ and a collection of BOREL functions $f_1, \ldots, f_n : \mathbb{R} \to \mathbb{R}$ such that $A_j = f_j(X)$ for all $j \in \{1, \ldots, n\}$.

The proof of this "theorem on the generator" (of rings of bounded self-adjoint operators) basically follows from the spectral theorem, see p. 312 in [AG81]. It goes back to Ref. 93 in that book and is mentioned already in Sec. 10 of Ch. II of [vN18]. For $\dim \mathscr{H} < \infty$ the proof is simple and can be found in § 84 of [Hal87].

*Example:* Derivation of iv) from i) to iii):
For $A = E = E^2$ is $\Theta(E) = \Theta(E^2) \overset{ii)}{=} \Theta(E)\Theta(E) = \big(\Theta(E)\big)^2$, hence $\Theta(E)(\omega) \in \{0,1\}$ for all $\omega \in \Omega$. The claim for a general $A$ then follows from its spectral resolution combined with i) and iii).

***Proof*** of the B–KS theorem (following [Ho01])
It is enough to suppose $3 \leq \dim \mathscr{H} < \infty$. The proof is by contradiction. Let me assume that a mapping $\Theta$ with the two properties exists and consider $\mu(E) := \Theta(E)(\omega)$, at a *fixed* $\omega \in \Omega$, for arbitrary projections $E$ on $\mathscr{H}$. By the above consequences i) and ii) it follows that $\mu$ is a probability measure on $\mathscr{P}(\mathscr{H})$ only taking values 0 and 1, see the above example. However, by GLEASON's theorem, see Section 4, there is a state $W$ acting on $\mathscr{H}$ such that $\mu(E) = \mathrm{Tr}(WE)$ for all $E$. Since the last trace belongs to the *open* unit interval $]0,1[$ for suitable $E$, we end up with a contradiction. □

*Remark* The formulation of the original B–KS theorem is somewhat involved and the proof given here, though rather short and elegant, is not elementary because it is based on the above lemma and on GLEASON's theorem. Moreover, the contextuality does not show up explicitly.

Fortunately, there is a simple version of the theorem with a proof for $\dim \mathscr{H} \geq 4$, which is spectacularly simple, direct, and transparent with regard to contextualiy. It is due to MERMIN [Mer93], see also the review [BCGKL22]. It should find its way into the textbooks.

MERMIN's version of the B–KS theorem

If $\dim \mathscr{H} \geq 3$, then there is always a set $\{A_1, A_2, \ldots, A_n\}$ of finitely many self-adjoint operators acting on $\mathscr{H}$, for which it is *impossible* to assign to it a set $\{\alpha_1, \alpha_2, \ldots, \alpha_n\}$ of corresponding real (eigen)values such that all functional relations between the *commuting* operators in $\{A_1, A_2, \ldots, A_n\}$ also hold for the assigned values.

***Proof*** (for $\dim \mathscr{H} = 4$, following [Mer93])

- Let me recall the two-spin system from Sections 3.2 and 3.3 with the total HILBERT space
$$\mathscr{H} \cong \mathbb{C}^2 \otimes \mathbb{C}^2 \qquad (\dim \mathscr{H} = \dim \mathbb{C}^2 \cdot \dim \mathbb{C}^2 = 2 \cdot 2 = 4)$$
- Each observable $A = A^*$ on $\mathscr{H}$ is a real linear combination of the 16 product observables
$$\mathbb{1} \otimes \mathbb{1},\ S^\gamma \otimes \mathbb{1},\ \mathbb{1} \otimes S^\delta,\ S^\gamma \otimes S^\delta \qquad (\gamma, \delta \in \{x,y,z\}).^{22}$$
- Especially, let me consider nine different product observables, arranged as entries of a $3 \times 3$ matrix (or "magic square", [BCGKL22]):

$$\begin{array}{ccc|c} S^x \otimes \mathbb{1} & \mathbb{1} \otimes S^x & S^x \otimes S^x & \mathbb{1} \otimes \mathbb{1} \\ \mathbb{1} \otimes S^y & S^y \otimes \mathbb{1} & S^y \otimes S^y & \mathbb{1} \otimes \mathbb{1} \\ S^x \otimes S^y & S^y \otimes S^x & S^z \otimes S^z & \mathbb{1} \otimes \mathbb{1} \\ \hline \mathbb{1} \otimes \mathbb{1} & \mathbb{1} \otimes \mathbb{1} & -\mathbb{1} \otimes \mathbb{1} & \end{array}$$

[22] To simplify the notation, $\mathbb{1}$ stands here and in the next section, unlike in Section 3, for the unit operator $\mathbb{1}_1$ resp. $\mathbb{1}_2$, when it appears as a left resp. right tensor factor, that is, when it refers to the HILBERT space of the first resp. second spin. This reference also applies to the three spin components $S^x$, $S^y$, and $S^z$.

- Facts:
  a) Each row and each column contains only pairwise *commuting* observables. In this sense each row and each column constitutes a "physical context", that is, a set of observables whose values could, in principle, be jointly measured. Therefore each observable can be considered to belong to two different contexts.
  b) The (usual) product from the three observables in the *right column* gives $-\mathbb{1}\otimes\mathbb{1}$. In contrast, the products from the respective three observables in the two other columns gives $\mathbb{1}\otimes\mathbb{1}$. The latter also holds for the products in each of the three rows.
  c) By the assumption of *functional consistency* the identities of b) imply the assigned (eigen)values $-1$ resp. 1. However: the *product of all nine values* is then $-1$ by the column identities, whereas it is 1 by the row identities. Contradiction! □

*Remark* In the case $\dim\mathscr{H}\geq 5$ it suffices to consider a 4-dimensional subspace. Unfortunately, the proof in the case $\dim\mathscr{H}=3$ remains more complicated [Per02].

## 6 Correlation Inequalities of BELL and Others

*Attention please:*

In the following theorem, unlike in previous sections, the commutator of two operators $X$ and $Y$ is defined as $[X,Y]:=XY-YX$, that is, without the factor $\mathrm{i}/\hbar$.

> **Theorem** (BELL 1964, CLAUSER–HORNE–SHIMONY–HOLT 1969; TSIRELSON 1980)
>
> Let $A,B,C,D$ be four self-adjoint operators on an at least 4-dimensional HILBERT space $\mathscr{H}$ with the properties $A^2=B^2=C^2=D^2=\mathbb{1}$ and $[A,C]=[A,D]=[B,C]=[B,D]=\mathbb{0}$. Moreover, consider the BELL-type operator $K:=A(C+D)+B(C-D)$ and the expectation functional $\langle\cdot\rangle:=\mathrm{Tr}(W\cdot)$ induced by a state $W$ on $\mathscr{H}$.
>
> Then one has:
>
> i) $K^*=K$ and $\mathbb{0}\leq K^2=4\,\mathbb{1}-[A,B][C,D]$
> ii) $\langle K^2\rangle\leq 8$ (TSIRELSON inequality)
> iii) If $\langle[A,B][C,D]\rangle\geq 0$, then $\langle K^2\rangle\leq 4$ (B–CHSH inequality).

***Proof*** i) By $(X+Y)^*=X^*+Y^*$, $(XY)^*=Y^*X^*$, and explicit squaring.

ii) The subadditivity (or triangle inequality) and submultiplicativity (or multiplicative inequality) of the uniform operator norm, see Explanation 3) in Section 2.2, gives:

$$\langle K^2\rangle\leq\|K^2\|\overset{i)}{\leq}\|4\,\mathbb{1}\|+\|[A,B]\|\|[C,D]\|\leq 4+(1+1)(1+1)=8\,.$$

iii) Obviously: $\langle K^2\rangle\overset{i)}{=}4-\langle[A,B][C,D]\rangle\leq 4\,.$ □

*Remarks:*

- By the positivity of general variances, ii) resp. iii) implies $|\langle K\rangle|\leq 2\sqrt{2}$ resp. $|\langle K\rangle|\leq 2$.
- For the above simple proof and proofs of extensions see [T/C80, La87, Stre07]. One should note the generality of the above result. Neither "stochastic independence" nor "locality" is assumed, whatever the latter may mean. The interpretation of the theorem is as simple as its proof. It merely states that non-commutativity can make a difference

in obtaining upper bounds on quantum expectation values. In the present case, the two (anti-self-adjoint) commutators $[A,B]$ and $[C,D]$ commute with each other, which implies that they have a common eigenbasis in $\mathscr{H}$. Nevertheless, it is not obvious how to measure the observable $K^2$ or even $K$ itself in realizable experiments [Per02, LeB06], but see [RZBB94].

- Calling $\langle K^2 \rangle \leq 4$ the "B–CHSH inequality" is justified, because it also holds, when $A,B,C,D$ are interpreted as four (commuting) *classical* random variables taking values $\pm 1$ and the angular brackets $\langle \cdot \rangle$ as the expectation with respect to an (unknown) *arbitrary* joint probability measure/distribution for these values, as in the HV-model of B–CHSH [$K^2 = 4$ implies $K = \pm 2$ for all $2^4 = 16$ realizations], see the second paper re-printed in [Bel04] and [CHSH69].

Application: Spin correlations in the singlet state

- QM system: Again composed of two spins with $\mathscr{H} = \mathscr{H}_1 \otimes \mathscr{H}_2$, $\mathscr{H}_1 = \mathscr{H}_2 \cong \mathbb{C}^2$
- Two arbitrary spin components on $\mathscr{H}_1$ resp. $\mathscr{H}_2$ as observables on $\mathscr{H}$ in terms of four *unit* vectors $\vec{a}, \vec{b}, \vec{c}, \vec{d} \in \mathbb{R}^3$:

$$\boxed{A := \vec{a} \cdot \vec{S} \otimes \mathbb{1},\ B := \vec{b} \cdot \vec{S} \otimes \mathbb{1},\ C := \mathbb{1} \otimes \vec{c} \cdot \vec{S},\ D := \mathbb{1} \otimes \vec{d} \cdot \vec{S}\,.}$$

Obviously, they satisfy the assumptions of the above theorem.
- *Singlet state* on $\mathscr{H}$ [in manifestly spin-rotation invariant form]:

$$\boxed{W^- := \frac{1}{4}\left(\mathbb{1} \otimes \mathbb{1} - \left(\vec{S} \otimes \mathbb{1}\right) \cdot \left(\mathbb{1} \otimes \vec{S}\right)\right)} \mathrel{\widehat{=}} |\Phi^-\rangle\langle\Phi^-|\,.$$

It is entangled with completely mixed reduced states, see Section 3.2. Moreover, it satisfies

$$(W^-)^2 = W^- \quad \text{and} \quad (\vec{S} \otimes \mathbb{1} + \mathbb{1} \otimes \vec{S})^2 W^- = \mathbb{0}\,.$$

It is therefore a pure eigenstate of the squared total-spin vector with eigenvalue $0 = 4 \cdot 0(0+1)$, which justifies its name. Finally, it is the (unique) ground state of the HAMILTONian $H := \mathrm{v}(\vec{S} \otimes \mathbb{1}) \cdot (\mathbb{1} \otimes \vec{S}) := \mathrm{v}\sum_{\gamma \in \{x,y,z\}} S^\gamma \otimes S^\gamma$, coupling the two spins isotropically and anti-ferromagnetically: $HW^- = -3\mathrm{v}W^-$, $\mathrm{v} > 0$.
- Combining the four observablen $A,B,C,D$, as defined above, to a [special] $K$ as in the theorem and computing its expectation value in the singlet state $W^-$ gives explicitly in terms of four correlation coefficients:

$$\begin{aligned}
\langle K \rangle &= \langle AC \rangle + \langle AD \rangle + \langle BC \rangle - \langle BD \rangle \\
&= \kappa_{A,C} + \kappa_{A,D} + \kappa_{B,C} - \kappa_{B,D} \\
&= -\vec{a} \cdot \vec{c} - \vec{a} \cdot \vec{d} - \vec{b} \cdot \vec{c} + \vec{b} \cdot \vec{d}\,.
\end{aligned}$$

- If all four unit vectors $\vec{a}, \vec{b}, \vec{c}, \vec{d}$ are *coplanar* and aligned as in the subsequent figure, then one finds $\langle K \rangle = -\sqrt{8}$. Combining this result with the TSIRELSON inequality gives $8 = \langle K \rangle^2 \leq \langle K^2 \rangle \leq 8$. Hence, $W^-$ is an eigenstate of this particular $K$ with eigenvalue $-\sqrt{8}$ and yields a maximum violation of the classic(al) B–CHSH inequality.

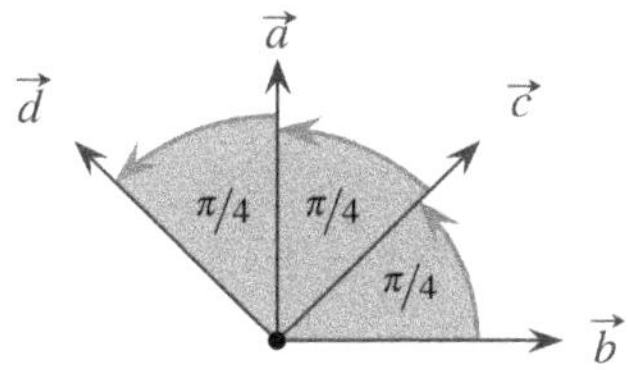


Four coplanar unit vectors with
$\vec{a} \| \vec{c} + \vec{d}$ and $\vec{b} \| \vec{c} - \vec{d}$

*Remarks:*

- When one computes $\langle K \rangle$ not with the singlet state, but with an arbitrary *non-entangled* state, one finds $|\langle K \rangle| \leq 2$ for all unit vectors $\vec{a}, \vec{b}, \vec{c}, \vec{d} \in \mathbb{R}^3$, in agreement with the classical HV-model interpretation of the B–CHSH inequality [Au07]. Conversely, WERNER has constructed *entangled* (mixed) states with $|\langle K \rangle| \leq 2$ for all unit vectors [Wer89].
- Given these mathematical results, what is their relevance for entangled states occurring in real experiments? BELL was motivated by D. BOHM's 1951 spin-singlet simplification of the famous "Gedankenexperiment" by EINSTEIN–PODOLSKY–ROSEN in 1935 to point out that correlations between two *spacelike separated* quantum systems which are in an entangled state due to interactions in the past may be stronger than is possible from any classical mechanism. A real-world, but not practical, example [Per02, LeB06] is the (rare) leptonic decay [Hus24] of an electrically neutral pion $\pi^0$ (with spin 0) into an electron $e^-$ and a positron $e^+$ flying along a straight line in opposite directions. Their different locations are, hence, associated with different individual HILBERT-space factors. In 1982, A. ASPECT and coworkers set up an equivalent scenario in a basement laboratory in Orsay near Paris and found that the correlation statistics of coincidence-measurement outcomes on pairs of polarized photons agrees with the TSIRELSON inequality, but in general not with the *classical* B–CHSH inequality [ADR82]. Later such so-called BELL tests, for example by A. ZEILINGER and coworkers [WJSWZ98] and by other (big) groups like [Chr-etal13] and the three ones mentioned in [Mil16], have confirmed that the probabilistic predictions of QM cannot be described by classical HV-models which are local (or locally realistic) in the sense of BELL, see [Bel04, BCPSW14, BG16]. Some authors use this fact to claim that the QM itself is genuinely non-local. I do not side with these authors, because the QM is clearly local in the sense that for two dynamically uncoupled systems, even if in an entangled state, no measurement or force on the second system affects the physical behavior of the first. Nevertheless, in an entangled state the systems may be spatially correlated or "causually non-separable", even over huge distances. For various inspiring discussions, see the 7th paper re-printed in [Bel04], FARIS in [W95], and [Mer93, As07, Stre07, HS10, Wis14a, Wis14b, BG16, Pe17, Gr20, Cab21, BCGKL22, Wer23]. It now seems that *contextuality* is largely responsible for the structural differences between classical and quantum [DF22].
- To summarize, quantum entanglement and quantum contextuality are genuinely non-classical concepts. These days they have become key resources for quantum information, communication, and computation. Trying to describe them by classical concepts is – quoting R.F. STREATER – like trying to model non-EUCLIDean geometry by using figures of a strange shape in EUCLIDean space. It can't be done.

I'll leave the last word to Mark TWAIN
(in his memoir *Life on the Mississippi*, published 1883):

*"There is something fascinating about science. One gets such wholesale returns of conjecture out of such a trifling investment of fact."*

## 7 Bibliography

### Selected popular books


[BH24] J. BAGGOTT and J.L. HEILBRON, *Quantum Drama: From the Bohr–Einstein Debate to the Riddle of Entanglement*, Oxford University Press, Oxford, 2024.

[Bha21] A. BHATTACHARYA, *The Man from the Future: The Visionary Life of John von Neumann*, Penguin Books, UK, 2021.

[Ga85] G. GAMOW, *Thirty Years that Shook Physics: The Story of Quantum Theory*, Dover, New York,1985 [re-publication of the edition in 1966 by Doubleday, New York].

[Gr04] B. GREENE, *The Fabric of the Cosmos: Space, Time, and the Texture of Reality*, Knopf Doubleday, New York, 2004.

[Hei69] W. HEISENBERG, *Der Teil und das Ganze: Gespräche im Umkreis der Atomphysik*, Piper, München, 1969 [in German].

[Hei71] W. HEISENBERG, *Physics and Beyond: Encounters and Conversations*, Harper & Row, New York, 1971 [translation of [Hei69] by A.J. POMERANS].

[Ka85] M. KAC, *Enigmas of Chance: An Autobiograhy*, Harper & Row, New York, 1985.

[Kl19] K. KLEINKNECHT, *Einstein and Heisenberg: The Controversy Over Quantum Physics*, Springer, Cham, 2019.

[Ku09] M. KUMAR, *Quantum: Einstein, Bohr and the Great Debate About the Nature of Reality*, Icon Books, London, 2009.

[R22] C. ROVELLI, *Helgoland: The Strange and Beautiful Story of Quantum Physics*, Penguin Books, UK, 2022.

[S11] J.S. STRICKLAND, *Weird Scientists: The Creators of Quantum Physics*, Lulu Press, 2011.

[Zei23] A. ZEILINGER, *Dance of the Photons: From Einstein to Quantum Teleportation*, newest edition, Penguin Books, UK, 2023.


### Selected textbooks and monographs


[AG81] N.I. AKHIEZER and I.M. GLAZMAN, *Theory of Linear Operators in Hilbert Space, Vol. I*, Pitman, Boston, Massachusetts, 1981.

[Au07] J. AUDRETSCH, *Entangled Systems: New Directions in Quantum Physics*, Wiley-VCH, Weinheim, 2007.

[Bal98] L.E. BALLENTINE, *Quantum Mechanics: A Modern Development*, 2nd ed., World Scientific, Singapore, 1998.

[Bau96] H. BAUER, *Probability Theory*, de Gruyter, Berlin, 1996.

[Bay81] G. BAYM, *Lectures on Quantum Mechanics*, 9th printing with revisions, Benjamin/Cummings, Reading, Massachusetts, 1981.

[Bel04] J.S. BELL, *Speakable and Unspeakable in Quantum Mechanics*, 2nd ed., Cambridge University Press, Cambridge, 2004.

[BG16] M. BELL and S. GAO, editors, *Quantum Nonlocality and Reality: 50 Years of Bell's Theorem*, Cambridge University Press, Cambridge, 2016.

[LeB06] M. LE BELLAC, *Quantum Physics*, Cambridge University Press, Cambridge, 2006.

[BC81] E.G. BELTRAMETTI and G. CASSINELLI, *The Logic of Quantum Mechanics*, Addison-Wesley, Reading, Massachusetts, 1981.

[BZ06] I. BENGTSSON and K. ŻYCZKOWSKI, *Geometry of Quantum States*, Cambridge University Press, Cambridge, 2006.

[BEH08] J. BLANK, P. EXNER, and M. HAVLÍČEK, *Hilbert Space Operators in Quantum Physics*, 2nd ed., Springer, Dordrecht, 2008.

[Boh93] A. BOHM, *Quantum Mechanics: Foundations and Applications*, 3rd ed., Springer, New York, 1993.

[BGL95] P. BUSCH, M. GRABOWSKI, and P.J. LAHTI, *Operational Quantum Physics*, Springer, Berlin, 1995

[BLPY16] P. BUSCH, P. LAHTI, J.-P. PELLONPÄÄ, and K. YLINEN, *Quantum Measurement*, Springer, Switzerland, 2016.

[CFZ19] T.L. CURTRIGHT, D.B. FAIRLIE, and C.K. ZACHOS, *A Concise Treatise on Quantum Mechanics in Phase Space*, World Scientific, Singapore, 2013/2019.

[d'Esp03] B. D'ESPAGNAT, *Veiled Reality: An Analysis of Present-Day Quantum Mechanical Concepts*, Westview Press, Boulder, Colorado, 2003.

[Far06] W.G. FARIS, editor, *Diffusion, Quantum Theory, and Radically Elementary Mathematics*, Princeton University Press, Princeton, 2006.

[FH10] R.P. FEYNMAN and A.R. HIBBS, *Quantum Mechanics and Path Integrals*, emended edition by D.F. STYER, Dover, Mineola, New York, 2010 [based on the edition in 1965/2005 by McGraw-Hill, New York].

[GP90] A. GALINDO and P. PASCUAL, *Quantum Mechanics I*, Springer, Berlin, 1990.

[GP91] A. GALINDO and P. PASCUAL, *Quantum Mechanics II*, Springer, Berlin, 1991.

[dG17] M. DE GOSSON, *The Wigner Transform*, World Scientific, London, 2017.

[GS20] S.J. GUSTAFSON and I.M. SIGAL, *Mathematical Concepts of Quantum Mechanics*, latest edition, Springer, Berlin, 2020.

[Hal87] P.R. HALMOS, *Finite-Dimensional Vector Spaces*, Springer, New York, 1987 [re-publication of the 2nd ed. in 1958 by Van Nostrand, Princeton, New Jersey].

[Hay17] M. HAYASHI, *Quantum Information Theory: Mathematical Foundation*, 2nd ed., Springer, Berlin, 2017.

[Ho01] A.S. HOLEVO, *Statistical Structure of Quantum Theory*, Springer, Berlin, 2001.

[Hug89] R.I.G. HUGHES, *The Structure and Interpretation of Quantum Mechanics*, Harvard University Press, Cambridge, Massachusetts, 1989.

[I95] C.J. ISHAM, *Lectures on Quantum Theory: Mathematical and Structural Foundations*, Imperical College Press, London, 1995.

[LévB90] J.-M. LÉVY-LEBLOND and F. BALIBAR, *Quantics: Rudiments of Quantum Physics*, North-Holland, Amsterdam, 1990.

[LL01] E.H. LIEB and M. LOSS, *Analysis*, 2nd ed., Amer. Math. Soc., Providence, Rhode Island, 2001.

[LoP77] J.L. LOPES and M. PATY, editors, *Quantum Mechanics, A Half Century Later: Papers of a Colloquium at the University Louis Pasteur, Strasbourg, May 2–4, 1974*, Reidel, Dordrecht, Holland, 1977.

[Lu85/87] G. LUDWIG, *An Axiomatic Basis for Quantum Mechanics, Vol. 1: Derivation of Hilbert Space Structure, Vol. 2: Quantum Mechanics and Macrosystems*, Springer, Berlin, 1985/87.

[Mas90] P.R. MASANI, *Norbert Wiener: 1894–1964*, Vita Mathematica: Vol. 5, Birkhäuser, Basel, 1990.

[Mit98] P. MITTELSTAEDT, *The Interpretation of Quantum Mechanics and the Measurement Process*, Cambridge University Press, Cambridge, 1998.

[Mor17] V. MORETTI, *Spectral Theory and Quantum Mechanics: Mathematical Foundations of Quantum Theories, Symmetries and Introduction to the Algebraic Formulation*, 2nd ed., Springer, Cham, 2017.

[vN32] J. VON NEUMANN, *Mathematische Grundlagen der Quantenmechanik*, Springer, Berlin, 1932 [in German].

[vN18] J. VON NEUMANN, *Mathematical Foundations of Quantum Mechanics*, new edition by N.A. WHEELER of the translation of [vN32] by R.T. BEYER in 1949, Princeton University Press, Princeton, 2018.

[NC10] M.A. NIELSEN and I.L. CHUANG, *Quantum Computation and Information*, 2nd ed., Cambridge University Press, Cambridge, 2010.

[Per02] A. PERES, *Quantum Theory: Concepts and Methods*, Kluwer Academic Publishers, New York, 2002.

[Pet08] D. PETZ, *Quantum Information Theory and Quantum Statistics*, Springer, Berlin, 2008.

[Roe94] G. ROEPSTORFF, *Path Integral Approach to Quantum Physics*, Springer, Berlin, 1994.

[SN21] J.J. SAKURAI and J. NAPOLITANO, *Modern Quantum Mechanics*, 3rd ed., Cambridge University Press, Cambridge, 2021.

[Si05] B. SIMON, *Trace Ideals and Their Applications*, 2nd ed., Amer. Math. Soc., Providence, Rhode Island, 2005.

[SS08] J. STOLZE and D. SUTER, *Quantum Computing: A Short Course from Theory to Experiment*, 2nd ed., Wiley-VCH, Weinheim, 2008.

[Stre07] R.F. STREATER, *Lost Causes in and beyond Physics*, Springer, Berlin, 2007.

[Stro08] F. STROCCHI, *An Introduction to the Mathematical Structure of Quantum Mechanics: A Short Course for Mathematicians*, 2nd ed., World Scientific, Singapore, 2008.
[T14] G. TESCHL, *Mathematical Methods in Quantum Mechanics: With Applications to Schrödinger Operators*, 2nd ed., Amer. Math. Soc., Providence, RI, 2014.
[V07] V.S. VARADARAJAN, *Geometry of Quantum Theory*, 2nd ed., Springer, New York, 2007.
[W95] D. WICK, *The Infamous Boundary: Seven Decades of Controversy in Quantum Physics*, Appendix: W.G. FARIS, *Probability in Quantum Mechanics*, Birkhäuser, Boston, Massachusetts, 1995.
[Ya00] D. YAFAEV, *Scattering Theory: Some Old and New Problems*, Lecture Notes in Math., Vol. 1735, Springer, Berlin, 2000.
[Zeh07] H.D. ZEH, *The Physical Basis of the Direction of Time*, 5th ed., Springer, Berlin, 2007.

## Articles

[Ac10] L. ACCARDI, *Quantum probability: New perspectives for the laws of chance*, Milan J. Math. **78**, 481–502 (2010).
[Ad00] A.M. ADAM, *Farewell to certitude: Einstein's novelty on induction and deduction, fallibilism*, J. Gen. Philos. Sci. **31**, 19–37 (2000).
[AL70] H. ARAKI and E.H. LIEB, *Entropy inequalities*, Commun. Math. Phys. **18**, 160–170 (1970).
[As07] A. ASPECT, *To be or not to be local*, Nature **446**, 866–867 (2007).
[ADR82] A. ASPECT, J. DALIBAR, and G. ROGER, *Experimental tests of Bell's inequalities using time-varying analyzers*, Phys. Rev. Lett. **49**, 1804–1807 (1982)
[Bec75] W. BECKNER, *Inequalities in Fourier analysis*, Ann. of Math. **102**, 159–182 (1975).
[BB–M75] I. BIAŁYNICKI-BIRULA and J. MYCIELSKI, *Uncertainty relations for information entropy in wave mechanics*, Commun. Math. Phys. **44**, 129–132 (1975).
[BJLR17] A. BLUM, M. JÄHNERT, C. LEHNER, and J. RENN, *Translation as heuristics: Heisenberg's turn to matrix mechanics*, Stud. Hist. Philos. Mod. Phys. **60**, 3–22 (2017).
[Bo24] M. BORN, *Über Quantenmechanik.*, Z. Physik **26**, 379–395 (1924) [in German].
[Bo26] M. BORN, *Quantenmechanik der Stoßvorgänge.*, Z. Physik **38**, 803–827 (1926) [in German].
[BHJ26] M. BORN, W. HEISENBERG, and P. JORDAN, *Zur Quantenmechanik. II.*, Z. Physik **35**, 557–615 (1926) [in German].
[BJ25] M. BORN and P. JORDAN, *Zur Quantenmechanik.*, Z. Physik **34**, 858–888 (1925) [in German]
[BLM04] K. BRODERIX, H. LESCHKE, and P. MÜLLER, *Continuous integral kernels for unbounded Schrödinger semigroups and their spectral projections*, J. Funct. Anal. **212**, 287–323 (2004).
[BCPSW14] N. BRUNNER, D. CAVALCANTI, S. PIRONIO, V. SCARANI, and S. WEHNER, *Bell nonlocality*, Rev. Mod. Phys. **86**, 419–478 (2014).
[BCGKL22] C. BUDRONI, A. CABELLO, O. GÜHNE, M. KLEINMANN, and J.-Å. LARSSON, *Kochen–Specker contextuality*, Rev. Mod. Phys. **94**, 045007, 62pp. (2022).
[BLW14] P. BUSCH, P. LAHTI, and R.F. WERNER, *Quantum root-mean-square error and measurement uncertainty relations*, Rev. Mod. Phys. **86**, 1261–1281 (2014).
[Cab21] A. CABELLO, *Bell non-locality and Kochen–Specker contextuality*, Found. Phys. **51**, 61,13pp. (2021).
[Car76] N.D. CARTWRIGHT, *A non-negative Wigner-type distribution function*, Physica **83A**, 210–212 (1976).
[Chr-etal13] B.G. CHRISTENSEN et al., *Detection-loophole-free test of quantum nonlocality, and applications*, Phys. Rev. Lett. **111**, 130406, 5pp. (2013).
[CHSH69] J.F. CLAUSER, M.A. HORNE, A. SHIMONY, and R.A. HOLT, *Proposed experiment to test local hidden-variable theories*, Phys. Rev. Lett. **23**, 880–884 (1969).
[CKM85] R. COOKE, M. KEANE, and W. MORAN, *An elementary proof of Gleason's theorem*, Math. Proc. Camb. Philos. Soc. **98**, 117–128 (1985).
[CGTVV95] A. CUCCIOLI, R. GIACHETTI, V. TOGNETTI, R. VAIA, and P. VERRUCCHI, *The effective potential and effective Hamiltonian in quantum statistical mechanics*, J. Phys. Condens. Matter **7**, 7891–7938 (1995).
[DM36] M. DELBRÜCK and G. MOLIÈRE, *Statistische Quantenmechanik und Thermodynamik.*, Abhandl. Preuß. Akad. Wiss., phys.-math. Klasse, Nr.1, 1–42 (1936) [in German].
[D25] P.A.M. DIRAC, *The fundamental equations of quantum mechanics.*, Proc. Roy. Soc. (London) A **109**, No. 752, 642–653 (1925).
[DF22] A. DÖRING and M. FREMBS, *Contextuality and the fundamental theorems of quantum mechanics*, J. Math. Phys. **63**, 072103, 20pp. (2022).

[DJ13] A. DUNCAN and M. JANSSEN, *(Never) Mind your p's and q's: Von Neumann versus Jordan on the foundations of quantum theory*, Eur. Phys. J. H **38**, 175–259 (2013).

[Ei19] A. EINSTEIN, *Induktion und Deduktion in der Physik.*, Berliner Tageblatt, 4. Beiblatt, S.1, 25 Dezember 1919 [in German]. English translation: The Digital Einstein Papers, Vol. 7, The Berlin Years: Writings, 1918-1919, Doc. 28 (English translation supplement) pp. 108–109. For another translation, with discussion, see [Ad00].

[Er-etal12] J. ERHART et al., *Experimental demonstration of a universally valid error-disturbance uncertainty relation in spin measurements*, Nat. Phys. **8**, 185–189, Corrigenda: 349, 634 (2012).

[Fal70] H. FALK, *Inequalities of J.W. Gibbs*, Am. J. Phys. **38**, 858–869 (1970).

[Far78] W.G. FARIS, *Inequalities and uncertainty principles*, J. Math. Phys. **19**, 461–466 (1978).

[FK86] R.P. FEYNMAN and H. KLEINERT, *Effective classical partition functions*, Phys. Rev. A **34**, 5080–5084 (1986).

[FiLe93] H. FINK and H. LESCHKE, *Phase-space- and other representations of quantum mechanics: A unifying approach*, pp. 416–419 in: *Classical and Quantum Systems: Foundations and Symmetries*, ed. by H.D. DOEBNER, W. SCHERER, and F. SCHROECK, JR., World Scientific, Singapore, 1993.

[FoSi97] G.B. FOLLAND and A. SITARAM, *The uncertainty principle: A mathematical survey*, J. Fourier Anal. Appl. **3**, 207–238 (1997).

[FL12] R.L. FRANK and E.H. LIEB, *Entropy and the uncertainty principle*, Ann. Henri Poincaré **13**, 1711–1717 (2012).

[GT85] R. GIACHETTI and V. TOGNETTI, *Variational approach to quantum statistical mechanics of nonlinear systems with applications to sine–Gordon chains*, Phys. Rev. Lett. **55**, 912–915 (1985).

[Gl57] A.M. GLEASON, *Measures on the closed subspaces of a Hilbert space*, Journal of Mathematics and Mechanics (now: Indiana Univ. Math. J.) **6**, 885–893 (1957).

[Gr20] R.B. GRIFFITHS, *Nonlocality claims are inconsistent with Hilbert-space quantum mechanics*, Phys. Rev. A **101**, 022117, 12pp. (2020).

[HS10] N. HARRIGAN and R.W. SPEKKENS, *Einstein, incompleteness, and the epistemic view of quantum states*, Found. Phys. **40**, 125–157 (2010).

[Heg85] G.C. HEGERFELDT, *Noncommutative analogs of probabilistic notions and results*, J. Funct. Anal. **64**, 436–456 (1985).

[Hei25] W. HEISENBERG, *Über quantentheoretische Umdeutung kinematischer und mechanischer Beziehungen.*, Z. Physik **33**, 879–893 (1925) [in German].

[Hei27] W. HEISENBERG, *Über den anschaulichen Inhalt der quantentheoretischen Kinematik und Mechanik.*, Z. Physik **43**, 172–198 (1927) [in German].

[Hir57] I.I. HIRSCHMAN, JR., *A note on entropy*, Amer. J. Math.**79**, 152–156 (1957).

[Hol00] G. HOLTON, *Werner Heisenberg and Albert Einstein*, Physics Today **53** (7), 38–42 (2000).

['t Hoo24] G. 'T HOOFT, *The hidden ontological variable in quantum harmonic oscillators*, Front. Quantum Sci. Technol. **3**, 1505593, 6pp. (2024).

[Hub70] A. HUBER, *An inequality for traces and its application to extremum principles and related theorems in quantum statistical mechanics*, pp. 37–73 in: *Methods and Problems of Theoretical Physics – In honour of* R.F. PEIERLS, ed. by J.E. BOWCOCK, North-Holland, Amsterdam, 1970.

[Hud74] R.L. HUDSON, *When is the Wigner quasi-probability density non-negative?*, Rep. Math. Phys. **6**, 249–252 (1974).

[Hus24] T. HUSEK, *Neutral-pion decay into an electron-positron pair: a review and update*, Phys. Rev. D **110**, 033004, 12pp. (2024).

[Ja77] J.M. JAUCH, *The quantum probability calculus*, pp. 39–62 in [LoP77].

[Ke27] E.H. KENNARD, *Zur Quantenmechanik einfacher Bewegungstypen.*, Z. Physik **44**, 326–352 (1927) [in German].

[KS67] S. KOCHEN and E.P. SPECKER, *The problem of hidden variables in quantum mechanics*, Journal of Mathematics and Mechanics (now: Indiana Univ. Math. J.) **17**, 59–87 (1967).

[La87] J.L. LANDAU, *On the violation of Bell's inequality in quantum theory*, Phys. Lett. A **120**, 54–56 (1987).

[LeW89] H. LESCHKE and S. WONNEBERGER, *Phase-space aspects of quantum partition functions*, pp. 480–497 in: *Path Integrals from meV to MeV, Bangkok, 9–13 January 1989*, ed. by V. SA-YAKANIT et al., World Scientific, Singapore, 1989.

[Lév86] J.-M. LÉVY-LEBLOND, *Correlation of quantum properties and the generalized Heisenberg inequality*, Am. J. Phys. **54**, 135–136 (1986).

[Lü51] G. LÜDERS, *Über die Zustandsänderung durch den Meßprozeß*, Ann. Physik, 6th series, **8**, 323–328 (1951) [in German].

[Mer93] N.D. MERMIN, *Hidden variables and the two theorems of John Bell*, Rev. Mod. Phys. **65**, 803–815 (1993) and its re-print with minor corrections and additions, arXiv:1802.10119 [quant-phys], 27pp. (2018).

[Mil16] J.L. MILLER, *Three groups close the loopholes in tests of Bell's theorem*, Physics Today **69** (1), 14–16 (2016).

[Moy49] J.E. MOYAL, *Quantum mechanics as a statistical theory*, [Math.] Proc. Camb. Philos. Soc. **45**, 99–124 (1949).

[Oz84] M. OZAWA, *Quantum measuring processes of continuous observables*, J. Math. Phys. **25**, 79–87 (1984).

[Oz03] M. OZAWA, *Physical content of Heisenberg's uncertainty relation: limitation and reformulation*, Phys. Lett. A **318**, 21–29 (2003).

[Oz19] M. OZAWA, *Soundness and completeness of quantum root-mean-square errors*, npj Quantum Information **5**, 1, 8pp. (2019).

[Pa25] W. PAULI, *Über den Zusammenhang des Abschlusses der Elektronengruppen im Atom mit der Komplexstruktur der Spektren.*, Z. Physik **31**, 765–783 (1925) [in German].

[Pa26] W. PAULI, *Über das Wasserstoffspektrum vom Standpunkt der neuen Quantenmechanik.*, Z. Physik **36**, 336–363 (1926) [in German].

[Pe17] H.S. PERLMAN, *Quantum mechanics is incomplete but it is consistent with locality*, Found. Phys. **47**, 1309–1316 (2017).

[RZBB94] M. RECK, A. ZEILINGER, H.J. BERNSTEIN, and P. BERTANI, *Experimental realization of any discrete unitary operator*, Phys. Rev. Lett. **73**, 58–61 (1994).

[Rob29] H.P. ROBERTSON, *The uncertainty principle*, Phys. Rev. **34**, 163–164 (1929).

[Rob30] H.P. ROBERTSON, *A general formulation of the uncertainty principle and its classical interpretation*, Phys. Rev. **35**, APS-conference abstract 40, p. 667 (1930).

[Rob34] H.P. ROBERTSON, *An indeterminacy relation for several observables and its classical interpretation*, Phys. Rev. **46**, 794–801 (1934).

[Sc26a] E. SCHRÖDINGER, *Quantisierung als Eigenwertproblem; (Erste Mitteilung.)*, Ann. Physik, 4th series, **79**, 361–376 (1926) [in German].

[Sc26b] E. SCHRÖDINGER, *Über das Verhältnis der Heisenberg–Born–Jordanschen Quantenmechanik zu der meinen;*, Ann. Physik, 4th series, **79**, 734–756 (1926) [in German].

[Sc30] E. SCHRÖDINGER, *Zum* HEISENBERG*schen Unschärfeprinzip.*, Sitzber. Preuß. Akad. Wiss. **XIX**, phys.-math. Klasse, 296–303 (1930) [in German].

[Sc99/08] E. SCHRÖDINGER, *About Heisenberg uncertainty relation*, arXiv:quant-ph/9903100v3, 10pp. (2008) [translation with annotations of [Sc30] by A. ANGELOW and M.-C. BATONI]. See also: Bulg. J. Phys. **26**, 193–203 (1999).

[Stre00] R.F. STREATER, *Classical and quantum probability*, J. Math. Phys. **41**, 3556–3603 (2000).

[Sy65] K. SYMANZIK, *Proof and refinements of an inequality of Feynman*, J. Math. Phys. **6**, 1155–1156 (1965).

[T/C80] B.S. TSIRELSON (=CIREL'SON), *Quantum generalizations of Bell's inequalities*, Lett. Math. Phys. **4**, 93–100 (1980).

[WJSWZ98] G. WEIHS, T. JENNEWEIN, C. SIMON, H. WEINFURTER, and A. ZEILINGER, *Violation of Bell's inequality under strict Einstein locality conditions*, Phys. Rev. Lett. **81**, 5039–5043 (1998).

[Wer89] R.F. WERNER, *Quantum states with Einstein–Podolsky–Rosen correlations admitting a hidden-variable model*, Phys. Rev. A **40**, 4277–4281 (1989).

[Wer23] R.F. WERNER, *Bloß ungenau oder falsch? – Laborsprache und verborgene Variablen*, pp. 93–112 in: *Unbestimmt und relativ? – Das Weltbild der modernen Physik*, ed. by H. FINK and M. KUHLMANN, Springer, Berlin, 2023 [in German].

[WF19] R.F. WERNER and T. FARRELLY, *Uncertainty from Heisenberg to today*, Found. Phys. **49**, 460–491 (2019).

[Wey27] H. WEYL, *Quantenmechanik und Gruppentheorie.*, Z. Physik **46**, 1–46 (1927) [in German].

[Wig32] E.P. WIGNER, *On the quantum correction for thermodynamic equilibrium*, Phys. Rev. **40**, 749–759 (1932).

[Wis14a] H.M. WISEMAN, *The two Bell's theorems of John Bell*, J. Phys. A: Math. Theor. **47**, 424001, 31pp. (2014).

[Wis14b] H.M. WISEMAN, *Bell's theorem still reverberates*, Nature **510**, 467–469 (2014).

[Zac07] C.K. ZACHOS, *A classical bound on quantum entropy*, J. Phys. A: Math. Theor. **40**, F407–F412 (2007).

[Zeh70] H.D. ZEH, *On the interpretation of measurement in quantum theory*, Found. Phys. **1**, 69–76 (1970).

# Chapter 3

# Quantum Evolution as the Square Root of the Master Equation

J.M.J. van Leeuwen (deceased)

February 2, 2026

Instituut-Lorentz, Universiteit Leiden,
Niels Bohrweg 2, 2333 CA Leiden, The Netherlands.

**Abstract**

The analogy between the Quantum Evolution and that of the Master Equation is exploited. By stressing the stochastic nature of quantum evolution a number of conceptual difficulties in the interpretation of quantum mechanics are avoided.

## 1 Introduction

Quantummechanics is hundred years old. As with many revolutions the precise birthdate is not clear, but the definite breakthrough from classical mechanics to quantummechanics was the article by Heisenberg in 1925 [1]. In this paper it became clear that the quantum theory was not a set of rules added to classical mechanics but a complete new mechanics, which forced us to think differently about the world on micro level. Quantities like position and impuls of particles changed from having definite values to matrices or operators as we call them now. After the invention of classical mechanics by Newton, it is the greatest discovery in physics with the largest impact on our view on the nature of matter. With the subsequent probability interpretation of Born [2], the deterministic classical dynamics was replaced by a quantum stochastic process.

In particular the stochastic character of the quantum dynamics met resistence. Einstein objected against this aspect with " Der lieber Herr Gott würffelt nicht" and asked for an underlying deterministic theory. However such a deterministic foundation is not compatible with the structure of the Heisenberg-Born formulation of quantummechanics as von Neumann [3] showed. Even today one still struggles with the proper appreciation of the quantum process as such weird notions as the "many worlds" and "the wave function of the universe" demonstrate.

An informal poll [4] taken in July 1999 at a conference on quantum computation at the Isaac Newton Institute in Cambridge, England, suggests that the prevailing viewpoint is shifting. Out of 90 physicists polled, only eight declared that their view involved explicit wave-function collapse. Thirty chose "many worlds or consistent histories (with no collapse)". As stated in [4]: the picture is not clear: 50 of the researchers chose "none of the above or undecided". Rampant linguistic confusion may contribute to that large number. It is not uncommon for two physicists, who say that they subscribe to the

Copenhagen interpretation, for example, to find themselves disagreeing about what they mean.

In this note I will show that much of these misunderstandings can be avoided if one explores the analogy between the quantum process and a classical stochastic process, in particular the Brownian motion. From this analogy one gets a much better insight in the quantum process than the often-used notion of particle-wave duality. On the other hand it is also interesting to elaborate on the fundamental differences between the quantum and classical stochastic processes.

A link between quantummechanics and classical stochastic processes is not new. Already in 1933 Furth [5] investigated the similarities between the classical stochastic processes and quantummechanics, in particular the appearance of uncertainty relations in stochastic processes. However this paper did not get much attention and it is recently translated and re-issued by Peliti and Muratore-Ginanneschi [6]. While Furth constructs uncertainty relations for the stochastic diffusive process in analogy with the uncertainty relations in quantummechanics, in this note the opposite viewpoint is taken and quantummechanics is elucidated by its analogy with classical stochastic processes. In order to make the point let us start with the analogy between the Brownian motion and the Schrödinger Equation for a free particle.

## 2 Brownian motion and the Schrödinger Equation

Brownian motion is governed by the diffusion equation

$$\frac{\partial P(\mathbf{r})}{\partial t} = D\Delta P(\mathbf{r}, t), \tag{1}$$

where $P(\mathbf{r}, t)$ is the probability of finding the particle at position $\mathbf{r}$ at time $t$. $D$ is the diffusion coeffient of the Brownian particle. The equation needs an initial distribution $P(\mathbf{r}, 0)$, for the complete solution $P(\mathbf{r}, t)$. The typical solution starts with a delta function $\delta(\mathbf{r})$ in the origin. The corresponding $P(\mathbf{r}, \mathbf{t})$ is then given by

$$P(\mathbf{r}, t) = \frac{1}{(4\pi Dt)^{3/2}} \exp[-r^2/(4Dt)] \tag{2}$$

Note that the equation conserves probability

$$\int d\mathbf{r} P(\mathbf{r}, t) = 1. \tag{3}$$

The evolution equation is deterministic, the initial value of $P$ determines the $P(\mathbf{r}, t)$ at later times.

If the particle is observed to be at $\mathbf{r}'$ at time $t'$, the new starting point for the probability distribution is

$$P(\mathbf{r}, t') = \delta(\mathbf{r} - \mathbf{r}'). \tag{4}$$

Through the observation, the probability distribution "collapses" from a wide distribution to a delta function. The collapse is not something physical since the probability distribution $P(\mathbf{r}, t)$ is not a property of the Brownian particle, it only gives the probabilities for finding the position upon observation. There is no way to associate a velocity with the Brownian particle.

By fourier decomposition one sees that all the eigenvalues of the diffusion operator are negative except one. All the wave components with finite eigenvalues decay, the one with eigenvalue zero is stationary. To phrase the physics of the Brownian particle in a formulation akin to the quantum mechanical: the Brownian particle is in a state, which can be found by measurement of the position. By measuring the position again at a later time one finds a new state with a probability given by the solution of the Master Equation with the earlier state as starting point. So the Master Equation gives the transition probabilities from state to state. One can only measure displacements of the Brownian particle, it has no velocity.

Now compare this with a free particle obeying the Schrödinger Equation

$$\frac{\partial \Psi(\mathbf{r},t)}{\partial t} = \frac{i\hbar}{2m} \Delta \Psi(\mathbf{r},t), \tag{5}$$

The similarity between the equations (1) and (5) is striking, but the association of the probability with $\Psi$ is different. Not $\Psi$ but $|\Psi|^2$ is conserved

$$\int d\mathbf{r} |\Psi(\mathbf{r},t)|^2 = \int d\mathbf{r} |\Psi(\mathbf{r},0)|^2. \tag{6}$$

In spite of the formal correspondence of the Eqns. (1) and (5), one cannot just copy the solution Eq. (2) to $\Psi(\mathbf{r},t)$. The reason is that the resulting $\Psi$ is not normalizable in the sense of Eq. (6). It is interesting however to start with a near $\delta(\mathbf{r})$ of the form

$$\Psi(\mathbf{r},0) = \frac{1}{(\pi\alpha)^{3/4}} \exp(-r^2/2\alpha) \tag{7}$$

Taking the square of this function yields a $|\Psi|^2$ which is normalized to 1. For small $\alpha$ the probability distribution is confined to a narrow region around $r = 0$. Via fourier transforms one can solve Eq. (5) leading to

$$|\Psi(\mathbf{r},t)|^2 = \left(\frac{\gamma}{\pi}\right)^{3/2} \exp(-\gamma r^2), \tag{8}$$

with $\gamma$ given by

$$\gamma = \frac{\alpha}{\alpha^2 + \hbar^2 t^2/m^2}. \tag{9}$$

The larger $t$ the smaller $\gamma$ and the more the probability distribution is spread out over all space.

This example shows that the deterministic evolution equation of the two processes are the same, but involve different domains: the classical Brownian motion refers to real positive probability numbers, while the quantum Schrödinger equation operates in the plane of complex probability amplitudes. The common aspect is that an observation interrupts the probability evolution and provides a new starting point for the probability distribution.

It may seem that the similarity between the Brownian motion and the evolution of a quantum free particle is an incident. In the next section we show that a much more general similarity exist between the Master Equation and the Schrödinger Equation.

# 3 The Master and Schrödinger Equation

The Master Equation provides the stochastic evolution of variable $X$. It is defined by the transition rates $W(X|X')$ of state $X'$ to $X$. The probability $P(X,t)$ on state $X$ then obeys the equation

$$\frac{\partial P(X,t)}{\partial t} = \sum_{X'} [W(X|X')P(X',t) - W(X'|X)P(X,t)] = \sum_{X'} H_{X,X'} P(X',t), \tag{10}$$

The first term gives the gain to state $X$ and the second term the loss from state $X$. The equation is deterministic i.e. the probability at any later time $t$ is determined by the probability at an earlier time (e.g. $t = 0$). Somewhat more precise: $P(X,t)$ gives the conditional probability to find the system in state $X$ at time $t$ given the probability distribution $P(X,0)$ at time $t = 0$. In this discrete formulation the time evolution results from a matrix operation on the distribution. The matrix $H_{X,X'}$ is called *stochastic* since the sum over the columns vanishes, as (10) shows. This property guarantees the conservation of probability

$$\sum_X P(X,t) = \sum_X P(X,0), \tag{11}$$

implying that one of the eigenvalues of $H_{X,X'}$ is zero; the others must have a negative real part, otherwise the probability distribution blows up.

The general Schrödinger Equation has the form

$$i\hbar \frac{\partial \Psi(X,t)}{\partial t} = \mathcal{H}\,\Psi(X,t), \tag{12}$$

where $\Psi$ is the wavefunction (state) of the system. This is also a deterministic equation, since the initial value of the wavefunction $\Psi(X,0)$ determines the behavior at later times. The hamiltonian operator $\mathcal{H}$ is hermitian. For classical systems there are strict rules of construction of the Hamiltonian from the Hamilton function by replacement of position and impuls by operators. Eq. (12) yields the conservation law

$$\int dX |\Psi(X,t)|^2 = \int dX |\Psi(X,0)|^2, \tag{13}$$

which is interpreted as conservation of probability.

More generally one sees $\Psi$ as a state $|\Psi\rangle$ and $\Psi(X)$ as the innerproduct of the eigenstate of the operator $X$ and $\Psi$

$$\Psi(X) = \langle X|\Psi\rangle. \tag{14}$$

This makes the transformation to other representations, e.g. the momentum representation, possible. The probability $P(X)$ of finding the value $X$ in state $\Psi$ is given by

$$P(X) = |\langle X|\Psi\rangle|^2. \tag{15}$$

After measuring $X$ the system is in the state $|\Psi\rangle = |X\rangle$ where $X$ is the result of the measurement. Thus the quantum process has an evolution equation and a rule for the probabilities. This defines a stochastic process. The important difference with the Master Equation is that the probability is the absolute square of the $P(X,t)$ that obeys the evolution equation.

# 4 Similarities and Differences between the Processes

The quantum problem is much richer than the stochastic process of the Master Equation. In the Appendix the correspondence is worked out between a Brownian particle in a harmonic force field and the groundstate of a quantum particle in a harmonic well. Only the groundstate distribution can be matched with a solution of Master Equation. The excited states of the quantum particle in the harmonic potential have no counterpart in the solutions of the Master Equation. Another fundamental difference is that the complex value of the probability amplitude allows for a solution where this is not possible for the Brownian motion, which makes the phenomenon of tunneling possible in quantummechanics.

One can view the Schrödinger equation as the square root out of the Master Equation. Square roots are not innocent though. To mention a few profound differences: the wave function is complex and the time evolution is unitary and not decaying. So while a plane wave decays in the Master Equation, it evolves via a phase factor in the Schrödinger Equation.

As the Schrödinger equation is linear one can make superpositions of states, which gives an association with waves. Therefore $\Psi$ is called the wave function, a somewhat unfortunate name, probability amplitude would have been better.

The analogy with stochastic processes is even stronger for many particles. Probabilities for more than one particle are naturally defined in the product space of the particles. The probability on a pair of particles is a function in a six-dimensional space. Similarly the wavefunction of two particles is a function in a six-dimensional space. This makes the analogy with waves even more obscure. For two well-separated particles a new twist of quantummechanics comes into play. Due to superposition, the probability distribution does not have to be the product of the probabilities on the particles, what one would expect for independent systems. So funny correlations (entanglement) between the particles may arise, which cannot occur in the classical Master Equation.

# 5 The two-slit experiment

We take one of famous experiments in quantummechanics, the two-slit experiment, in order to illustrate the differences between the Master Equation and the Schrödinger Equation. An electron is fired at a barrier with two open slits. After passing through the slits, the electron is detected at a screen. For the full experiment two detection devices are registrating through which of the slit the electron passes. These detection devices can be turned off and on.

According to the Master Equation the probability to arrive at a point at the screen is the sum of the probabilities to pass via one of the slits. Both paths give the product of the probability to arrive at the slit times the probability to go from the slit to the screen. Since probabilities are essentially positive all points on the screen have a positive probability for detecting the electron. If the electron is detected at one of the slits the probability to reach the screen is the probability of that path (and not the sum of both).

However, the screen displays an interference pattern, since in quantum mechanics one must add the wave function and not the absolute squares. The interference pattern is reminiscent of coherent light passing through the two slits. No wonder that the analogy with waves is invoked and that the Schrödinger Equation is seen as a *wave equation*. An unfortunate misnomer, suggesting that $\Psi$ is a wave of a field, as Schrödinger originally thought. The analogy is misleading, since one can also detect the passing of the electron through the slits. The outcome is that the electron is always detected as passing through one of the slits, never through both at the same time. Detection of the electron at one of the slits destroys the interference pattern. A wave, on the contrary, interferes with itself, since it is half present at both slits. The result for the electron is not strange if one realizes that quantum evolution is a stochastic process. So if the electron is found at one slit one must continue with that new initial value for the wavefunction.

# 6 Classical versus Quantum

A probabilistic description is meaningful if the probabilities have a non-zero variance. If the variance shrinks the probability distribution becomes a prediction. Consider the trajectory of a charged elementary particle in a cloud chamber. The condensation trail that the particle leaves behind, can be seen as a repeated measurement of the position of the particle. Each droplet that is formed due to the passage of the charge is a measurement of its position. The trail of droplets is wide with respect to the size of the particle. So the probability distribution of the position of the particle shows a large variance. If we look on the other hand to the trajectory of the moon, the quantum variance is negligble. One can predict the trajectory of the moon many years ahead.

There is a gradual transition from quantum to classical behaviour. It is not sharp as in a phase transition but it is a cross-over. Sufficiently far away from the cross-over region the distinction between micro and macro is clear.

In the quantummechanical description there are parameters (involving Planck's constant $h$) that tell whether the system behaves classically or quantummechanically. An example is the de-Boer parameter $\Lambda$ for noble gases

$$\Lambda = \frac{h}{\sigma\sqrt{m\epsilon}}, \tag{16}$$

telling whether a noble gas is classical like argon or quantum mechanical like helium. Here $\sigma$ is the range of the hard core, $m$ the mass of the particles and $\epsilon$ the depth of the interaction well.

There is no use to associate a quantum distribution to a macroscopic object. Invoking the wave function of the universe makes no sense and wrongly suggests that the time evolution of the this wave function has a relation to how the universe actually develops.

# 7 The measurement process

Contrary to measurements of macroscopic objects, where the variables have definite values, the measurement of a fluctuating quantum observable deserves some commentary. Before discussing this question it is good to note that there are few measurements that probe the wavefunction in a fundamental way as e.g. in the two-slit experiment. The most striking successes of quantummechanics are the prediction of the properties of matter: the possible elements in the periodic table, the covalent bonds, magnetism and superconductivity, the existence of bosons and fermions and their profound implications. All these properties do not require an interpretation of the wavefunction or a theory of the measurement. For instance the energy levels of the electrons in an atom are not determined by measuring the energy but by inspection of the absorption spectrum.

There are three types of interaction: between two classical systems, between two quantum systems and between a quantum system and a classical system. The interaction between classical systems is controlled by the classical equations of motion, the interaction between two quantum systems is controlled by the Schrödinger Equation for the combined state of the two quantum systems. The interaction of a quantum system with a classical system gives a projection of the quantum state according to the Born rules. If it also leads to a detectable changes in the classical system one has a measurement and one can infer from the change in the classical system the state to which the quantum system is projected. So a measuring device is a macroscopic instrument which changes its status upon interaction with the microscopic probe. It is important that the combination, observable and measuring device, are isolated from the environment otherwise the measurement is spoiled by external noise.

Measuring whether the electron in the two-slit experiment has passed one slit requires a macroscopic change in the status of the measuring device. If the device is fluctuating as a microscopic object it is useless as a measuring device. Measuring a quantum observable requires that the interaction of the observable with the device is capable of changing the status of the device. The trail of an elementary particle in the cloud chamber is a prototype of the interaction of a quantum particle with a macroscopic system. Each droplet generated by the elementary particle is a measurement of its position. The thickness of the trail is a measure for the variance of the position of the particle. It is magnified by the growing of the droplets. Such magnification is often helped by a metastable state of the device in which small effects can have large consequences. It is of great importance that the probability distribution of the electron does not depend on the details of the measuring device but only on the state of the electron.

# 8 Why is the Master Equation so easy to accept ?

There is little discussion on the status of the Master Equation. It is clearly an approximate description of the physical process. One usually knows the underlying mechanical theory and the Master Equation is invoked because the underlying mechanics is too difficult. Another aspect showing the approximate character of the equation is that the temporal evolution is irreversible. All the eigen modes of the Master Equation decay, except the mode corresponding to the eigenvalue zero, which yields the stationary state. Although the macro world is obviously irreversible, the micro equations have to be time reversal invariant. The Master Equation applies to the mesoscopic behaviour: it is less detailed than the micro description but it still contains the fluctuations around the macroscopic variables.

The probabilities obey the standard rules of classical probability theory. In particular the Kolmogorov theorem is obeyed, which states that the probability to get from the initial state $X_0$ at $t_0$ to the final state $X_2$ at $t_2$ equals the sum over the probabilities of all paths via intermdiate states $X_1$ and time $t_1$. In the two-slit experiment, the probability for the electron to reach the screen, is the sum of the probabilities to pass through the slits and from there on to the screen. This is in line with our intuitive notion of probabilities.

# 9 Why is quantummechanics so difficult to accept?

The first clash between the two heroes of the old quantum theory, Einstein and Bohr, concerned the two-slit experiment. Einstein considered quantummechanics incomplete since the electron sometimes behaved like a particle (when passing through the slits) and sometimes as waves (when interfering at the screen). The answer of Bohr was rather clumsy using particle-wave duality and the uncertainty relations for the interaction between the electron and the measuring device. Viewing the quantum process as a stochastic process with the rules of forming the probabilities, the answer is clear and unambiguous. Since Bohr was the major opponent of Einstein the Born rules came to be known as the Kopenhagen interpretation. But Bohr was, with his notion of particle-wave duality and later of complimentarity, not the ideal spokesman of the Born rules for the probability.

The stochastic character of quantum evolution is unavoidable, but hard to swallow for a mind that is grown up with causality in the form: every event has a cause. This does not hold in quantum mechanics. If the electron passes in the two-slit experiment through the upper slit, there is no reason why it has chosen the upper and not the lower slit. The reasoning of Bohr, which led him to the notion of complementarity is a dead-end street. One simply has to accept that the outcome is intrinsically unpredictable. Von Neumann [3] already showed that the structure of quantummechanics does not permit hidden variables.

The dispute also triggered the question "what happens during the measurement?" "Is the Schrödinger Equation still valid?". All what happens is a transition of the state of the object to the eigenstate of the observable, corresponding to the found value. One can call that a collapse of the wave function. A similar question for the Master Equation has the non-controversial answer: yes, since collapse only means a new start for the probability distribution. So the same answer holds in quantummechanics: yes it collapses, but there is nothing physical going on. Just another starting point for the probabilities: the eigenstate of the measured quantity.

Still the Schrödinger Equation is given by some people the status as the exact mechanics of the wave function. Van Kampen [7] aptly summarizes his opinion in his fourth theorem (with a slight modification):

Theorem IV: *Whoever endows the wave function with more meaning than is needed for computing transition probabilities, is responsible for the consequences.*

The extreme interpretation of the wavefunction as a physical field governing the evolution of any system is the idea of the wavefunction of the universe. One should keep in mind that quantummechanics is only tested in isolated systems and that only in isolated systems meaningful calculations can be carried out. The marriage of the quantum evolution with the probabilistic rules leads to the idea of many worlds. Every time a choice is made a new world is created. All other possible choices create simultaneously parallel universes. Apart from the fact that this idea is useless for calculations there is also no trace of influences of the many world on each other.

The psychology of the status of the Schrödinger Equation is understandable. quantummechanics was developped to replace the classical deterministic mechanics and in fact the quantum equations are mostly constructed from the classical equations by replacing variables as position and momentum by operators in a well defined way. Later on there appeared operators without a classical analog such as spin, (which Dirac derived by studying the square root out of the Klein-Gordon equation!). As the hamiltonian is the basic ingredient in the evolution of the wavefunction, all the experience with hamiltonians, such as conservation laws, can be exploited in the Schrödinger Equation.

Stochastic processes were, so far, known to be approximate. One knew the underlying mechanistic motion of the Brownian particle. Moreover the Master Equation cannot be fundamental, since it is not time reversal invariant, while a fundamental theory has to have this property. In fact in quantummechanics the situation is opposite: the Schrödinger Equation is time reversal invariant and there cannot be an underlying deterministic mechanics.

For a number of people this is no satisfactory answer. They hope to get information on the measurement by studying the Schrödinger Equation for the combined system of object and measuring device. This is a mathematical interesting exercise, but it will not give a clue which of the possible answers is realized. For the probability distribution of the measured value one does not need to know the role of the measuring device since the probability distribution is given by the state of the observable.

While the discussion between Bohr and Einstein becomes irrelevant once the stochastic nature of the quantum process is accepted, the second attack on quantummechanics of Einstein, Podolsky and Rosen [8] is much more inspiring. It involves a system of two particles which are initial close together and interacting and then fly apart. By measuring on one of them one can get information on the other particle without the possibility that the first signals to the second how it has to behave. The issue originally refered to the question whether one could associate "elements of reality" to a system without actually measuring these. Later Einstein objected against the "spoky interaction at the distance". The conclusion is that quantummechanics, relativity and locality do not go together.

In the later variant of the "EPR paradox" by Bohm [9] one constructs, "EPR pairs", a singlet state of two particles one with spin up and one with spin down

$$|\psi\rangle = \frac{1}{2}(|1\uparrow, 2\downarrow\rangle + |1\downarrow, 2\uparrow\rangle) \tag{17}$$

After the construction of the pairs the constituing particles 1 and 2 are separated far apart. Then the spins are measured. There is a 50% probability that one finds particle 1 to have spin up or down and the same holds for particle 2. Inspecting the records after the measurements one finds that if 1 is up, 2 is down or vice versa. The distance is of the particles is so large and the times of measurement so close that no signal from 1 to 2 is possible to tell how 2 has to react on the measurement of 1. This is Einstein's "spoky interaction at a distance". If we replace "interaction" by "correlation" it sounds already less suggestive. The particles 1 and 2 are indeed correlated or entangled in the singlet state.

The discussion remained purely theoretical till Bell [10] made the distinction between classical and quantum probabilities more explicit. He constructed an inequality holding for all possible classical correlations and showed that in special cases quantum correlations violate this classical criterion. This was confirmed by the experiments of Aspect [11]. Presently one is searching to exploit the features of entanglement. It cannot be used as a means to transmit information (if so John Bell hoped that they would call it the "Bell Telephone"). But it can be used as a non-breakable encryption and as a check on eaves dropping. Also for the operation of the quantum computer entanglement is an essential feature.

The unitary evolution of the wave function of a quantum system is denoted as *conservation of information*. This becomes a big point in astrophysics: can quantum information escape from a black hole? As soon as the word information is used, the discussion becomes a bit vague. The sort of information carried by the wavefunction is nothing else than a means to calculate probabilities.

We take again the two-slit experiment as illustration. One would think that by measuring the passage at the slit, information is gained on the electron. On the other hand information on the electron is destroyed since its wave function "collapses". I presume that the proponents of the "conservation of information" will argue that the combined system of electron and measuring device evolves unitarily, so total information is conserved. But by unitary evolution the combined system remains forever in a superposition of possible outcomes, while in reality only one happens. So the sort of information carried by the unitary wavefunction of the combination is at odds with the reality.

Suppose that we did a two slit experiment with *classical* particles, in which not only the arrival distribution at the screen is observed, but also through which slit the particle has passed. Then one has two options to calculate the distribution at the screen: use the information of detection at the slits or ignore it. Both lead to the same result. Using it, one can calculate two sub-distributions related each to a passage at one of the slits. The sub-distributions contain more information. Their sum yields, by Kolmogorov's theorem, the same as calculated without the knowledge of passing.

The situation in quantummechanics is different. Measuring both passing through the slit and arrival at the screen, gives no interference pattern. So, if the passing through the slits is measured but ignored (may be someone else measures them), one draws the wrong conclusion about the probability distribution at the screen. This is the clue why eaves dropping shows up in transmission of quantum information and not in classical information. It prompts me to formulate the following theorem:

Theorem IVa: *Anyone ignoring information provided by measurements, draws wrong conclusions.*

# 10 Conclusion

It is remarkable that even Einstein, one of the founding fathers of Brownian motion theory, had such a trouble accepting the stochastic nature of quantummechanics. Originally he set up a stochastic interpretation, but later got convinced that "God does not play dice", on which Bohr replied: "How do you know what God does?"

The quantum evolution equation for the probability amplitude together with the Born rules for calculating the probabilities define a stochastic process. Acknowledging this, Quantum Dynamics is crystal clear. There is no room for an underlying deterministic mechanics.

Maybe the lingering quarrels with quantummechanics have to do with the traditional introduction for freshmen, which follows the historical part from the old quantum theory of Bohr, via particle-wave duality to the Schrödinger Equation. The latter has the advantage that it is easier to visualize as a wave equation similar to the equation for electromagnetic waves. The more fundamental approach of Heisenberg remains out of sight for a while in the education. This is putting the students on the wrong track, one should make students familiar with stochastic processes before teaching them quantummechanics.

This advise is against the present trend where the introduction to quantummechanics is already placed in the high-school curriculum. Indeed quantummechanics and relativity are the most spectacular theories in physics and it is tempting to teach them in an early stage in the curriculum. The price of this approach is that the students get the impression that quantummechanics is a mystery, while the rules of quantummechanics are clear. As consolation one then offers them the quote from Feynman [12]: "nobody understands quantum mechanics".

# 11 Postscript

Editor's note: Sadly, Professor J.M.J. van Leeuwen passed away in Dec 2024. I have decided to include here his final response to one of the referee's objections which the reader may find useful.

The author's reply to the referee.

Comparing classical stochastic processes with quantum evolution one finds correspondences and differences. The correspondence is that both involve a stochastic process i.e. random motion. The difference is the association of the probability distribution with the process. In the classical case there is an equation for the evolution of probability, while in the quantum case there is an equation for the amplitude of the probability (wave function). The absolute square gives the probability. In both cases the evolution of probability is deterministic. The referee finds the fact that the probability amplitude is in general complex, an overwhelming difference which makes any comparison between the quantum and classical case futile. Indeed, this is a very important difference. Moreover, the quantum probability amplitude gives the probability distributions of all observables. Focusing so strongly on the (deterministic) evolution of the probability amplitude, the referee does not appreciate the fact that the quantum case is also stochastic. His question: "how does a quantum state, which is a superposition of two eigenstates, become one of the two through measurement?", misses the stochastic nature of the quantum process. In the classical case the comparable question "how does the Brownian particle know that it is here and not there" has the trivial answer: sometimes it is here and sometimes there. Likewise, the superposition of quantum states gives sometimes one state and sometimes the other, according to the statistics dictated by the probability amplitude. In this light "wave function" and "collapse of the wave function" give the wrong impression, where "probability amplitude" and "restart of the probability amplitude" are much more appropriate. This may seem only a question of semantics, but acknowledging the stochastic character of the quantum process makes the Bohr-Einstein discussion irrelevant just as notions such as "the wave function of the universe" and "the many worlds interpretation" are.

**Ackowledgement**

I am indebted to critical comments of Michael Nauenberg, Henk Hilhorst, Gerard Nienhuis and Dennis Dieks.

# A A particle in a harmonic potential

A specific example of the Master Equation is the Brownian particle. The state of the particle is the position $\mathbf{r}$. When the transition probabilities are confined to small steps, the Master Equation can be written in the form of a Fokker-Planck equation. Introduce the step

$$\mathbf{s} = \mathbf{r} - \mathbf{r}' \tag{18}$$

and write the transition probability as

$$W(\mathbf{r}|\mathbf{r}') = W(\mathbf{r}'; \mathbf{s}). \tag{19}$$

Using this notation the Master Equation gets the form

$$\frac{\partial P(\mathbf{r},t)}{\partial t} = \int ds\, W(\mathbf{r}-\mathbf{s};\mathbf{s})P(\mathbf{r}-\mathbf{s},t) - P(\mathbf{r},t)\int ds\, W(\mathbf{r};-\mathbf{s}). \tag{20}$$

The assumption is that $W$ and the probability $P$ depend smoothly on the first argument $\mathbf{r}$. Then we may expand the first term in the Master Equation with respect to $\mathbf{r}$, with keeping the second argument in its full glory

$$W(\mathbf{r}-\mathbf{s};\mathbf{s})P(\mathbf{r}-\mathbf{s},t) = W(\mathbf{r};\mathbf{s})P(\mathbf{r},t) - [\mathbf{s}\cdot\nabla]W(\mathbf{r};\mathbf{s})P(\mathbf{r},t) + \cdots \tag{21}$$

Inserting this expansion in the Master Equation (20), the first term in the expansion cancels the second term in the Master Equation. Thus the Master Equation transforms into the Fokker-Planck equation

$$\frac{\partial P(\mathbf{r},t)}{\partial t} = -\nabla\cdot\mathbf{a}_1(\mathbf{r})P(\mathbf{r},t) + \frac{1}{2}\nabla\nabla\cdot\mathbf{a}_2(\mathbf{r})P(\mathbf{r},t). \tag{22}$$

where the $\mathbf{a}_i(\mathbf{r})$ are the jump moments

$$\mathbf{a}_i(\mathbf{r}) = \int ds\, \mathbf{s}^i\, W(\mathbf{r};\mathbf{s}). \tag{23}$$

The simplist jump moments are

$$\mathbf{a}_1 = -a_1\mathbf{r}, \qquad \mathbf{a}_2(\mathbf{r}) = 2D\,\mathbf{I}. \tag{24}$$

The first jump moment favors the jumps towards the origin, the more the further out the particle is. It represents the harmonic force on the particle. The second jump moment takes as lowest order the isotropic distribution. This gives the Fokker-Planck equation the form

$$\frac{\partial P(\mathbf{r},t)}{\partial t} = a_1\nabla\cdot\mathbf{r}\,P(\mathbf{r},t) + D\Delta P(\mathbf{r},t). \tag{25}$$

The stationary state distribution has the form

$$P(\mathbf{r}) = \left(\frac{\kappa}{2\pi}\right)^{3/2}\exp(-\kappa r^2/2). \tag{26}$$

with $\kappa$ given by

$$\kappa = a_1/D. \tag{27}$$

The Schrödinger Equation for a particle in an harmonic potential $b\,r^2/2$ reads

$$i\hbar\frac{\partial\Psi(\mathbf{r},t)}{\partial t} = -\frac{\hbar^2}{2m}\Delta\Psi(\mathbf{r},t) + \frac{1}{2}b\,r^2\Psi(\mathbf{r},t). \tag{28}$$

The stationary states are the energy eigenstates obeying the equation

$$E\Psi(\mathbf{r}) = -\frac{\hbar^2}{2m}\Delta\Psi(\mathbf{r}) + \frac{1}{2}b\,r^2\Psi(\mathbf{r}). \tag{29}$$

The groundstate is again a Gaussian

$$\Psi_0(\mathbf{r}) = \left(\frac{\lambda}{2\pi}\right)^{3/4} \exp(-\lambda r^2/4), \tag{30}$$

with the relations

$$E_0 = \frac{3\hbar^2\lambda}{4m} \qquad b = \frac{\hbar^2\lambda^2}{4m}. \tag{31}$$

The second relation gives the value of $\lambda$ in terms of the harmonic force parameter $b$. If we match the groundstate distribution with the probability distribution of the stationary state of the Fokker-Planck Equation

$$\Psi_0(\mathbf{r})^2 = P(\mathbf{r}), \tag{32}$$

we have to set $\kappa = \lambda$ and have to choose $a_1$ as

$$a_1 = D\kappa = D\lambda = D\left(\frac{4mb}{\hbar^2}\right)^{1/2}. \tag{33}$$

# References


[1] W. Heisenberg, Z. Phys. **33** (1925), 879-893.

[2] M. Born, Z. Phys. **37** (1926).

[3] J. von Neumann, "Mathematical Foundations of quantummechanics", Princeton University Press 1955.

[4] M. Tegmark, and J. A. Wheeler, Scientific American **284** 75.

[5] R. Furth, Z. Phys. **81** (1933), 143-162.

[6] L. Peliti and P. Muratore-Ginanneschi, arXiv:2006.03740v1 [physics.hist-ph] 5 Jun. 2020.

[7] N. G. van Kampen, Physica A, **153**, (1988) 97-113.

[8] A. Einstein, B. Podolsky and N. Rosen Phys. Rev. **47** (1935), 777–780.

[9] D. Bohm, Phys. Rev. **85** (1952), 180.

[10] J. Bell, Physics **1** (1964), 195-290.

[11] A. Aspect, Nature 398 (1999), 6724.

[12] R. P. Feynman, The Messenger Lectures, 1964, MIT.

# Chapter 4

## A centennial reappraisal of Heisenberg's Quantum Mechanics with a perspective on Einstein's Quantum Riddle.

Tuck C. Choy*

*Center for Space, Time and the Quantum, Parc Maraveyre Bât. 1, 13260, Cassis, Bouches du Rhône, France and Departmento de Fisica-CIOyN, Universidad de Murcia, Murcia 30071, Spain.*

## Abstract

Heisenberg's breakthrough in his July 1925 paper that set in motion the development of Quantum Mechanics through subsequent papers by Born, Jordan, Heisenberg and also Dirac (from 1925 to 1927) is reexamined through a modern lens. In this paper, we shall discuss some new perspectives on (i) what could be *the* guiding intuitions for his discoveries and (ii) the origin of the Born-Jordan-Heisenberg canonical quantisation rule. From this vantage point we may get an insight into Einstein's Quantum Riddle [1–3] and a possible glimpse of what might come next after the last hundred years of Heisenberg's quantum mechanics.

(This paper is dedicated to the celebration of one hundred years of quantum mechanics, on the anniversary of Heisenberg's founding paper on the subject in July 1925, delivered by the author at the IQSA 2025 Conference at Tropea in Calabria, Italy from 30 June to 4th July 2025).


## I. INTRODUCTION

Unlike earlier discoveries in theoretical physics prior to 1925, Quantum Mechanics was the result of the efforts of more than one man, namely: Heisenberg, Born, Jordan, Dirac and Schrödinger. It was indeed Heisenberg who made the first and major step in the discovery of Quantum Mechanics, his priority and legacy is therefore unquestionable [4]. However, had he not made that step, it is clear from the historical developments that it would have been Schrödinger who would claim priority, though his perspective using the de Broglie matter wave mechanical approach, though ubiquitous, had turned out to be incorrect eventually. Some physicists famously Weinberg [6] and also Fermi [7], considered Heisenberg's paper of July 1925 incomprehensible because it was clouded in mystery as to why he made the assumptions that he did and what motivated the steps in his thinking. In 2004 Aitchison et al [8] made an attempt to provide an "understanding" of Heisenberg's July 1925 paper by filling in the apparent gaps in the calculational details (some omitted by Heisenberg in his 1925 paper) thereby making a reconstruction of this landmark paper. While their effort is instructive and may be useful for the teaching of advanced courses in quantum mechanics, and may have relieved Weinberg's and Fermi's "incomprehensibility" objection, it has done little in my opinion to elucidate the "magical" part of Weinberg's observations.

* tuckvk3cca@gmail.com

Although one concur with Aitchison et al [8] that it may not in fact be possible to "render completely comprehensible the mysterious processes" whereby physicists "gain new insights about nature" through a phenomenal breakthrough, nevertheless there are some useful clues that are worth exploring. This is so, especially as it is now a century after the development of the subject, many of whose stones had been laid as early as the period from 1925 to 1927 and are in need of scrutiny under a modern lens. By doing so we will not only make the process of discovery less "magical" and plausible, but in this case we may also gain some new insight, one missed during the last century, that might lead us closer to a solution for Albert Einstein's famous Quantum Riddle, by which he meant: what are "the principal reasons behind the quanta?" [1, 9]. The words Quantum Riddle were coined by Einstein himself (*not* Lande [1]) as early as 1923[10] but the spirit of the riddle has been in his writings and correspondence even earlier since 1916[1] until his death in 1954 with numerous interpretations as to its meaning [10]. For this paper we shall stick closely to Lande's [1] version, in fact explicitly defining here the quantum riddle as the origin for the Born-Jordan-Heisenberg quantization rule eqn(1) from which everything else quantum follows [11]. This is the aim in this paper. Even with this limited aim, this paper has found a number of fascinating conclusions and speculations that will hopefully motivate further research for the next hundred years.

## II. HEISENBERG'S DISCOVERIES - A SUMMARY

The central result of Heisenberg's landmark paper of July 1925 [12] as later clarified by Max Born and Pascual Jordan [13], was the quantum rule:

$$[\hat{q}, \hat{p}\,] = i\hbar\hat{1} \;, \tag{1}$$

which Born considered a postulate of the new quantum theory and even today is still accepted by many as such or as an empirical rule beyond that of the old Bohr-Sommerfeld quantization rule. In modern language the left hand side is the famous commutator product of two non-commuting Hermitian operators (aka matrices in the early days) while the right hand side is the imaginary constant $i\hbar$ multiplied by the unit operator. As we shall see later, one can call the LHS a closure operator. Heisenberg in his July 1925 paper [12] was able to obtain only the diagonal part of this expression (not knowing anything about matrices at that time, which was all the more remarkable). Within a matter of days of receiving Heisenberg's paper, Born immediately realised that it represented a matrix product, conjectured that the off-diagonal elements are zero and recruited the assistance of Pascual Jordan to settle the issue (which they did) and produced their more comprehensive paper which formed the foundation of quantum mechanics as we know it today [13]. Born considered his discovery to be so important and fundamental that he had it engraved on his tomb stone [4, 14]. It is perhaps interesting to speculate as to the reason why he did this, apart from the fact that Born was obviously very proud of his discovery (which he had said so himself in a commentary of his July 1925 letter to Einstein [15]). Bertrand Russell who contributed a forward to the book [15] had mentioned that both men were "brilliant and humble", so pride or ego could not have been the answer or the only answer. A little history may throw some light on this. Archimedes was considered perhaps the greatest scientist and mathematician in antiquity. He had the result of his favourite theorem that the volume *or* surface area of a sphere circumscribed in a cylinder is 2/3 that of the volume *or* surface area of the cylinder (first published in 225 BC) made into a sculpture with inscriptions in his tomb stone which was later discovered by the Roman statesman and philosopher Marcus Tullius Cicero (106-43 BC). Archimedes basically issued a challenge to future generations: find another way of proving this that is better than mine [16]. As is well known a formal proof was not available until the advent of integral calculus in the 17th century. Max Born in one's opinion issued a similar challenge, for though critical of Einstein's objections to the Copenhagen interpretation in his later years, he was well aware of Einstein's Quantum Riddle and had to remind future generations that it should not be forgotten. Born apparently continued to toy with the fundamental significance of his commutation rule until the war in collaboration with Klaus Fuchs according to Bernstein [17].

We are now jumping ahead of the story so let us back track a little. In July 1925 Heisenberg had no idea about matrices, no concept of what we now know as the basic formalism of quantum mechanics. His "magical" discovery essentially came from an assignment set to him by Born, to find this new mechanics which Born already knew must exist as early as 1924, to which he first put the name Quantum Mechanics in print[18]. Born also suggested in regular discussions with his two assistants that they focus on the transition amplitudes and look for "some kind of symbolic multiplication" of these quantities that would form the new mechanics [19] . It is worth reading this paper [18] again to find out why he thought so.

Heisenberg too must have studied this paper before he took off to Helgoland. In addition he had his own paper with Kramers written in January 1925 [20] and his earlier discussions with Bohr on the same subject of dispersions and transition amplitudes [21] in the summer of 1922. So by the time he left for Helgoland he had this knowledge with him and his strategy was clear too. Rather than trying out his new ideas on the atomic orbits of the Bohr hydrogen atom model which had caused him difficulties, he chose a simpler model: the one dimensional anharmonic oscillator. He also knew from his discussions with Bohr that the harmonic oscillator model might be too simple as transitions there are all evenly spaced harmonics just as in the classical case, an issue he had mentioned to Bohr in 1922 in relation to the quadratic Stark effect, which caused Bohr some discomfort [21]. On this he was perhaps overly cautious or super prescient (see section V later) and he made a fundamental error to postulate a theory based entirely on *observable* quantities among which he considered the particle co-ordinates not to be. As noted by van der Warden, this was an error but a 'fruitful error' [22] as it caused him to focus on developing methods for calculating only those quantities which were directly accessible at that time. Heisenberg's theory was developed on the basis of a number of assumptions that we now summarise with some condensation of the ideas and updated notes [23].

(1) In the atomic range, classical mechanics is no longer valid.

(2) Any new theory must satisfy Bohr's Correspondence principle i.e. in the limit of large quantum numbers, it must agree with classical mechanics. An important device used to satisfy this was the trick of replacing differential quotients in classical formulae by difference quotients. This trick originates from Born [18] and has been used by Kramers and Heisenberg [12].

(3) Heisenberg's hypothesis. Heisenberg felt that the difficulties encountered at that time were due to the failure of the kinematics underlying the mechanics but *not* its laws which remain the same as in the classical theory. Hence the equations of motion such as Newton's law: $\ddot{x} = f(x)$ should remain the same, except that the kinematic quantity $x$ must be re-interpreted. This was perhaps the single most powerful insight of his paper, it was bold and it was super intuitive and could be considered "magical". Later we shall have more to say as to how he could have been so convinced. It remains today as his most fundamental quantization axiom; however a more thorough scrutiny is long overdue. In one's opinion, too much focus has been given to the Uncertainty Principle and Entanglement in

the last hundred years without a thorough examination of this hypothesis which underlies the foundation of quantum mechanics.

(4) Knowing that the classical Fourier series: for a periodic quantity[24] $x(n,t) = \sum_\alpha a_\alpha(n)e^{i\alpha\omega_n t}$ is only likely to be valid for large quantum numbers he set about to make the transition to quantum quantities by replacing the classical amplitudes $a_\alpha(n)$ by $a_\alpha(n,n-\alpha)$ and the classical resonance frequencies $\omega_n$ by quantum transitional frequencies $\omega(n,n-\alpha)$. This was a major breakthrough on his part, though it had been earlier suggested by Born [18].

(5) From the assumption (4), Heisenberg then deduced that the quantities concerned have certain 'multiplication features'. He showed how this is done for $x^2(t)$ and also $x(t)y(t)$ which is later recognised by Born as matrix multiplication. To do so, he was guided by the Ritz spectral combination rule: $\omega(n,n-\alpha)+\omega(n-\alpha,n-\beta) = \omega(n,n-\beta)$ but had to invoke the Sommerfeld-Wilson quantum rule. Unfortunately this makes his theory and the later papers by Born, Jordan and himself somewhat ad-hoc and semi-empirical [13] as they were derived essentially from the old quantum theory, in particular Planck's radiation formula. We will return to this point later.

(6) To make further progress he must now obtain a formula to generalise the old quantum rules of Bohr and Sommerfeld. This uses the trick given by (2) above as well as the matrix multiplication rule. He effectively derived the diagonal elements of the commutator eqn(1) and, using a correspondence formula obtained by Born [18], showed that it is equivalent to the Thomas-Kuhn sum rule of dispersion [12]. Heisenberg seemed to intuitively realise that his "algebra" is incomplete or unspecified without this rule eqn(1). One can call this his closure hypothesis.

As the details of Heisenberg's paper and other commentaries now exist, we need not repeat them here [8, 25]. Instead we shall focus on items (3), (5) and (6). However in the seminal paper of 1925 [13], the Born-Jordan rule eqn(1) was only discovered as an extension of Heisenberg's earlier work [12] from the old quantum theory. This was somewhat ad-hoc as mentioned above and was therefore not rooted in very strong theoretical foundations which were, indeed not possible at that time.

Let us first consider item (3) above, as to why Heisenberg felt that it was the kinematics which was at fault in trying to develop a new form of mechanics. First he had by then spent a great deal of time with Bohr and almost everyone at that time was familiar with the success of the Bohr's model for the H atom. Bohr did not use any fanciful ideas, no new supersymmetric fields or gauge boson field models etc to quote a common present day parlance. Bohr used the Rutherford model and plain old Newtonian mechanics albeit with two new ingredients: the Einstein-Planck energy frequency relation and a postulate for the notorious quantum jumps between stationary states, a concept retained till this day. Next Heisenberg had spent some time talking with Einstein, and he must have convinced himself that something like special relativity was at work here. Einstein told him that it is theory that decides what can be observed, so he decided to pursue a theory based on so called observables. Now special relativity was a modification of Newtonian theory based on a change of kinematics, not dynamics. So Heisenberg, then only 23 must also have convinced himself to do the same. So take Newton's law and make some kinematic changes to discover the new mechanics. The question was what kinematic changes? His knowledge about relativity must be as good as that of his contemporary Wolfgang Pauli, who at the young age of 21 had already written a full monograph on relativity covering both the special and general theories. He might even have tried generalising his observables from vectors to tensors or even dyadics and things like that which he might have found in his textbooks, but failed in the attempt. So he chose to use the Fourier series and knowing the coefficients and Ritz law he came up with non-commuting objects that lead to a spectrum that agreed with experiments, for by that time the vibrational spectrum of atoms and molecules was well known. In addition he has now a way to calculate transition amplitudes. Could he have done better? One may think the answer to this question is yes, if he was familiar with the French literature [26]. The reason is that between the time of Newton and Einstein, a most significant theory of dynamics had been found which has deep implications to the present day. This was the treatise: Traite de Dynamique [27] published by the French physicist and mathematician Jean-Baptiste le Rond d'Alembert who preceded Hamilton and Lagrange by defining the force of inertia and incorporating it into the principle of virtual work and so extending it from statics to dynamics. Pars [28] called d'Alembert's principle the fundamental law of mechanics, superseding the laws of Newton. The implications of d'Alembert's principle for mechanics and especially quantum mechanics are deeply profound. Unfortunately for the last hundred years this connection has rarely been made or fully explored. For d'Alembert, dynamical motion is an equilibrium condition once we incorporate the force of inertia. Now Heisenberg could have better understood why he must introduce new kinematic objects in quantum mechanics, if he had realised that it is required by d'Alembert's principle. In classical mechanics, d'Alembert's principle can be stated in terms of the applied force and the inertial force in an equilibrium condition using the principle of virtual work, *subject to harmony with given kinematic constraints* [29]. For N bodies this is usually written in the form:

$$\sum_{n=1}^{N}(\mathbf{F}_n - m_n\mathbf{A}_n)\cdot\delta\mathbf{R}_n = 0\ , \tag{2}$$

where $\mathbf{F}_n$ is the external applied force, $\mathbf{A}_n = \ddot{\mathbf{R}}_n$ is the acceleration and $\delta\mathbf{R}_n$ is the virtual displacement of each particle $n$. So what Heisenberg is intuitively saying is that d'Alembert's principle in the form eqn(2) is untenable. He could and may have tried other extensions of the equation to tensors, dyadic, quaternion and other objects. In the modern context most of these attempts would have been fruitless if he had indeed made such attempts. This is because they are inadvertently local hidden variable theories that we now know are doomed to failure. Instead Heisenberg opted to choose his quantisation procedure, in the non-relativistic case, by an extension to objects which he later found out through Born are some kind of matrices, nowadays known to be in fact complex Hermitian operators. So we put hats on all the old classical quantities to indicate this:

$$\sum_{n=1}^{N}(\hat{\mathbf{F}}_n - m_n\hat{\mathbf{A}}_n)\cdot\delta\hat{\mathbf{R}}_n = 0\ . \tag{3}$$

This is a deep and powerful statement, it says that with these objects, d'Alembert's equilibrium principle is once again restored. That, one must think is *the fundamental postulate* of quantum mechanics. While we could find no direct evidence that Heisenberg's intuition was guided by d'Alembert's principle, it is not inconceivable that he had been influenced by Sommerfeld's lectures and therefore his "magical" hypothesis item (3) was something quite natural for him. Obviously with such a view the measurement postulates that Dirac later had to propose and von Neumann adopted, including the difficulties with concepts of quantum reality of EPR, wavefunction collapse and so on could have been quite different, if not avoidable. Naturally one can pursue this quantum d'Alembert's equilibrium principle even more deeply. For sure one should not think it will lead us back to the classical equilibrium conditions d'Alembert envisaged at his time and in the words of Einstein " the Lord is subtle", but with equilibrium one will immediately see fluctuations, see later. To pursue this Class I(a) problem [11] would take us too far afield so for now let us return to points (5) and (6). Having decided that he must create new quantum objects which he had shown to be non-Commutative, which greatly disturbed him, he must have realised that the algebra is incomplete and worse even inconsistent without setting up some rules. Consider a simple example: let C=AB and D=BA. Then Y=ABA can be written as Y=CA or Y=AD. To be consistent clearly this requires CA=AD and similarly DB=BC. As a matter of fact, in 1928 Hermann Weyl caught on to this and to avoid other issues with the Born-Jordan-Heisenberg quantisation rule eqn(1), such as the trace anomaly [30], he solved this consistency issue by proposing the quantisation rule: $AB = e^{i\alpha}BA$, since named after him, with A and B being unitary operators and $\alpha$ a real number phase factor [31]. Heisenberg's concerns must have disturbed him so much that he went about finding out how to do the algebra using the technique as mentioned earlier of Born's in item(4) above, finishing with the diagonal part of the quantum rule eqn(1) via items (5) and (6). As noted earlier this approach is ad-hoc but his procedure worked and the results are in agreement with experiment though it may appear to be magical or even mysterious. Einstein for one complained that " the theory says a lot but does not really bring us any closer to the secret of the 'old one' " [32]. For the rest of this paper, we shall show how Heisenberg's motivations can be put on a more fundamental basis, that may perhaps have been more palatable to Einstein.

## III. A RE-FOUNDATION OF QUANTUM MECHANICS - DIRAC QUANTIZATION

As d'Alembert's principle involves an inertial force that is of a polygenic character, the usual practice is to perform an integration by parts [33] and convert d'Alembert's variational principle into Hamilton's least action principle involving only scalar functions. Note that this process is only valid classically for monogenic applied forces and holonomic constraints [29]. We shall assume the same after quantisation. Note also it is only after this process is carried out in classical mechanics that the concepts of virtual and actual paths for the particles become possible. In configuration space the latter can be shown to be determined by Jacobi's principle [29], of great importance in Schrödinger's Wave Mechanics. Clearly in the quantisation of the d'Alembert's principle as presented here, these concepts now no longer have the same meaning. Nevertheless when Born and Jordan [13] realised that the objects Heisenberg had introduced were matrices they did try to propose a variational principle to justify the quantum Hamilton equations in their analysis, which was unfortunately severely inadequate and subsequently ignored in their three men paper [34]. Therefore eqn(1) remains an ad-hoc rule which Born had always considered a postulate [4, 5, 14], although not explicitly considered as an axiom in von-Neumann's axiomatic foundations. It was left to Dirac [35, 36] who in 1926 tried to lay a more fundamental theoretical foundation by proposing his more general quantum Poisson bracket quantisation condition from first principles. However Dirac, in spite of being a native French speaker, was unlikely to have been influenced by d'Alembert's principle; he never cited d'Alembert's principle in any of his works. He was a Hamiltonian man, being influenced in his own words, by Whittaker's "Analytical Dynamics" which contained no mention of d'Alembert's principle[37, 38]. Dirac's rule was:

$$\{\hat{u}, \hat{v}\} = -\frac{i}{\hbar}[\hat{u}, \hat{v}] \ , \tag{4}$$

where the operator Poisson bracket $\{\hat{u}, \hat{v}\}$ on the LHS must be defined as:

$$\{\hat{u}, \hat{v}\} = (\frac{\partial \hat{u}}{\partial \hat{q}} \circ \frac{\partial \hat{v}}{\partial \hat{p}} - \frac{\partial \hat{u}}{\partial \hat{p}} \circ \frac{\partial \hat{v}}{\partial \hat{q}}) \ , \tag{5}$$

by which the Born-Jordan quantisation rule eqn(1) is a corollary, (taking $\hat{u} = \hat{q}$ and $\hat{v} = \hat{p}$) . Here we retain the $\circ$ to denote a symmetrised product, (ignored by Dirac), whose details we will not go into here [39], except that it is symmetric in the exchange of the multiplicands. The Dirac rule eqn(4) was initially obtained through correspondence arguments in the case of large quantum numbers [35, 36]. He was obviously dissatisfied with this approach that he used to propose his quantum law, since in his seminal book of 1931 he proposed a separate theoretical argument, that detached the correspondence principle. Moreover he was unclear what eqn(4) actually meant. Rather than a mathematical equality he seemed to later settle on a corresponding identity or a replacement rule or algorithm for quantisation. Nevertheless this was a fundamentally important advance and his argument was ingenious. Any student of his classic book[40] should be familiar with it and it was based on algebraic manipulations of a four operator Poisson bracket using the Leibniz rule, nowadays associated with dynamical Lie algebraic group properties [40]. Appendix 1 also gives an alternative proof of this important rule. With this "derivation", Dirac canonised his quantum condition eqn(4) as the foundation of quantum mechanics and then went on to show that the Born-Jordan rule eqn(1) can be obtained from it. In 1961, in a letter to van der Warden, Dirac confessed that he had obtained his condition in 1926 first because he had expected "some kind of connection between the new mechanics and Hamilton mechanics". It is unclear however why Dirac was satisfied with his 1931 derivation of his rule [40] because its connection between quantum mechanics and Hamiltonian mechanics contains some logical gaps, which were never really clarified in 1931 nor in fact up to the present day. Perhaps he had searched all his life and failed, or perhaps he felt the problem is sufficiently solved so that he felt no longer bothered by it. In particular, starting from Poisson brackets, the Leibniz rule provides logically only a one way proof of the relation: $\{\hat{u}, \hat{v}\} \Rightarrow= \frac{1}{i\hbar}[\hat{u}, \hat{v}]$ . Owing to this he continued to discuss this rule as a correspondence rule or quantisation rule but not as an equality, let alone an identity. Closer examination shows that his "derivation" can also be viewed as a consistency condition between multi operator quantum Poisson bracket algebra with the Lie algebra of quantum commutators and thus cannot on its own constitute a fundamentally complete derivation of canonical quantisation [40], see Appendix 1. We shall later also discuss other consistency conditions by which we can extend our arguments to quantum Lagrange brackets, a subject that appeared to have been missed in the quantum literature for nearly a hundred years. Later authors such as Groenewold [41] also discovered inconsistencies in Dirac's scheme which nowadays go by the name of "quantization obstructions", technical difficulties we shall not go into in this work.

In this section we shall first fill the logical gaps in Dirac's canonical quantisation scheme and show that it can actually be derived in the reversed way, i.e. $[\hat{u}, \hat{v}] \Rightarrow i\hbar\{\hat{u}, \hat{v}\}$. In his 1963 Scientific American article, Paul Dirac [42] proposed as a major class 1 problem the question: " How can one form a consistent picture behind the rules for the present quantum theory?", perhaps his own way of framing Einstein's Quantum Riddle. Understanding the origin of these rules may help put us closer to the "consistent picture" he so desired. Recall that in classical mechanics, Hamilton's variational principle requires that *both* p and q can and must be varied independently to obtain the Hamilton equations of motion,(see for example Lanczos [43] ). This procedure is a unique consequence of the Legendre transformation that takes one from the Lagrangian $L$ to the Hamiltonian function $H = p\dot{q} - L$. In quantum mechanics this independence of variation is no longer possible due to non-commutativity and the requirement of rules to define the algebra. To restore this independence, we shall now write (restricting to one dimension without loss of generality) the quantum action as:

$$\hat{S} = \int \Big(\hat{p}d\hat{q} - \hat{H}(\hat{p}, \hat{q})dt - \hat{\lambda}\hat{F}(\hat{u}, \hat{v})dt\Big). \tag{6}$$

Strictly speaking a term like $\hat{p}d\hat{q}$ should be replaced by its symmetrised form $\frac{1}{2}(\hat{p}\ d\hat{q} + d\hat{q}\ \hat{p})$ and there will be ordering issues in defining the Hamiltonian $\hat{H}(\hat{p}, \hat{q})$ which however will not affect our arguments here [39]. The kinematic constraint function $\hat{F}(\hat{u}, \hat{v})$ will be further discussed below, where $\hat{u}(\hat{p}, \hat{q})$ and $\hat{v}(\hat{p}, \hat{q})$ are arbitrary functions at this stage. We must emphasize here that as soon as the classical Hamiltonian action is replaced by the quantum one with operators then due to non-commutativity a relationship must exist between the operators as mentioned earlier to ensure consistency and algebraic closure [44]. The variation is no longer constraint free, even for a non-interacting particle as in the classical case. Here $\hat{\lambda}(\hat{u}, \hat{v})$ is the Lagrange multiplier and we have assumed it as a function of $\hat{u}$ and $\hat{v}$ for simplicity, but in fact it need not be [48]. The reader familiar with the theory of Lagrange multipliers will recall that in general $\hat{\lambda}$ need not be determined, hence the name undetermined multiplier and in the case of a vanishing derivative of the constraint (see below), it is in fact undeterminable. Now there are in fact many ways to choose the function $\hat{F}(\hat{u}, \hat{v})$ that will satisfy the Heisenberg hypothesis item (3) as we shall see. One way as alluded to by Dirac [44], is that quantisation can be specified either by a rule using either the commutator or the anti-commutator. Another is the Weyl rule as discussed above [31]. Although $\hat{u}(\hat{p}, \hat{q})$ and $\hat{v}(\hat{p}, \hat{q})$ are arbitrary functions at the moment, later their specific choice will determine the quantisation rule. The inclusion of a constraint function $\hat{F}(\hat{u}, \hat{v})$ in the action is what's missing in the original variational principle of Born and Jordan's paper in 1925 [13]. Also why was the specific choice $\hat{u} = \hat{q}, \hat{v} = \hat{p}$ necessary as a quantization rule in eqn(4) was never given by Dirac [40] (1931), other than that it yields the corollary Born-Jordan rule eqn(1). Once again, note that the algebra cannot be closed or consistent, without specifying the constraint and any application of the variational principle must respect that,; see statement in italics before eqn(2). So what can Heisenberg's hypothesis (see item (3) above) that Newton's law will continue to hold, tell us about the unknown constraint function $\hat{F}(\hat{u}, \hat{v})$. Is it always true, or indeed was Heisenberg just a lucky man? As can be easily seen, a straightforward application of the variational calculus will now provide us with the equations of motion (see for example Landau and Lifshitz [45] ):

$$\dot{\hat{q}} = \frac{\partial \hat{H}}{\partial \hat{p}} + \hat{\lambda}\frac{\partial \hat{F}}{\partial \hat{p}} \text{ and } \dot{\hat{p}} = -\Big(\frac{\partial \hat{H}}{\partial \hat{q}} + \hat{\lambda}\frac{\partial \hat{F}}{\partial \hat{q}}\Big). \tag{7}$$

For Heisenberg's hypothesis to be valid, the canonical equations of motion must remain unchanged. Therefore we require:

$$\hat{\lambda}\frac{\partial \hat{F}}{\partial \hat{p}} = 0 \textit{ and } \hat{\lambda}\frac{\partial \hat{F}}{\partial \hat{q}} = 0. \tag{8}$$

However there are many ways in which this can be satisfied. As an example consider the Weyl quantisation rule [31] in which the constraint function $\hat{F}(\hat{u},\hat{v})$ now takes the form:

$$\hat{F}(\hat{q},\hat{p}) = e^{is\hat{q}}e^{it\hat{p}} - e^{-i\hbar st}e^{it\hat{p}}e^{is\hat{q}} = 0, \tag{9}$$

in which eqn(1) must still hold implicitly for it to be valid. Weyl's rule is nowadays better known as a special case of the Baker–Campbell–Hausdorff formula, which has a long history [46]. Here it is straightforward using Leibniz's rule to show that eqn(8) is now satisfied by way of the partial derivatives vanishing. However we must be clear that this is not the object of our exercise, that is to insert the known quantum rules as the constraint function and derive quantum mechanics. This would at best only give us a bootstrap theory. Instead we want to exploit the constraint function as a means to find out more about Albert Einstein's Quantum Riddle [1] or in his own words: "If only I knew which little screw the Lord applies here" [2]. An alternative could also be that $\hat{\lambda}$ is in fact zero in which case the derivatives of $F$ do not matter. We shall look further into all these factors later; for now we shall follow Dirac's [44] specification and assume that the constraint can be given as some unknown function of the commutator alone (plus perturbations which we shall discuss later, and in view of Heisenberg's hypothesis *must* be *very small* ) i.e. $\hat{F}([\hat{u},\hat{v}])$, see later. This in turn requires:

$$\hat{\lambda}\hat{F}'\frac{\partial}{\partial \hat{p}}[\hat{u},\hat{v}] = 0, \tag{10}$$

and:

$$\hat{\lambda}\hat{F}'\frac{\partial}{\partial \hat{q}}[\hat{u},\hat{v}] = 0. \tag{11}$$

For $\hat{\lambda} \neq 0$ (undetermined multiplier) and $\hat{F}' \neq 0$ i.e. differentiability, we now have:

$$\frac{\partial}{\partial \hat{p}}[\hat{u},\hat{v}] = 0, \tag{12}$$

and:

$$\frac{\partial}{\partial \hat{q}}[\hat{u},\hat{v}] = 0. \tag{13}$$

Now eqn(12) and eqn(13) imply that for the specified choice of $\hat{u}$ and $\hat{v}$ that we must take to define the quantisation rule, the commutator $[\hat{u},\hat{v}]$ cannot be a function of $\hat{p}$ nor of $\hat{q}$. It must be a constant matrix operator, but we have not yet proven that it must be diagonal, nor what specified choice to make for $\hat{u}$ and $\hat{v}$ . To proceed further we need to invoke the Poisson algebra eqn(5). Then we must have for any arbitrary $\hat{u}$ and $\hat{v}$ :

$$\frac{\partial}{\partial \hat{p}}[\hat{u}, \hat{v}] \ = \{\hat{q}, [\hat{u}, \hat{v}]\} \ = 0, \tag{14}$$

and:

$$\frac{\partial}{\partial \hat{q}}[\hat{u}, \hat{v}] \ = -\{\hat{p}, [\hat{u}, \hat{v}]\} \ = 0. \tag{15}$$

Now we are nearly there, for under any canonical transformation that takes $\hat{p} \to \hat{P}$ and $\hat{q} \to \hat{Q}$ *and* must leave the canonical equations of motion eqn(7) intact, *and* indeed also eqn(14) and eqn(15), we require that the brackets:

$$\{\hat{q}, [\hat{u}, \hat{v}]\}_{\hat{p},\hat{q}} = \{\hat{Q}, [\hat{u}, \hat{v}]\}_{\hat{P},\hat{Q}} \tag{16}$$

and

$$\{\hat{p}, [\hat{u}, \hat{v}]\}_{\hat{p},\hat{q}} = \{\hat{P}, [\hat{u}, \hat{v}]\}_{\hat{P},\hat{Q}}. \tag{17}$$

Here I use the notation of Goldstein [47] where the subscripts indicate the canonical variables associated with the bracket. Eqn(17) implies that we must have the necessary condition (Lemma 1):

$$[\hat{u}, \hat{v}]_{\hat{p},\hat{q}} = [\hat{u}, \hat{v}]_{\hat{P},\hat{Q}}. \tag{18}$$

There are only two brackets with this property namely Poisson $\{u, v\}$ or Lagrange $(u, v)$ [47]. However because the commutators satisfy the Jacobi identity, as do the Poisson brackets (see Appendix 1) while the Lagrange brackets do not, we finally arrive at Dirac's Quantum Condition see eqn(19) below. Note that this is not yet a quantisation rule which is why we must be careful to distinguish between what is a quantum condition and what is a quantum rule. The equation:

$$[\hat{u}, \hat{v}] = \kappa\{\hat{u}, \hat{v}\}, \tag{19}$$

where here $\kappa = i\hbar$ by dimensional analysis for Hermitian operators, *is* a quantum condition. Now since this proof of equality goes in the direction: $[\hat{u}, \hat{v}] \to \kappa\{\hat{u}, \hat{v}\}$ i.e. it is a proof of necessity. We also need Dirac's [40] proof (see also Appendix 1) which goes in the opposite direction $[\hat{u}, \hat{v}] \leftarrow \kappa\{\hat{u}, \hat{v}\}$, as a proof of sufficiency. We can now conclude that Dirac's famous Quantum Condition [40] is in fact an identity or equivalence:

$$[\hat{u}, \hat{v}] \equiv \kappa\{\hat{u}, \hat{v}\}. \tag{20}$$

Mathematically what we have proven here is that the commutator bracket Lie algebra and the Dirac-Poisson bracket algebra (subject to operator ordering requirements which we cannot go into here) are isomorphisms. That is not all. Further to eqn(14) and eqn(15) and comments following them, there can be only *one* choice for the unique quantisation rule, namely: $\hat{u} = \hat{q}$ and $\hat{v} = \hat{p}$, or vice versa, that satisfies Heisenberg's hypothesis (item 3). Thus we arrive at the non-commuting canonical Born-Jordan-Heisenberg quantisation rule eqn(1), as the fundamental rule of canonical quantisation. Note that the commutating relations $[\hat{q}, \hat{q}\ ] = [\hat{p}, \hat{p}\ ] = 0$ corresponding to the choices $\hat{u} = \hat{q}\ ,\ \hat{v} = \hat{q}$ and $\hat{u} = \hat{p}\ ,\ \hat{v} = \hat{p}$ respectively also follow from our arguments and are trivial here. Note also that in evaluating the Poisson bracket to obtain the quantum rule eqn(1) using eqn(20), the latter now collapses into an equality, in fact a weak equality $\approx$ and becomes a secondary constraint, in the terminology of Dirac [48]. It is impossible to prove eqn(1) from right to left; physically such a general statement would also be absurd. However this does not diminish its status as a fundamental physical law, with tribute to Max Born who recognised it straightway in 1925.

Before moving on, we shall merely state that it is straightforward to generalise the above results to the case of boson field theory. The standard approach well known since 1929, see for example Heisenberg [49] and later Schiff [50], Dirac [48], is to generalise all derivatives to functional derivatives, and the quantisation rule for fields becomes:

$$[\hat{\psi}(x), \hat{\pi}(x')] = \kappa\ \delta(x - x'), \tag{21}$$

where $\kappa$ is in general a complex dimensional constant ($= i\hbar$ for bosons) and $\hat{\pi}(x)$ is the canonical field momentum operator, all other commutators between canonical field variables being zero. This is usually written in the standard form:

$$[\hat{\psi}(x), \hat{\psi}^\dagger(x')] = \ \delta(x - x'), \tag{22}$$

using the following result: $\hat{\pi}(x) = i\hbar\hat{\psi}^\dagger(x)$ from the Lagrangian field density [49, 50], see also eqn(56) below.

## IV. LAGRANGE BRACKET QUANTISATION

In this section we shall replace the constraint function $\hat{F}$ in eqn(6) by a function $\hat{G}$ which is explicitly written as a function of an anti-commutator.

$$\hat{S} = \int \Big(\hat{p}d\hat{q} - \hat{H}(\hat{p},\hat{q})dt - \hat{\lambda}\hat{G}([\hat{u},\hat{v}]_+)dt\Big), \tag{23}$$

where $[\hat{u},\hat{v}]_+ = \hat{u}\hat{v} + \hat{v}\hat{u}$. Then all our previous arguments follow up to eqn(18), by replacing commutators with anti-commutators. Since anti-commutators and also the Lagrange brackets do not satisfy the Jacobi identity, we must now modify the Dirac Quantum Condition as:

$$[\hat{u},\hat{v}]_+ = \bar{\kappa} P_{\hat{u}\hat{v}}(\hat{u},\hat{v}), \tag{24}$$

where $\bar{\kappa}$ is a different complex dimensional constant that now depends on the choice for $\hat{u}$ and $\hat{v}$ and $P_{ab} = 1$ for the identity permutation of a,b and -1 for odd ones and $(\hat{u},\hat{v})$ is the Lagrange bracket:

$$(\hat{u},\hat{v}) = \big(\frac{\partial\hat{q}}{\partial\hat{u}} \circ \frac{\partial\hat{p}}{\partial\hat{v}} - \frac{\partial\hat{p}}{\partial\hat{u}} \circ \frac{\partial\hat{q}}{\partial\hat{v}}\big) \ . \tag{25}$$

The factor $P_{ab}$ is required here as the Lagrange bracket $(u,v)$ is antisymmetric in the exchange of $u \leftrightarrow v$ while the anti-commutator is symmetric. The one way proof $[\hat{u},\hat{v}]_+ \rightarrow \kappa P_{\hat{u}\hat{v}}(\hat{u},\hat{v})$ is now obtained in a similar way. However, the reverse proof $[\hat{u},\hat{v}]_+ \leftarrow \kappa P_{\hat{u}\hat{v}}(\hat{u},\hat{v})$ is less straightforward, Lagrange brackets do not satisfy Leibniz's rule, and Dirac's arguments (see Appendix 1) do not apply. The proof requires super Jacobi identities and it is provided in the Appendix 2. Eqn(20) now becomes a new isomorphism:

$$[\hat{u},\hat{v}]_+ \equiv \bar{\kappa} P_{\hat{u}\hat{v}}(\hat{u},\hat{v}). \tag{26}$$

Eqn(26) is a fundamentally new result of this paper; it modifies the Born-Jordan-Heisenberg rule eqn(1) to:

$$[\hat{q},\hat{p}\,]_+ = \hbar\hat{1} \ . \tag{27}$$

There is now no imaginary $i$ in this quantum rule as the anti-commutator of two Hermitian operators is Hermitian . There does not seem to be any application for this rule in regular physical systems of particles as far as I am aware, for Hermitian operators $\hat{q}$ and $\hat{p}$ that must be nilpotent according to eqn(26) are null. However, extending to field theory where the field operators are non-Hermitian, it can be shown to provide a direct justification for the Jordan and Wigner's 1928 hypothesis [51] for fermions in quantum field theory. Lagrange bracket quantisation provides a fundamental theoretical basis for the construction of a quantum field theory of fermions, ending a near century old puzzle for fermion fields as to the origin of their anti-commutation relations, see Appendix 3. We can now further state a theorem regarding the mixing of brackets:

Theorem I:

Just as in classical mechanics, the mixing of Poisson and Lagrange brackets in a mathematical expression, which upon quantisation become commutators and anti-commutators is allowed, see Appendix 3. However mixing of quantization rules is forbidden.

A simple counter example suffices. Consider again ABC [44] and assuming this is allowed i.e. $ABC = [A, B]C + BCA = i\hbar C + BCA$ using the quantum rule eqn(1). Then also $ABC = [A, B]_+C - BCA = \bar{\kappa}C - BCA$ by the quantum rule eqn(27). By simple manipulation we have: $2BCA = (\bar{\kappa} - i\hbar)C$. Now taking determinants and using $det(ABC) = detA\ detB\ detC$ we now have $2detA\ detB = (\bar{\kappa} - i\hbar)$. For Hermitian matrices, the LHS is always real while the RHS is complex, $\bar{\kappa}$ being real. We have a contradiction, hence by *reductio ad absurdum* the theorem is proved.

To conclude this section, we shall make some remarks about Lagrange brackets. Lagrange brackets have been historically displaced by Poisson brackets in quantum mechanics. In celestial mechanics and astronomy, at least in the days before electronic computing, Lagrange brackets are of far more value, especially in perturbational calculations [52]. A revival of Lagrange brackets in quantum mechanics would be of great importance. Lagrange brackets and therefore anti-commutators do not form a Lie Algebra.

## V. BEYOND HEISENBERG

Let us start with a quick review of Heisenberg's quantum mechanics in terms of Poisson operator algebra. We first write down the total Hamiltonian $\hat{H}_T$ (symbolically) as :

$$\hat{H}_T = \hat{H} + \hat{\lambda} \circ \hat{F}, \tag{28}$$

where $\hat{F}$ is the mysterious " Quantum Red October" function [53] (for want of a better name) and $\hat{\lambda}$ is the equally illusive (undetermined) Lagrange multiplier operator, which we shall call its ghost. The time evolution of any dynamical operator $\hat{G}$ is given as usual by:

$$\dot{\hat{G}} = \{\hat{G}, \hat{H}_T\} = \{\hat{G}, \hat{H}\} + \{\hat{G}, \hat{\lambda} \circ \hat{F}\} = \{\hat{G}, \hat{H}\} + \hat{\lambda} \circ \{\hat{G}, \hat{F}\} + \{\hat{G}, \hat{\lambda}\} \circ \hat{F}. \tag{29}$$

We leave the symbol $\circ$ as a multiplier to remind the reader that in general a symmetrization procedure is required in order that Leibniz's rule can be used in the second equation to expand the Poisson brackets above, although we will not consider those exotic Hamiltonian systems where such a procedure is required in this paper [30, 39]. The last term in eqn(29) can now be dropped. Now Heisenberg's hypothesis:

$$\dot{\hat{G}} = \{\hat{G}, \hat{H}\} \tag{30}$$

is equivalent to:

$$\hat{\lambda} = 0 \ or \ \{\hat{G}, \hat{F}\} = 0. \tag{31}$$

In particular if the constraint is time independent, which implies that the quantum rule such as eqn(1) is also time-independent, then we must have $\dot{\hat{F}} = 0$ which requires from eqn(29) that $\{\hat{F}, \hat{H}\} = 0$ i.e, both $\hat{F}$ and $\hat{H}$ must commute. In this case we must also have:

$$\dot{\hat{\lambda}} = \{\hat{\lambda}, \hat{H}\} + \hat{\lambda} \circ \{\hat{\lambda}, \hat{F}\} = 0 \Rightarrow \{\hat{\lambda}, \hat{H}\} = 0 \ and \ \{\hat{\lambda}, \hat{F}\} = 0, \tag{32}$$

by eqn(29). The last requirement is redundant if the operators are (i) Hermitian and (ii) non-degenerate, but we cannot assume that, as commutativity is in general non transitive. $\hat{\lambda}$ and $\hat{F}$ can both be Hermitian i.e. are observables or anti-Hermitian and therefore are not observables, but the Hamiltonian $\hat{H}$ and total Hamiltonian $\hat{H}_T$ must be Hermitian. Nevertheless, we can now define a class of operators that satisfy Heisenberg's hypothesis, which we shall call the Heisenberg class. For a given Hamiltonian $\hat{H}$, the set of mutually commuting operators $\hat{H}$, $\hat{\lambda}_j$ and $\hat{F}_j$, $j = 1, 2, 3...$ will define a (Heisenberg) class of quantum mechanical systems, all having the same equation of motion albeit with apparently different quantum rules such as eqn(1) and eqn(9). In short there are many $\hat{F}_j$ functions and many ghosts. However for the same initial conditions i.e. a prior prepared state, the evolved states are therefore all identical. We shall not examine the details of this algebra further here, which is rather rich. For example, it is easy to see by successive evolutions using a pair $\hat{\lambda}_i$, $\hat{F}_i$ and another pair $\hat{\lambda}_j$, $\hat{F}_j$ that we have (see Appendix 4):

$$\Delta^2\hat{G} = \delta\hat{G}' - \delta\hat{G}'' = \hat{\gamma}_i\hat{\epsilon}_j\{\hat{G},\{\hat{F}_i,\hat{F}_j\}\} + \{\hat{\gamma}_j,\hat{\epsilon}_i\}\{\{\hat{G},\hat{F}_j\},\hat{F}_i\}. \tag{33}$$

Hence unless the $\hat{\lambda}$ ghost algebra is commutative i.e. $\{\hat{\gamma}_j,\hat{\epsilon}_i\} = 0$, the $\hat{F}_j$ do not even form a group i.e. $\{\hat{F}_i,\hat{F}_j\} = \hat{F}_l$. Eqn(33) vanishes for the Heisenberg class which forms a commutative group. As noted earlier this class is commutative, hence its members are first class primary constraints [38] and do not lead to changes of state. This appears to be related to Herman Weyl's[31] observation way back in 1928 that the quantum rule eqn(1) forms an Abelian group in *ray* space.

For now we shall merely look at the evolution of the quantum bracket $\hat{\chi} = [\![\hat{u},\hat{v}]\!]$ which could be the commutator or anti-commutator. From eqn(29) we have:

$$\dot{\hat{\chi}} = \{\hat{\chi},\hat{H}\} + \hat{\lambda}\circ\{\hat{\chi},\hat{F}\} = 0, \tag{34}$$

for time-independent evolution of the bracket. Clearly as discussed earlier in section III, Heisenberg's hypothesis is equivalent to $\{\hat{\chi},\hat{H}\} = 0$ *and* $\{\hat{\chi},\hat{F}\} = 0$. The latter is true *if* $F$ is only a function of $\hat{\chi}$ as assumed in section III. Failure of eqn(34) will lead to perturbations and $\hat{\chi}$ is no longer a constant of motion, with consequences for entanglement and the uncertainty principle.

### A. Schwinger's quantum rule

The following discussions should be read as symbolic only, as there are some complex mathematical issues that we cannot discuss here. First we shall make a Legendre transformation back to the Lagrangian:

$$\hat{L} = \hat{p}\dot{\hat{q}} - \hat{H}_T = \hat{p}\dot{\hat{q}} - \hat{H} - \hat{\lambda}\circ\hat{F} = \hat{L}_0 - \hat{\lambda}\circ\hat{F}, \tag{35}$$

with $\hat{L}_0$ the unconstrained Lagrangian. Now applying the variational principle to the action $\hat{S} = \int \hat{L}dt$ we now have:

$$\delta\hat{S} = 0 \Rightarrow \delta\hat{S}_0 = \delta\hat{W}, \tag{36}$$

where $\hat{S}_0 = \int \hat{L}_0 dt$ and $\hat{W} = \int(\hat{\lambda}\circ\hat{F})dt$. Eqn(36) is an extension of Schwinger's variation principle [54]. To see this, consider calculating the expectation value of an operator $\hat{\Omega}$ from eqn(36):

$$<\hat{\Omega}\delta\hat{S}_0> = <\hat{\Omega}\delta\hat{W}> = -<\delta\hat{\Omega}\hat{W}>, \tag{37}$$

where the latter follows via a (symbolic) integration by parts, or a variational principle: $\delta < \hat{\Omega}\hat{W} >=< \delta\hat{\Omega}\hat{W} > + < \hat{\Omega}\delta\hat{W} >= 0$ . Now Schwinger's variational principle can be written as: $< \delta\hat{\Omega} >= -\frac{i}{\hbar} < \hat{\Omega}\delta\hat{S}_0 >$, a result that can also be obtained by Feynman's path integrals . Eqn(37) identifies $\hat{W}$ as similar to a Quantum Anomaly factor well known in quantum field theory.

### B. Hunt for the Red October?

This is of course an interesting question. Clearly the $\hat{F}_i$'s are objects of the universe, it is perhaps one way to explain why our universe is a quantum universe. In his later years, Heisenberg proposed the notion that Quantum Mechanics could fail as the energy involved in interactions increases. Dirac on the other hand conjectured that because the fine structure constant $\alpha = \frac{e^2}{\hbar c}$ is dimensionless approximately $\frac{1}{137}$, he voted that $\hbar$ can be got rid of in a better theory[42]. A convenient hypothesis may be the following:

$$\hat{F}(\hat{p}, \hat{q}) = [\hat{p}, \hat{q}] + i\hbar\hat{1} + A\hat{p} + B\hat{p}^2 - C\hat{p}^3 - D\hat{p}^4 - A'\hat{q} + B'\hat{q}^2 + C'\hat{q}^3 + D'\hat{q}^4 - ... \quad (38)$$

We eliminate cross-terms to avoid symmetrization problems [39], but a casual look at this naive hypothesis of terms in the total Hamiltonian $\hat{H}_T$ already reveals some maybe interesting cosmological features. Is the $A$ constant which gives a background velocity related to the expansion of the universe? Is the B constant a renormalisation of the particle mass related to Mach's Principle? Is the $A$' constant (which gives a background acceleration) related to the universe's accelerated expansion like a cosmic Stark effect? Is the $B$' constant a cosmic harmonic force, while $C$' and $D$' are cosmic anharmonic forces and so on? As noted earlier in section II, Heisenberg was very concerned about anharmonic forces in his founding paper [12]. From the empirical perspective, current technology can produce quantum dots with single electrons trapped in quantum wells. These are ideal candidates to search for the $\hat{F}_i$'s. Without doing more number crunching, we cannot tell how big a challenge this enterprise would be. With the fabrication of millions, perhaps even billions of quantum dots, subtle effects beyond the Heisenberg class such as depicted in eqn(38) may be detectable? This is an aspiration for the future. Interestingly in 1999 much controversy arose and still exists from the announcement of results of the Karl Popper experiment by Korean physicists Y-H. Kim and Y. Shih [55] then at Cornell. Popper's experiment, first proposed in 1934 in his book Logik der Forschung (German edition), was a predecessor of EPR which like the latter does not involve spin, unlike most subsequent studies inspired by Bohm's book of 1954. In 1935, after the now famous Einstein-Poldolsky-Rosen (EPR) paper was published, Einstein wrote a long letter to Popper criticising his experiment; this caused Popper to abandon his proposal. Einstein, Poldolsky and Rosen did not cite Popper's experiment in their celebrated EPR paper. As Poldolsky was responsible for most of its writing this may have been an oversight [56]. Einstein and Poldosky in fact fell out after the EPR paper. Einstein's hand written letter and its translation can be found in a report of an interview Popper gave in Paris just a few years before his death to Marie-Christine Combourieu in 1992 [57]. In this letter, according to Popper, there was no definition of "reality" as in the EPR paper. Sadly Popper died before the announcement of the results of the first realisation of his experiment, made possible only by advancement in parametric down-conversion and co-incident photon counting techniques down to single photons [55]. The surprising result, still disputed, is that it seemed to agree with Popper's prediction that was designed to refute the Copenhagen interpretation of Quantum Mechanics. Without becoming involved in the controversy, it would be interesting to research if the Y-H. Kim and Y. Shih [55] experiment or a variant of it could be used to detect features beyond Heisenberg's quantum mechanics. In any case refinement of these experiments will surely take place over the next hundred years.

## VI. D'ALEMBERT AND EINSTEIN'S QUANTUM RIDDLE

Let's now return to d'Alembert's principle eqn(3). In classical mechanics "harmony with given kinematic constraints" is generally built into the equation by the use of generalised co-ordinates appropriate to the constraint, see remarks before eqn(2). However in the expression for our quantum d'Alembert's principle (see eqn(39)) below we can introduce a quantum force $\hat{\mathbf{\Phi}}_n$ into eqn(3) as an additional inertia force:

$$\sum_{n=1}^{N}(\hat{\mathbf{F}}_n - m_n\hat{\mathbf{A}}_n - \hat{\mathbf{\Phi}}_n)\cdot\delta\hat{\mathbf{R}}_n = 0\ . \tag{39}$$

The recovery of the force $\hat{\mathbf{\Phi}}_n$ from $\hat{F}_n$ is by no means trivial even if we know the latter such as our hypothesis eqn(38), considering the complicated algebra generated by eqn(38). On the contrary, *if* we hypothesize the force we can reconstruct the function, subject to certain assumptions such as that the force $\hat{\mathbf{\Phi}}_n$ is monogenic and is related to the function $\hat{F}_n$ in the usual way [29]. In 1946 the Hungarian physicist Imre Fényes seemed to have been the first to deduce the Schrödinger equation from classical stochastic mechanics [58, 59]. This was further taken up in 1966 by the mathematician Edward Nelson [60] who provided a mapping of quantum mechanics to a classical stochastic model, albeit for Hamiltonians with quadratic momentum terms only. He obtained *both* the time-independent and time-dependent Schrödinger equations, effectively removing all the operator hats in eqn(39) with the force $\hat{\mathbf{\Phi}}_n$ postulated as a random Langevin force which defines a Markov process. The equivalence is striking and has since motivated a lot of research into stochastic models of quantum mechanics in spite of various limitations. Mapping from classical stochastic models to matrix algebra, already well known for Markov processes, see for example van Kampen [61], is relevant to quantum mechanics. In fact this has been constructed for the Fényes-Nelson stochastic dynamics model by Davidson in 1979 [62]. Unfortunately none of the above authors started from the d'Alembert's principle which is deeper and in the quantum form eqn(39) is a statement akin to a *detailed balance* principle. In particular in the classical case, the constraint force $\hat{\mathbf{\Phi}}_n$ must do no work for all virtual displacements [29] and this must be strictly maintained upon quantization in the operator form. Moreover in accordance with Ehrenfest's theorem, the average $< \hat{\mathbf{\Phi}}_n \cdot \delta\hat{\mathbf{R}}_n >= 0$, which implies that this force can only generate states in the Hilbert space that must be mutually orthogonal. Understanding this will throw light on the origin of quantum fluctuations and non-locality, our aspired solution for the full class I problem. It is unclear at this stage what else can be learnt from more sophisticated classical stochastic models about the Heisenberg Group $\hat{\chi}$'s let alone going beyond in the pursuit to mimic quantum mechanics. Nevertheless these studies are useful illustrations for the interpretation of quantum mechanics and the attack on the measurement problem, see for example Weinberg [63]. In the next hundred years, we may finally be able to understand quantum mechanics, contrary to Richard Feynman's claim.

## VII. CONCLUSION

*Raffiniert ist der Herr Gott, aber boshaft ist Er nicht.* translated as: Subtle is the Lord, but malicious he is not - Albert Einstein.

In this paper we have given an appraisal of Heisenberg's great discovery of 1925 which laid the foundation of quantum mechanics. We have looked at some logical gaps in the foundation left by the early pioneers, notably Heisenberg, Born, Jordan, Dirac and Wigner. We have provided a response to Born's challenge and Einstein's riddle as to the origin of the fundamental Born-Jordan-Heisenberg quantum rule. We have extended our arguments from bosons to fermions by the introduction of Lagrange bracket quantization, which was missed by Dirac and have finally examined what could lie beyond Heisenberg's quantum mechanics. In this we are led to consider if the fundamental quantum constraint function $\hat{F}_j$ which subtly underlies the basic quantum rule eqn(1) is in fact a cosmological object. Indeed if this is so, it will have major implications for any application of quantum mechanics to cosmological studies, such as black holes and dark matter etc. Over the last hundred years there have been many attempts at going beyond Heisenberg's quantum mechanics. The author is not competent to provide even a glancing review of the wide subject. However there are a few that should be mentioned. In 1979, Tom Kibble suggested a program to 'geometrize' quantum mechanics [64]. This builds upon earlier works by mathematicians such as Bertram Kostant and Jean-Marie Souriau in the 1970s which use a generalization of Hilbert space and a symplectic structure as in classical mechanics. An excellent textbook on geometrical quantization which appeared in the last decade of the 20th century is given by Woodhouse [65]. It is an active field of mathematical research which aims to extend quantum mechanics to treat non-linear relativistic systems, and " rewrite quantum mechanics in a form better suited to unificaton with general relativity" [64], an objective that remains elusive. The same is true of other "deformation" theories of geometrical quantization which date back to the early $\star$ product phase space formulation of quantum mechanics by Groeneweld [41] and Moyal [66]. These techniques have nevertheless been embraced in recent years by quantum chemists because of their classical-like phase space approach and computational convenience, see for example Zachos et al [67]. In recent decades too there have also been sophisticated generalizations of the Markovian master equation approach due in particular to Göran Lindblad [68], see for example Breuer and Petruccione [69]. Though originally developed to deal with open systems with non-unitary evolution due to the environment, it has now also attracted particle physicists to the field, such as Weinberg [63]. I do not know if any of these studies are related to the ideas presented here but certainly there is much work for future generations to consider.

## VIII. PERSONAL TRIBUTE AND DEDICATIONS

This work is a tribute to my many teachers, friends and colleagues, many of whom have since departed to paradise. Over the years, I have had the good fortune to learn from them, and to share with them my own journey towards an understanding of quantum mechanics. Alfred Lande taught me to question the foundations of quantum mechanics at a very young age at 21. John Valatin taught me all I needed to know about superconductivity and that it does not matter if a problem has been solved, the only thing that does matter is *how much* of it has been solved. Derek Martin taught me all I had to know about classical and quantum solid state physics. John Charap taught me all about relativistic quantum mechanics. As he was a student of Paul Dirac then I must have got most of it all from the horse's mouth. Rudolph Peierls, our neighbour down the street at Oxford during my Harwell days, taught me a lot about physics, including his work with Landau on the measurability of quantum fields which angered Bohr. Wim Caspers shared with me the great joy of finding exact solutions to quantum many body problems, one of my favourite was the exact solution of the Thirring model. I still remember how stunned I was when Walter Thirring gave a seminar in our department and we talked, I was just a student then. Geoffrey Sewell taught me a lot about C* algebra and quantum statistical mechanics but failed to convince me to become a mathematician. Cyril Domb was like a father to many of us in our early graduate student days. Nicolaas van Kampen introduced me to the subtleties of Lagrangian constraints in classical and quantum mechanics at my first Dutch summer school in 1980 at Enschede. Tom Kibble taught me a lot about quantum field theory, he was a true gentleman and an inspiration while I was at Imperial College. Moses Blackman, another student of Max Born's and Heisenberg was his PhD examiner, had his office next to mine at Imperial College. It was on a Sunday afternoon in 1983 when he met me in my office and announced that he had just finished his article on the history and physics of the magnetism of lodestones [71]. He told me something about it and that I must read it. Later that week, he passed away at 91. Many would know that it was Heisenberg and his quantum mechanics that provided the first solution to the millennia old puzzle for the origin of ferromagnetism. Walter Kohn a student of Julian Schwinger was a personal friend and mentor through one of my difficult times. He taught me a lot about quantum condensed matter theory; our friendship and mutual correspondence lasted until his death in 2016. Marshall Stoneham took me into

Harwell in 1988 and we started a long collaboration until his death in 2010. Peter Schofield, Alan Lidiard and Norman March were old friends and mentors, they gave me many fond memories at Harwell. It was there that I regularly met with Nevil Mott who said he enjoyed talking to me even though my accent was funny. Roger Elliott was a wonderful friend who took the trouble to care and wrote to me during my difficult time after I have left Monash for Taiwan. Sam Edwards a great friend who hosted me for high table at Caius College several times, we had such fun together on other occasions. The multi-talented Peter Landsberg influenced me greatly, he was one of my final year undergraduate examiners who supported me firmly even though I did not know it then. His astounding career as a mathematician, semi-conductor physicist and pioneering worker on the foundation of quantum mechanics taught me how boring life would be without all three. Sadly, some of my colleagues have also passed on, some far too early. Among them was Paul Kirkman (1986), recently Jill Bonner (2021), Henk Blöte (2022) and J.M.J. van Leeuwen (2024).

## IX. ACKNOWLEDGEMENT

I am grateful to Hajo Leschke for correcting some references, Dr Andrew Smith for his dedicated proofreading, Dr Aris Alexopoulos for useful comments and the late J.M.J. van Leeuwen for review and comments on my earlier version of this manuscript.

## X. APPENDIX 1- PROOFS OF DIRAC'S QUANTUM CONDITION

Dirac's sufficiency proof of his quantum condition [40] was an ingenious argument based on four operators $\hat{u}_1$,$\hat{v}_1$,$\hat{u}_2$ and $\hat{v}_2$. These are all functions of the two canonical variables $\hat{p}$,$\hat{q}$. Although not stated by Dirac explicitly we can now see that since there are only two canonical variables $\hat{p}$,$\hat{q}$ and the $\hat{u}$,$\hat{v}$'s must be related to them via canonical transformations, then the four variables $\hat{u}_1$,$\hat{v}_1$,$\hat{u}_2$ and $\hat{v}_2$ must satisfy certain consistency relations between them when we generalise from classical variables to quantum operators. To do so Dirac exploited the Leibniz's rule for Poisson brackets: $\{uv, w\} = u\{v, w\}+\{u, w\}v$ and $\{u, vw\} = v\{u, w\} + \{u, v\}w$. Dirac then argued that if we have four operators in a Poisson bracket: $\{\hat{u}_1\hat{u}_2, \hat{v}_1\hat{v}_2\}$, we have effectively two ways to evaluate this bracket. Equating the results of these two evaluations, Dirac obtained essentially:

$$\{\hat{u}_1, \hat{v}_1\}[\hat{u}_2, \hat{v}_2] = [\hat{u}_1, \hat{v}_1]\{\hat{u}_2, \hat{v}_2\}. \tag{40}$$

If this relation must hold for $\hat{u}_1$,$\hat{v}_1$ independent of $\hat{u}_2$,$\hat{v}_2$, it is sufficient that:

$$\{\hat{u}_1, \hat{v}_1\} = \kappa^{-1}[\hat{u}_1, \hat{v}_1] \text{ and} \{\hat{u}_2, \hat{v}_2\} = \kappa^{-1}[\hat{u}_2, \hat{v}_2], \tag{41}$$

with $\kappa = i\hbar$, which is Dirac's famous Quantum Condition [40]. We shall now provide an alternative proof of this famous condition using only three operators instead of four. In this exercise we shall start with the operator Jacobi identity which is easily proved for any non-commutating set of operators:

$$[\hat{a}, [\hat{b}, \hat{c}]] + [\hat{c}, [\hat{a}, \hat{b}]] + [\hat{b}, [\hat{c}, \hat{a}]] = 0. \tag{42}$$

Since we have already a one way proof of the Dirac quantum condition eqn(19), we now have two ways to replace some of the commutators in eqn(42) namely:

$$\{\hat{a}, [\hat{b}, \hat{c}]\} + \{\hat{c}, [\hat{a}, \hat{b}]\} + \{\hat{b}, [\hat{c}, \hat{a}]\} = 0, \tag{43}$$

or alternatively:

$$[\hat{a}, \{\hat{b}, \hat{c}\}] + [\hat{c}, \{\hat{a}, \hat{b}\}] + [\hat{b}, \{\hat{c}, \hat{a}\}] = 0, \tag{44}$$

and they must be consistent. Now subtracting equations (43) and (44), we now require that:

$$\{\hat{a}, [\hat{b}, \hat{c}]\} = [\hat{a}, \{\hat{b}, \hat{c}\}], \tag{45}$$

etc, for each of the remaining two cyclic permutations. We can now see that $\{\ ,\ \} = \kappa^{-1}[\ ,\ ]$ is a sufficient condition for eqn(45) to be satisfied for by applying it we have $\kappa^{-1}[\hat{a}, [\hat{b}, \hat{c}]]$ on the LHS while on the RHS we have $[\hat{a}, \kappa^{-1}[\hat{b}, \hat{c}]]$ proving equality. Alternatively a direct expansion of eqn(45) now gives the equations:

$$\{\hat{a}, \hat{b}\hat{c} - \hat{c}\hat{b}\} = \{\hat{a}, \hat{b}\hat{c}\} - \{\hat{a}, \hat{c}\hat{b}\} = \{\hat{a}, \hat{b}\}\hat{c} + \hat{b}\{\hat{a}, \hat{c}\} - \{\hat{a}, \hat{c}\}\hat{b} - \hat{c}\{\hat{a}, \hat{b}\}, \tag{46}$$

using Leibniz rule for the LHS and:

$$[\hat{a}, \{\hat{b}, \hat{c}\}] = \hat{a}\{\hat{b}, \hat{c}\} - \{\hat{b}, \hat{c}\}\hat{a}, \tag{47}$$

for the RHS. We now have an equation written only entirely in terms of Poisson brackets:

$$\{\hat{a}, \hat{b}\}\hat{c} + \hat{b}\{\hat{a}, \hat{c}\} + \{\hat{b}, \hat{c}\}\hat{a} = \hat{a}\{\hat{b}, \hat{c}\} + \{\hat{a}, \hat{c}\}\hat{b} + \hat{c}\{\hat{a}, \hat{b}\}. \tag{48}$$

Again a sufficiency condition for this to hold is of course Dirac's famous Quantum Condition written in the form: $\{\hat{a}, \hat{b}\} = \kappa^{-1}[\hat{a}, \hat{b}]$, for all the Poisson brackets as easily verified.

## XI. APPENDIX 2- PROOF OF LAGRANGE BRACKET QUANTUM CONDITION

In a similar way to Appendix 1, we shall use brackets involving three operators. We start with the following operator Jacobi identity in terms of commutators [ , ] and anti-commutators $[\ ,\ ]_+$ which is easily proved for operators:

$$[[\hat{a},\hat{b}]_+,\hat{c}]] + [[\hat{b},\hat{c}]_+,\hat{a}]] + [[\hat{c},\hat{a}]_+,\hat{b}]] = 0. \tag{49}$$

Next we need a super or graded Jacobi identity from classical mechanics involving both Poisson brackets { , } and Lagrange brackets ( , ) [47] which can also be proved for operators, subject to symmetrization conditions [30, 70]:

$$\{(\hat{a},\hat{b}),\hat{c}\} + \{(\hat{b},\hat{c}),\hat{a}\} + \{(\hat{c},\hat{a}),\hat{b}\} = 0. \tag{50}$$

From Appendix 1 we can replace the Poisson brackets with commutators using the Dirac Condition:

$$[(\hat{a},\hat{b}),\hat{c}] + [(\hat{b},\hat{c}),\hat{a}] + [(\hat{c},\hat{a}),\hat{b}] = 0. \tag{51}$$

Equations eqn(49) and eqn(51) are now strikingly similar. We now multiply eqn(51) by a dimensional constant $\bar{\kappa}$ to ensure that they are dimensionally equivalent. Then they will be identical if the following holds:

$$\bar{\kappa}[(\hat{a},\hat{b}),\hat{c}] = [[\hat{a},\hat{b}]_+,\hat{c}], \tag{52}$$

for each of the cyclically permutated terms. A sufficiency condition is the equality $(\hat{a},\hat{b}) = \bar{\kappa}^{-1}[\hat{a},\hat{b}]_+$ with $\bar{\kappa}$ an appropriate dimensional constant. This completes the proof.

## XII. APPENDIX 3- PROOF OF JORDAN-WIGNER QUANTUM CONDITION FOR FERMION FIELDS

The non-relativistic second quantized Schrödinger field is given by the Hamiltonian:

$$H = \int \Big[ (\hat{\psi}^{\dagger} \frac{-\hbar^2}{2m} \nabla^2 \hat{\psi}) + \hat{\psi}^{\dagger} V \hat{\psi} \Big] d\tau. \tag{53}$$

Upon integration by parts and setting boundary terms to zero, we then have:

$$H = \int \Big[ \frac{\hbar^2}{2m} (\nabla \hat{\psi}^{\dagger} \nabla \hat{\psi}) + \hat{\psi}^{\dagger} V \hat{\psi} \Big] d\tau. \tag{54}$$

We can now identify $\pi = -i\hbar\nabla\psi$ as the canonical field momentum for the Schrödinger field . Then the Lagrange bracket quantisation rule eqn(24) is now given by:

$$[\hat{\psi}, \hat{\pi}]_+ = i\hbar(\hat{\psi}, \hat{\pi}) = i\hbar\delta(x - x'). \tag{55}$$

Here we have a factor of $i$ (which amounts to a choice for an arbitrary global gauge as in the boson case) since $\psi$ is no longer a Hermitian operator field. Now from the Lagrangian field density $L$ it is well known that:

$$\hat{\pi} = -i\hbar\nabla\hat{\psi} = \frac{\partial L}{\partial\dot{\hat{\psi}}} = i\hbar\hat{\psi}^\dagger, \tag{56}$$

see for example Heisenberg [49] or Schiff [50]. With this convention we arrive immediately at the Jordan-Wigner field quantisation rule [51] for fermions in the usual form:

$$[\hat{\psi}, \hat{\psi}^\dagger]_+ = \delta(x - x'). \tag{57}$$

## XIII. APPENDIX 4- PROOF THAT $F_j$'S DO NOT FORM A GROUP UNLESS ITS GHOSTS ARE ABELIAN

The proof follows that of Dirac [38], who had only considered classical constraints and their Poisson brackets. Consider an evolution in $\hat{G}$ induced by a change in $\hat{\lambda}$. From eqn(29) we have:

$$\delta\hat{G} = \hat{\epsilon}_i\{\hat{G}, \hat{F}_i\}, \tag{58}$$

where $\hat{\epsilon}_i = \delta t(\hat{\lambda}_i - \hat{\lambda}_i')$ and we have dropped the $\circ$ reminder as understood. We can consider a further evolution induced by new $\hat{\gamma}_j$'s, so that:

$$\delta\hat{G}' = \hat{\epsilon}_i\{\hat{G}, \hat{F}_i\} + \hat{\gamma}_j\{\hat{G} + \hat{\epsilon}_i\{\hat{G}, \hat{F}_i\}, \hat{F}_j\}. \tag{59}$$

Now we can perform these evolutions in reverse order:

$$\delta\hat{G}'' = \hat{\gamma}_j\{\hat{G}, \hat{F}_j\} + \hat{\epsilon}_i\{\hat{G} + \hat{\gamma}_j\{\hat{G}, \hat{F}_j\}, \hat{F}_i\}. \tag{60}$$

Now taking the difference and making use of the Jacobi identity we shall arrive at eqn(33) after some algebra.

---

[1] "Albert Einstein and the Quantum Riddle" Am. J. Phys. **42** 459-464 (1974). This paper by Alfred Lande was published just two years before his death. The then Editor of the journal,

Edwin F. Taylor admitted that the paper had been through more than the usual number of reviewers, had received more than the usual spread of judgements and that the referees simply cannot agree on its appropriateness for publication. It was clearly his editorial decision to have it published anyway for "the enlightenment that our readers may obtain from it." The paper is an enlightening one, perhaps too ahead of its time and incorrectly perceived as a challenge to the Copenhagen school. It was clearly an incomplete work, important these days on the 'probability' aspect of quantum mechanics but lacking a development of the Schrödinger equation.

[2] See reference 1 in Lande [1].

[3] See letter no 52, in the "Born-Einstein Letters" MacMillan, (London) 1971.

[4] See for example Jeremy Bernstein in "Max Born and the quantum theory" Am. J. Phys. **73** 999-1008 (2005) and Erratum Am. J. Phys. **74** 160 (2006).

[5] Max Born, "The statistical interpretation of quantum mechanics", Nobel lecture, 11 Dec 1954.

[6] Steven Weinberg in *Dreams of a Final Theory*, Pantheon, New York 53-54, (1992).

[7] Mentioned by Malcolm Longair in "Quantum Concepts in Physics- An Alternative Approach to the Understanding of Quantum Mechanics" pg 207 Cambridge University Press Cambridge (2013). Unfortunately I was unable to locate the reference to Fermi provided by the author of this book, which might have been a printing error.

[8] Ian J.R. Aitchison et al "Understanding Heisenberg's "magical" paper of July 1925: a new look at the calculational details " Am. J. Phys. **72** 1370–1379 (2004)

[9] The great debates between Bohr and Einstein have been well documented, see for e.g. Pais [56], see also Niels Bohr: "Discussions with Einstein on Epistemological Problems in Atomic Physics" in "Albert Einstein: Philosopher-Scientist", by Paul Arthur Schilpp, Cambridge University Press, 1949. Over the last four decades or more this has been dominated by arguments about EPR, Bell and the nature of local realism etc, which to the author's mind are issues that pale or perhaps even moot by comparison with Einstein's fundamental Quantum Riddle.

[10] Michel Paty, "The Nature of Einstein's Objections to the Copenhagen Interpretation of Quantum Mechanics", Foundations of Physics, **23**(1) 183–204 (1993). See also "Einstein's Criticism of Quantum Mechanics" by Michel Paty, in The Oxford Handbook of the History of Quantum Intepretations, edited by Olival Freire Jr, et al, Chapter 12, 302-338, Oxford University Press, (Oxford) 2022.

[11] One may consider Einstein's Quantum Riddle as a Class I problem in the classification of Dirac [42]. In this paper we shall be focusing only on the origin of the Born-Jordan quantum rule eqn(1) which we shall dub a Class I(b) problem. One can merely allude to an answer for the Class I(a) problem of why we must have a Hilbert Space with Hermitian Operators representing observables for the canonical quantisation procedure. We state this as a necessary requirement for the validity of the quantum d'Alembert's principle as intuitively adopted by Heisenberg in 1925, see later. Further research is required on the Class I(a) problem to complete the Class I problem. For further remarks on this Class I problem see section VI later.

[12] W. Heisenberg, "Über quantentheoretische Umdeutung kinematischer und mechanischer Beziehungen " Z. Phys.**33** 879-893 (1925) reprinted in B.L. van der Waerden (editor), "Sources of Quantum Mechanics", vol V of classics of modern science series 261-276, Dover, New York 1968.

[13] Max Born and Pascual Jordan, "Zur Quantenmerchanik" Z. Phys.**34** 858-888 (1925) reprinted in B.L. van der Waerden (editor), "Sources of Quantum Mechanics", vol V of classics of modern science series 277-306, Dover, New York 1968.

[14] K. Schönhammer, "Göttingen and Quantum Mechanics", to be published in this volume in 2025.

[15] Max Born in "Born-Einstein Letters" pg 84, MacMillan, London 1971.

[16] A palimpsest stolen from the Greek Orthodox Church in the 20th century which surfaced in a 1998 auction revealed that Archimedes' first proof of the volume of the sphere in fact did not use any idea of limits of circumscribed polyhedrons as in his later works, but used his ingenious Method of Mechanical Theorems or the method of levers to first arrive at his result.

[17] See Bernstein [4] pg 1007. The author has approached both Jeremy Bernstein who could not recall where he obtained this information from and Frank Close who quite recently researched the life of Klaus Fuchs but found no notes on this collaboration, which of course might have disappeared since Fuchs went to prison for many years.

[18] Max Born "Quantum Mechanics" Zs. f. Phys. **26** 379-395 (1924) reprinted in B.L. van der Waerden (editor), "Sources of Quantum Mechanics", vol V of classics of modern science series 181-198, Dover, New York (1968).

[19] See B.L. van der Waerden (editor), "Sources of Quantum Mechanics", vol V of classics of modern science series 19-21, Dover, New York (1968)

[20] H.A. Kramers and W. Heisenberg, "Über die Streuung von Strahlen durch Atome" Z. Phys. **31** 681-710 (1925) reprinted in B.L. van der Waerden (editor), "Sources of Quantum Mechanics", vol V of classics of modern science series 223-252, Dover, New York (1968).

[21] See B.L. van der Waerden (editor), "Sources of Quantum Mechanics", vol V of classics of modern science series 21-22, Dover, New York (1968).

[22] See B.L. van der Waerden (editor), "Sources of Quantum Mechanics", vol V of classics of modern science series 33, Dover, New York 1968. In fact Heisenberg was not altogether wrong, the co-ordinates of a particle are *not* experimentally observable in the ultra-relativistic limit such as that of a photon or a neutrino due to limits imposed by *both* special relativity and the uncertainty principle. Heisenberg in 1938 returned to a theory based solely on transition amplitudes: the S matrix theory for elementary particles.

[23] See B.L. van der Waerden (editor), "Sources of Quantum Mechanics", vol V of classics of modern science series 28-29, Dover, New York (1968).

[24] The $n$ here is a quantum number borrowed from the Bohr model of the atom which specifies a particular orbit allowed by the old quantum rule of Bohr and Sommerfeld: $\oint pdq = J = nh$.

[25] E. MacKinnon "Heisenberg models and the rise of matrix mechanics"", Hist. Stud. Phys. Sci **8** 137-188 (1977).

[26] An examination of Heisenberg's PhD thesis on hydrodynamics revealed no references to the rich French literature on the subject but Heisenberg's supervisor Arnold Sommerfled's lecture notes in Munich (dating back to 1910) where he submitted his PhD did cover all the relevant French works of Lagrange, Bernoulli and d'Alembert, see Sommerfeld: Mechanics, Lectures on Theoretical Physics Vol 1, translated from the fourth German edition by Martin O. Stern, Academic Press, New York (1952) pg 59 to 62. Heisenberg himself did not seem to have much difficulty with the French language as he and Born penned an article in French at the 1927 Solvay Conference in Brussels. He had also taken part in several French discussions there including advancing the view (now a part of the Copenhagen dogma) that reality is only defined "at the moment of observation *after* an observer had *a prior* made a choice". See "Électrons et Photons" Reports and Discussions of the 1927 Solvay Conference in Brussels, Paris (1928).

[27] Jean Le Rond d'Alembert, "Traité de Dynamique, dans lequel les Loix de l'Equilibre et du Mouvement des Corps sont réduites au plus petit nombre possible, et démontrées d'une

maniére nouvelle, et oú l'on donne un Principe général pour trouver le Mouvement de plusieurs Corps qui agissent les uns sur les autres, d'une maniére quelconque", Paris (1743).

[28] L.A. Pars, "A Treatise on ANALYTICAL DYNAMICS", Heinemann, London (1965).

[29] Cornelius Lanczos, "The Variational Principles of Mechanics", 88–93, fourth edition, Dover, New York (1986).

[30] T. C Choy, "On the Dirac Quantisation rules and the trace anomaly", http://arxiv.org/abs/2007, 12025 (2021).

[31] Herman Weyl, "The theory of Groups and Quantum Mechanics", translated from German by H.P. Robertson, 272-276 Dover Inc., New York (1931).

[32] Letter from Einstein to Born, 4th Dec 1926, see Max Born in "Born-Einstein Letters" pg 91, MacMillan, London (1971). Neverthless Einstein proposed all three men for the Nobel prize in 1928.

[33] Cornelius Lanczos, "The Variational Principles of Mechanics", 111–115, fourth edition , Dover , New York (1986).

[34] M. Born, W. Heisenberg and P. Jordan, "Zur Quantenmerchanik II" Z. Phys.**35** 557-615 (1926) reprinted in B.L. van der Waerden (editor), "Sources of Quantum Mechanics", vol V of classics of modern science series 321-385, Dover, New York, 1968.

[35] K. Gottfried, "P.A.M. Dirac and the discovery of quantum mechanics", American Journal of Physics, **79**(3), 261-266 (2011).

[36] P.A.M. Dirac, "The Fundamental Equations of Quantum Mechanics" Proc. Roy. Soc. A **109**, 642-653 (1925) reprinted in B.L. van der Waerden (editor), "Sources of Quantum Mechanics", vol V of classics of modern science series 307-320, Dover, New York (1968).

[37] E.T. Whittaker, "A Treatise on Analytical Dynamics of Particles and Rigid Bodies", Cambridge University Press, second edition, Cambridge (1917). Whittaker made no mention of d'Alembert's Principle but started with Newton's laws and jumped straight into the Lagrange equations of motion following Lagrange's Méchanique Analytique of 1788. In the final chapter of his book, Whittaker introduces Hamilton's Principle of Least action without mentioning that it can be derived directly from d'Alembert's Principle and only a brief mention of the equation of virtual work in Hamilton's form.

[38] See B.L. van der Waerden (editor), "Sources of Quantum Mechanics", vol V of classics of modern science series 40-42, Dover, New York (1968).

[39] One method is via Weyl ordering, see for example: Z. Rakić and S. Prvanović, "Weyl ordering rule and new Lie brackets for quantum mechanics" https://www.researchgate.net/scientific-contributions/Zoran-Rakic-77122007 (2003). As noted by Schwinger[54], a simple symmetrization $A \circ B = \frac{1}{2}(AB + BA)$ for polynomials will only work if either $A$ or $B$ is less than cubic in $q$ or $p$, or both $A$ and $B$ are less than cubic in q, or in p. We do not discuss such exotic Hamiltonian systems in this paper.

[40] P.A.M. Dirac, "Principles of Quantum Mechanics", the international series of monographs in physics Oxford University Press, fourth edition (revised), Oxford 84-89 (1958). We shall later see that his derivation constitutes a consistency condition for four quantum operators since the dynamical problem has originally only two operators $\hat{p}$ and $\hat{q}$ and only one quantum condition can be defined for a one dimensional system. For Cartesian coordinates, the generalization to higher dimensions is straightforward.

[41] H.J. Groenewold,"On the principles of elementary quantum mechanics", Physica **12**, 405-460 (1946). Groenewold discovered that Dirac's quantisation rule eqn(4) and eqn(5) can lead to inconsistencies when applied to operators of order equal to or higher order than $\hat{p}^2\hat{q}^2$. It is now known that Weyl symmetrization of eqn(5) can remove such obstructions, see for example: Z. Rakić and S. Prvanović [39].

[42] P.A.M. Dirac, "The Evolution of the Physicists picture of Nature", Scientific American **208** (5) 45-53 May (1963).

[43] Cornelius Lanczos, "The Variational Principles of Mechanics", 161–172, fourth edition , Dover , New York (1986).

[44] P.A.M. Dirac, "Principles of Quantum Mechanics", the international series of monographs in physics Oxford University Press, fourth edition (revised), 28 Oxford (1958). If $AB \neq BA$ then for example $ABC = [A, B]C + BCA$ *or* $[A, B]_{+}C - BCA$ for algebraic closure. The term algebraic closure is loosely used here to mean that we can now establish a relation between $ABC$ and $BCA$ and thereby be able to reduce the algebra. So the commutaor $[a, b]$ or the anti-commutator $[a, b]_{+}$ is required in this case. Quantum algebraic closure is more complex than that, see text later. Dirac in 1925 also explored higher order product commutators $[A, B, C]$ and $[A, B, C, D]$ for closure[36] but he never pursued it further throughout his life as far as we know. Such products may well be useful for treating curved spaces and gravitational quantisation, a pursuit of his in his later years [48].

[45] L.D. Landau and E.M. Lifshitz, "Mechanics", third edition , 139–140 Pergamon Press, Oxford (1988) (with corrections). See also Lanczos *ibid* and Goldstein *ibid.*

[46] See for example R. Achilles and A. Bonfiglioli, "The early proofs of the theorem of Campbell, Baker, Hausdorff, and Dynkin", Archive for History of Exact Sciences. **66 (3)**, 295–358 (2012). Eqn(9) is only valid if *both* $\hat{p}$ and $\hat{q}$ commute with their commutator $[\hat{q}, \hat{p}]$. It can thus be developed into an infinite series of commutators within commutators etc, hence solely as a function of the commutator; see remarks before eqn(10) later.

[47] Herbert Goldstein, "Classical Mechanics", second edition, 397–405 Dover Inc., New York (1980).

[48] P.A.M. Dirac, "Lectures on Quantum Mechanics", 10-15 Dover, New York, (2001).

[49] Werner Heisenberg, "The Physical Principles of the Quantum Theory", translated by Carl Eckhart and Frank C. Hoyt, based on lectures by Heisenberg at the University of Chicago in 1929, 165–172 Dover Inc., New York (1949). In the preface written from Leipzig in 1930, Heisenberg wrote "But even today the physicist more often has a kind of faith in the correctness of the new principles than a clear understanding of them", valid even today.

[50] Leonard I. Schiff, "Quantum Mechanics", 329–350 first edition, McGraw Hill Inc., New York (1949).

[51] P. Jordan and E. Wigner, Zeits. f. Physik, **47** 631 (1928).

[52] William M. Smart, "Celestial Mechanics", Longmans, Green and Co , London (1953).

[53] The Red October is a fictitious stealth nuclear ballistic missle submarine in the novel "The Hunt for the Red October" by Tom Clancy 1984 subsequently made into a film by Hollywood in 1990, staring Sean Connery, Alec Baldwin, Scott Glenn and others. The author has pointed out a deep analogy between the Red October function and the Rayleigh dissipation function in classical mechanics which can be extended to Quantum Mechanics during the delivery of his talk at the conference in Tropea, Italy on 04 July 2025.

[54] See for example Julian Schwinger, "Quantum Mechanics - Symbolism of Atomic Measurements", edited by Berthold-Georg Englert, Springer, Berlin (2000).

[55] Y.-H. Kim and Y. Shih, "Experimental realization of Popper's experiment: violation of the uncertainty principle?". Foundations of Physics, 29 (12), 1849–1861 (1999).

[56] Abraham Pais, "Subtle is the Lord...", The Science and Life of Albert Einstein , paper back edition Oxford University Press, Oxford (1983).

[57] Marie-Christine Combourieu, "Karl R. Popper: About the EPR Controversy", Foundations of Physics, **22** (10) 1303-1323 (1992).

[58] I. Fényes, "A Deduction of Schrödinger Equation". Acta Bolyaiana. 1 (5): ch. 2 (1946).

[59] I. Fényes, "Zur wellenmechanischen Herleitung des statistischen Atommodells". Zeitschrift für Physik. 125 (4–6): 336–346, (1948).

[60] E. Nelson, "Derivation of the Schrödinger Equation from Newtonian Mechanics". Physical Review **150**(4), 1049-1085, (1966).

[61] N.G. Van Kampen, "Stochastic Processes in Physics and Chemistry", Elsevier (Amsterdam) (1980).

[62] M.P. Davidson, "The Origin of the Algebra of Quantum Operators in the Stochastic Formulation of Quantum Mechanics". Letters in Mathematical Physics **3** (5), 367–376, (1979). See also "Stochastic Mechanics Redux" by Mark Davidson , https://zenodo.org/records/17259543 (2025).

[63] Steven Weinberg, "Quantum Mechanics Without State Vectors", Phys. Rev. A. **90** (4), 42102 (2014).

[64] T.W.B. Kibble "Geometrization of Quantum Mechanics", Comm. Math. Phys. **65** 189-201, (1979).

[65] N.M.J. Woodhouse, "Geometric Quantization",Clarendon Press, Oxford (1991).

[66] Moyal, J. E. and Bartlett, M. S., "Quantum mechanics as a statistical theory". Mathematical Proceedings of the Cambridge Philosophical Society. **45** (1), 99–124 (1949)

[67] C.K. Zachos, D.B. Fairlie and T.L. Curtright, "Quantum Mechanics in Phase Space", World Scientific Series in 20th century physics **34**, 2005.

[68] See for example M, Daniel, "A short introduction to the Lindblad master equation". AIP Advances. **10** (2), 25106 (2020).

[69] H-P Breuer and F. Petruccione, "The Theory of Open Quantum Systems", Oxford Universoty Press, Oxford (2003).

[70] N.A. Lemos, " Short Proof of Jacobi's identity for Poisson brackets" Am. J. Phys. **68**(1) 88 (2000)

[71] M. Blackman, "The lodestone: a survey of the history and the physics", Contemporary Physics, **24** (4),319-331 (1983).

# Chapter 5

## Deriving Heisenberg's operator algebra from a classical model of stochastic mechanics

Mark Davidson*
*Spectel Research Corporation*
(Dated: 10/22/24)

A feature of quantum mechanics that distinguishes it from classical Newtonian mechanics is that momentum and position are described by non-commuting operators on a Hilbert space. It has long been known that certain diffusion theories provide stochastic models for Schrödinger's equation, together with a natural way to understand this non-commutative structure. This theory is revisited here. It suggests that the origin of quantum theory might be found in an algebraic extension of the geometry of space-time to complex numbers. It is a possible stepping stone to an emergent explanation for quantum mechanics and to a unification of classical and quantum physics.

### I. INTRODUCTION[1]

Heisenberg introduced the concept of a time-dependent and non-commuting algebra for dynamic variables like position and momentum in his landmark 1925 paper [1]. Stochastic mechanics (SM) provides a stochastic interpretation for quantum mechanics [2–4]. It also provides a toolkit of rigorous mathematical theorems in stochastic processes for proving theorems about quantum mechanics. This is particularly relevant to the path-integral approach [4–8]. Another interesting reason for studying it is that it can help in the search for an emergent explanation for quantum mechanics, as envisioned by Einstein [9] and many

* mdavid@spectelresearch.com
[1] INVITED PAPER DEDICATED TO THE CELEBRATION OF 100 YEARS OF QUANTUM MECHANICS ON THE ANNIVERSARY OF HEISENBERG'S FOUNDING PAPER ON THE SUBJECT IN JULY 1925, TO BE PUBLISHED IN A CELEBRATORY VOLUME IN JULY 2025 BY WORLD SCIENTIFIC PUBLICATIONS, SINGAPORE.

others [10–14]. It may also provide insight into quantum noise and vacuum fluctuations. Although stochastic mechanics has focused much more on the wave function formalism of Schrödinger than on the operator formalism of Heisenberg, there is an operator approach to generalized stochastic mechanics (GSM) that is closely related to the Heisenberg operator formalism in quantum mechanics [15–17]. Whereas SM provides a diffusion model for quantum mechanics with a unique value for the diffusion constant, GSM provides many diffusion models that all have different diffusion constants [17, 18]. Jaekel and Pignon generalized this result even further by allowing the diffusion constant to depend on position and time [19]. Position-dependent diffusion constants are a well-known phenomenon in condensed matter physics ([20], § 3.7), and GSM might find applications there as well. GSM is based on the mathematical framework of SM [2–4, 21], and it depends on a nonlinear transformation of Schrödinger's equation and a modification of Nelson's force equation in stochastic mechanics [16, 18]. The reason why any diffusion constant value can be used in GSM, as explained below, is that the combination of wave-function collapse combined with the Heisenberg uncertainty principle makes it probably impossible to directly measure the diffusion constant for a quantum particle. A review of wave-function collapse is given in [22]. This is a contentious subject as there are quantum interpretations that attempt to remove this collapse from physics such as Many Worlds, Decoherence, Bayesian, and Spontaneous Collapse interpretations. None of these has yet successfully predicted experimental outcomes which differ from those predicted by standard Copenhagen quantum mechanics. Therefore we can take them as effectively equivalent to wave-function collapse as far as the measurability of the diffusion parameter is concerned here. Should this change in the future, and this is certainly a possibility, then it may be possible to measure the diffusion parameter in GSM, and this would be great.

It is fair to say that GSM has been somewhat ignored by mathematicians in favor of a unique value for the diffusion constant. One reason for this is the action principle of Nelson [3, 23] that many have argued supports the claim of uniqueness of stochastic mechanics with the diffusion constant of $\hbar/2m$. But actually, it's possible to generalize the action principle to allow for a variable diffusion constant as is shown below.

Heisenberg introduced the wave-function collapse in 1927 [24]. It was made into a rigourous mathematical postulate by Von Neumann [25]. Dirac discussed the measurement process and the wave function collapse postulate in his 1930 book [26]. The desire to have maximum mathematical rigor in stochastic mechanics made it difficult to incorporate it, so it was mostly ignored in the SM literature. The measurement problem is still a fundamental problem of quantum theory [27]. Petroni and Morato [28] remarked on the lack of attention to this topic in the SM literature. The physical interpretation of SM or GSM can be taken to be similar to Bohmian mechanics, and good sources for this are [29–32]. Both theories are essentially non-local when multi-particle states are considered. Wave function collapse simply doesn't fit neatly into rigorous mathematical stochastic analysis. But theoretical interpretations of experimental results require it. At the very core of quantum theory is this mysterious collapse, and it suggests that the diffusion parameter cannot be measured in an experiment. There have been many efforts to demystify it. Many Worlds, Decoherence and Spontaneous Collapse are prime examples, but none of these efforts has achieved unanimous acceptance by physicists.

In this paper the non-commutative approach is reviewed, and it is shown how the usual Heisenberg algebra can be constructed. The analytic continuation to complex values for the diffusion constant can be interpreted as a diffusion in complex space (or space and time in a relativistic framework). This adds to a growing list of reasons that we should take the space-time to be a complex 4D manifold, where the real subspace plays a prominent role at the macroscopic level, but the complex embedding plays a large role when quantum mechanics is necessary.

There are similarities between the techniques used here and the path integral approach of Feynman in his thesis [33]. Feynman derives the commutation rules of quantum mechanics by time-ordering products of position and momentum at infinitesimally different times, and summing over real-valued paths with a complex weighting factor. This is the same technique we use here, but our weighting factor in GSM is positive definite, and it qualifies therefore as a true stochastic process. We find here that the commutator between position and momentum is real-valued for a real diffusion constant. Only by analytically continuing to a complex space-time can we obtain an imaginary commutator between position and momentum, whereas Feynman obtains an imaginary commutator for paths in real space-time. So GSM seems to favor the introduction of complex space-time into physics more strongly than the Feynman path integral method does. It's also obvious that one can generalize the Feynman path integral approach by using the same nonlinear gauge transformation of the Schrödinger equation used in deriving GSM.

Stochastic mechanics, like Bohmian mechanics, is a non-local theory because of wave-function collapse for entangled multi particle states. Moreover, since the drift term in GSM for multi particle states depends on the coordinates of all the particles, so that it is a function on configuration space, then such a system can be considered to be non-local even without considering collapse. There are good arguments that quantum mechanics is inherently non-local, and this doesn't only apply to hidden variable interpretations [32, 34–36]. Nelson eventually became very worried about this and recommended concentrating on field theory instead of wave function analysis of particle motion [37]. The non-locality problems of stochastic mechanics are no worse than those faced by Bohmian mechanics, and since a fair number of physicists accept Bohmian mechanics as a viable interpretation, they would probably find stochastic mechanics equally acceptable. See, for example, [30, 38].

One possible way of avoiding non-locality is if underlying the stochastic behavior is due to a hidden fully-deterministic theory, yet still a local one. An important example of this is 't Hooft's Planck scale cellular automata theory [14]. It was inspired by Conway's Game of Life simulation [39]. He has proposed that the universe is completely deterministic and local, and he has shown that such a theory is completely consistent with Bell's theorem, despite common beliefs to the contrary. For example, here is a quote from his book (page 10):

"Our models suggest that Einstein may have been right, when he objected the conclusions drawn by Bohr and Heisenberg. It may well be that, at its most basic level, there is no randomness in Nature, no fundamentally statistical aspect to the laws of evolution. Everything, up to the most minute detail, is controlled by invariable laws. Every significant event in our universe takes place for a reason, it was caused by the action of physical law, not just by chance. This is the general picture conveyed by this book.We know that it looks as if Bell's inequalities have refuted this possibility, in particular because we are not prepared to abandon notions of locality, so yes, they raise interesting and important questions that we shall address at various levels."

and on page 11, he writes

"We hope to inspire more physicists to investigate these possibilities, to consider seriously the possibility that quantum mechanics as we know it is not a fundamental, mysterious, impenetrable feature of our physical world, but rather an instrument to statistically describe a world where the physical laws, at their most basic roots, are not quantum mechanical at all. Sure, we do not know how to formulate the most basic laws at present, but we are collecting indications that a classical world underlying quantum mechanics does exist."

The key reason why 't Hooft's models do not lead to non-locality is because there is a loophole in the Bell arguments, namely that they require that some random decision be made, typically of a polarization orientation, as a critical part of the measurement. In the early years this loophole was called a "conspiracy against the experimenter", and was dismissed as preposterous. The modern term for this is superdeterminism. In a fully deterministic theory of the universe, the orientations of polarizers and the like would be determined by the initial conditions of the universe. 't Hooft does not endorse or write favorably about models of the de Broglie-Bohm pilot wave family, and that would include generalized stochastic mechanics. Nevertheless, it's certainly concievable that the motions of particles that emerge in a theory such as he is envisioning could behave as a stochastic diffusion as in GSM, at least as an approximation in some appropriate limit. Finding the algorithm that nature might use for this is a difficult problem. The stochastic motion of particles in stochastic mechanics might therefore be caused by a deterministic theory at the Planck scale. Although the motion of the whole universe is deterministic in this picture, if one were to just look at a single particle, then its motion might look like a diffusion. Consider, for example, a small test particle moving in a turbulent fluid. The whole system, fluid plus particle, is deterministic and is described by the Navier-Stokes equation, but the motion of the single particle would look like a stochastic diffusion. This is the sort of picture that we have in mind. Obviously it can be extended to any number of particles.

I am not fully committed to a local universe and superdeterminism, but it is a possibility. Another is that we are living in a complex space-time manifold that appears real to us because the events that we are aware of are happening on or near to a real subspace [40], but the space-time is overlayed with multiple Riemann sheets of multivalued fields. Generalized stochastic mechanics on complex manifolds provides a continuous path to the Heisenberg algebra by analytic continuation to imaginary diffusion constant. This requires embedding the theory in complex space-time. The Feynman path integral approach also derives the Heisenberg algebra, but it does so in real space-time [33]. So the Feynman path integral picture might be considered superior since it doesn't require a complex space-time to derive the Heisenberg algebra, but on the other hand it's not a true diffusion theory since the path integrals are weighted by a complex measure. For a true diffusion theory such as GSM we must analytically continue it to complex space-time to get the Heisenberg algebra. If this is how nature actually works, then GSM is giving us an important clue to it that the Feynman path integral approach does not give.

In this paper, we shall mainly use the Markov transition function, the forward and backward differential equations for it, the Chapman-Kolmogorov equations, and the multiple-time densities of generalized Brownian motion. The mathematics needed is therefore familiar to most physicists, and is sufficient to do calculations. We treat only the 1 dimensional case here for simplicity, but we use the $\nabla_x$ and $\triangle_x$ operator symbols to represent the first and second derivatives in 1D. For simplicity of notation, we shall refer to a Wiener process as having an arbitrary diffusion constant $\nu$, and denote it by $W(t,\nu) = \sqrt{2\nu}W(t)$, where $W(t)$ is the standard Wiener process with diffusion constant of $1/2$. When writing out operators that involve derivatives, we shall use the operator notation used in quantum mechanics. For example $\hat{\nabla} f(x) = \nabla f + f\hat{\nabla}$.

## II. ON MEASURING OR PLACING AN EXPERIMENTAL BOUND ON THE DIFFUSION CONSTANT OR THE DRIFT TERM IN STOCHASTIC MECHANICS

A widely accepted assertion about quantum mechanics is that all measurements ultimately boil down to positions measurements. It is then argued that a consequence of this is that if we can develop a stochastic model of quantum mechanics that correctly describes probability densities as they evolve in position and time, then this is sufficient to describe nature. For example, in ([41], § 16, p. 111) Nelson writes:

"It is a triviality that all measurements are reducible to position measurements, since the outcome of any experiment can be described in terms of the approximate position of macroscopic objects. Let us suppose that we observe the outcome of an experiment by measuring the exact position at a given time of all the particles involved in the experiment, including those constituting the apparatus. This is a more complete observation than is possible in practice, and if the quantum and stochastic theories cannot be distinguished in this way then they cannot be distinguished by an actual experiment. However, for such an ideal experiment stochastic and quantum mechanics give the same probability density $|\psi|^2$ for the position of the particles at the given time."

In a similar vein, Feynman and Hibbs ([42], § 5, 96) wrote:

"Indeed, all measurements of quantum-mechanical systems could be made to reduce eventually to position and time measurements (e.g., the position of a needle on a meter or the time of flight of a particle). Because of this possibility a theory formulated in terms of position measurements is complete enough in principle to describe all phenomenon."

Without this reasonable assertion, the path integral approach would not be equivalent to other interpretations of quantum mechanics, nor would Bohmian mechanics or stochastic mechanics. A formal mathematical investigation of the degrees of freedom in describing quantum position-equivalent nonlinear wave equations was performed by Doebner and Goldin [43]. GSM is possible because it exploits a particular nonlinear gauge transformation of Schrödinger's equation which is included in Doebner and Goldin's more general framework.

Quantum mechanics does not agree with experiments unless the postulate of wave-function collapse is included in some form or another. An ideal experiment is viewed in the Copenhagen interpretation as manufacturing a change in the quantum state succinctly described by a projection operator acting on the quantum Hilbert state vector. This projection imprints the result of a measurement onto the canvas of reality [25]. This is exploited routinely in quantum computers. When wave function collapse is combined with the Heisenberg uncertainty principle, it makes it seemingly impossible to measure the diffusion constant. This section is partially a response to an argument that the diffusion constant cannot be larger than $\hbar/2m$ without abandoning the "usual assumptions" ([3], § 14, p. 66). It is not an observable. The reason why is quite simple, but worth presenting here. For the purpose of this argument, I assume that the collapse as described by Von Neumann occurs. If it doesn't then the following argument is not correct, and there may be a way to measure the diffusion constant. Consider a non-relativistic particle without spin. Let us consider a real and positive diffusion constant in GSM. It's important to note that we are talking about GSM here, and not the general diffusion situation. GSM is very special in that the Schrödinger equation is the same for all values of the diffusion constant. The diffusion constant (in 1D for simplicity) is defined by the stochastic expectation

$$\nu = \frac{1}{2} \lim_{\epsilon \to 0} \left( \frac{(E(x(t+\epsilon) - x(t))^2)}{\epsilon} \right) \tag{1}$$

The expectation value here is the stochastic process expectation, and it's real-valued (for real $\nu$), unlike a quantum expectation which would be complex valued because of the non-commuting property of position operators at different times. For each sample path in the expectation $x(t)$ is a continuous real-valued function of t (almost surely). Now let's ask how we can measure this. We shall accept with Nelson and Feynman that we must make position measurements. We must measure the same particle in quick succession, first at time $t$ and then later at time $t + \epsilon$. We take that the particle is described by a wave function $\psi(x, t)$ undergoing unitary time evolution up until the first measurement is performed, and this can be described as a diffusion process up until that point defined by the standard form of stochastic differential equation as

$$dx = b(x, t, \nu)dt \; + \; dw(t, \nu) \tag{2}$$

where $b(\nu, x, t)$ is a sufficiently well-behaved function called the drift term, $w(t, \nu)$ is a Wiener function with diffusion constant $\nu$, and the second term is called the diffusion term [41]. The drift term depends on the wave function, and it also depends on what value of $\nu$ we are considering. If we precisely measure the particle position at time t, and then instantly the wave function collapses, the drift term will become huge after this measurement because of the Heisenberg uncertainty principle. The collapsed $\psi$ will be localized at a point x, but it will have a large spread in momentum, and so the drift term will dominate the diffusion term even during the short time delay $\epsilon$ to the next measurement. This situation persists no matter how small we make $\epsilon$, and consequently, we can never measure the diffusion constant. This was pointed out, for example, in [19]. In [3] it was argued that for large values of $\nu$, the diffusion term would dominate and could therefore be ruled out with a measurement, but this assumes that the $b$ term doesn't grow with increasing values of $\nu$, and this is exactly what happens in GSM, as the probability density in GSM is independent of $\nu$. After measuring the position at time t, the wave function collapses and evolves to time t+ε through the Schrödinger equation, which is independent of $\nu$. Therefore, the argument against very large values of $\nu$ does not apply to GSM. One might try to get around this situation with some sort of weak measurement [44] that might indirectly determine the diffusion constant without disturbing the wave function so much, but thus far there's no known way to do this. One can't even place a bound on the magnitude of $\nu$, as we have just shown. GSM is possible because of this inability. Nelson ([3], p.117) agrees that strictly speaking, the diffusion constant cannot be measured, but he suggests that a reasonable assumption is that the forward velocity $b$ right after a position measurement should be no bigger than $p/m$, where $p$ here denotes the uncertainty in the momentum of the particle after measurement as required by the Heisenberg uncertainty principle. This assumption is not satisfied in GSM, where b would also depend on the diffusion constant after a measurement because of (22), and for a large diffusion constant, it would be bigger than Nelson's suggested upper bound. So in this paper, we abandon the "usual assumptions" used to dismiss large diffusion constants as metaphysical. Thus, we are free to analytically continue the stochastic functions of GSM into the entire complex $\nu$ plane.

Note that the Heisenberg uncertainty principle alone, without collapse of the wave function, is not sufficient to prevent us from measuring the diffusion constant. A visualization of this comes from the bouncing drop experiments of Yves Coudet [45–47]. In these experiments we see a wave influencing a particle's motion as envisioned by de Broglie. The wave equation is not exactly the same as quantum mechanics, but it shows some aspects of quantum behavior. It's quite possible that these wave functions can in some configuration satisfy an approximate form of the Heisenberg uncertainty principle as far as the wave function alone is concerned, but nevertheless we can easily measure the position and momentum of the bouncing drops without affecting them so there's no collapse. We might be called super-observers because of this. If there were ways to measure a real electron without disturbing its wave function in this way, then the diffusion constant in GSM could be measured. It's conceivable that someday this might be possible, but this would be revolutionary. The analog of a quantum observer in the bouncing drop experiment would have to be an apparatus built from only bouncing drops that could somehow perform a measurement on other bouncing drops. This might lead to a wave-function collapse or change and to something like a Heisenberg uncertainty principle. The subject of noninvasive measurability in quantum mechanics has been researched intensively, and there is very strong evidence that it is impossible to achieve [48–50]. It might still be possible to infer a value for the diffusion constant in some special situations. In Section X below, several speculative candidate systems are discussed.

So one can construct a stochastic model for Schrödinger's equation with any positive diffusion constant whatsoever. Negative values of the diffusion constant could correspond to a time reversal or to a diffusion in complex space-time. If you use the value, $\nu = \hbar/2m$, you get standard stochastic mechanics. If you take the limit $\nu \to 0$ you get Bohmian mechanics, a result first shown by Shucker [51]. If you analytically continue $\nu$ to imaginary values using GSM, you get the Heisenberg algebra when $\nu = \pm i\hbar/2m$. This requires complex space-time which we discuss below. From a strictly empirical point of view all of these theories are equivalent, at least given our current understanding of measurement limitations. This non-uniqueness of stochastic mechanics reflects a kind of nonlinear gauge invariance of Schrödinger's equation, as proved in [18], which is a subset of [43]. Modern physics is happy with gauge invariance, and so striving to fix an unmeasurable gauge with an arbitrary mathematical bias might not be the wisest strategy. If any value of the diffusion constant is preferred, it is a purely imaginary value of $\pm i\hbar/2m$, for with this value the Heisenberg algebra is perfectly reproduced in GSM. Unlike the Feynman path integral approach though, such an imaginary value requires embedding the diffusion in a complex space-time manifold.

If we can't measure the diffusion constant if it is real-valued and positive, we certainly can't measure it if it is complex-valued because we have no way of measuring the imaginary part of the position of a particle even if we are embedded in a complex space-time. We can only observe the real part of a particle's position. Our macroscopic tools for measurement are all located on the real subspace of complex space-time, or at least that's the picture that is emerging. It's like we are in a higher dimensional flatland and not aware of it [52].

Similar arguments can be applied to the forward velocity $b$ and the backward velocity $b_*$. They cannot apparently be measured directly, either. It is remarkable how effectively quantum mechanics manages to cover its tracks in this way. Perhaps someday a way to get around these limitations will be found, and a measurement of these hidden parameters might be possible. But considerable time has passed since the original publications on GSM [16, 18], and so far no experiments have been performed or even proposed to actually measure the diffusion constant.

## III. A SHORT REVIEW OF STOCHASTIC MECHANICS WITH ARBITRARY DIFFUSION CONSTANT

Most of this can be found in [41], especially in § 13. We have two stochastic differential equations: a forward and a backward equation in 1D for simplicity, although the higher dimensions are straightforward.

$$dx(t,\nu) = b(x(t),t,\nu)dt + dw(t,\nu) \tag{3}$$

$$dx_*(t,\nu) = b_*(x_*(t),t,\nu)dt + dw_*(t,\nu) \tag{4}$$

$b(x,t,\nu)$ can be interpreted as the stochastic expectation value of the velocity just after the particle has passed through the point x, and $b_*(x,t,\nu)$ as the velocity expectation just before the particle passes through x. If the Markov trajectories were differentiable, then these would be equal. But they are not differentiable, and so the forward and backward velocities are different, and this leads to great richness and complexity in this theory. The existence of this pair as both Markov processes is assured by a theorem by Doob (see [41], § 13, p. 85). We mention it because the backward velocity $b_*$ plays are large role in creating a time symmetric diffusion which is necessary for a stochastic model of quantum mechanics, and otherwise we would have to make it a separate assumption. As Nelson points out, the usual approach to diffusion in Brownian motion, that is not time symmetric, is not suitable for time reversible stochastic mechanics. In Appendix A a list of formulas that are useful in stochastic mechanics are presented. The parameter $\nu$ is the time independent diffusion parameter defined by

$$\nu = \frac{1}{2}E\left((dw(t,\nu))^2\right)/dt \tag{5}$$

and $w(t,\nu)$ is a Wiener process. Although $b$ and $b_*$ depend on $\nu$, we shall not always include it in the argument list of these and other functions to simplify the notation.

Central to our discussion is the Markov transition density function. We follow the convention used in [15] such that the earlier time is to the right of the semicolon which is opposite to the convention in [3], so we have

$$P(x,t;y,s) = \lim_{dx\to 0}\frac{1}{dx}P(x(t)\in dx \mid x(s)=y), \quad t>s \tag{6}$$

The notation $P(A|B)$ means the probability of A conditioned by B. This transition function satisfies the Chapman-Kolmogorov equation (unless otherwise specified, the domain of integration is taken to be $-\infty$ to $+\infty$)

$$P(x,t;y,r) = \int P(x,t;z,s)P(z,s;y,r)dz, \ for\ times\ t > s > r \tag{7}$$

Continuity of paths leads to the limiting behavior

$$\lim_{t \downarrow s} P(x,t;y,s) \ = \ \delta(x-y) \tag{8}$$

where $\delta$ denotes the Dirac delta function-distribution. The transition function is normalized.

$$\int P(x,t;y,s)dx = 1 \tag{9}$$

The time reversed process has a different transition density function $P_*$, but it is simply related to $P$ provided that the density function is never zero by the formula from ([3], eqn. 6.2, p 35)

$$P_*(y,t;x,s) = P(y,t;x,s)\frac{\rho(x,s)}{\rho(y,t)}, \ t > s \tag{10}$$

which also satisfies a Chapman-Kolmogorov equation. We have for the probability density $\rho$:

$$\rho(x,t) = \int P(x,t;y,s)\rho(y,s)dy, \ t > s \tag{11}$$

$$\rho(y,s) = \int P_*(x,t;y,s)\rho(x,t)dx, \ t > s \tag{12}$$

In general this does not describe a stationary process, and therefore the time reversed process is different from the forward process, even though the Schrödinger equation is invariant under time reversal. The Markov transition function plays a role similar to the time evolution operator in quantum mechanics. We can calculate the probability densities for multiple times very easily by forming products of Markov transition functions ([3], eqn. 6.3, p 35).

It is customary to write the probability density as

$$\rho(x,t) = exp(2R(x,t)) \tag{13}$$

We define $S_N$ up to a function independent of x by

$$2\nu \frac{\partial}{\partial x} S_N(x,t,\nu) \equiv \frac{b(x,t,\nu) + b_*(x,t,\nu)}{2} \tag{14}$$

and similarly ([41], § 15)

$$2\nu \frac{\partial}{\partial x} R(x,t,\nu) \equiv \frac{b(x,t,\nu) - b_*(x,t,\nu)}{2} \tag{15}$$

and it follows that (in 3D this assumes that $curl(b) = 0$)

$$b(x,t) = 2\nu \nabla \left(R + S_N\right) \tag{16}$$

$$b_*(x,t) = 2\nu \nabla \left(-R + S_N\right) \tag{17}$$

The wave function will be written as

$$\psi(x,t) = exp(R(x,t) + iS_Q(x,t)) \tag{18}$$

In stochastic mechanics, $S_N = S_Q$ and $\nu = \hbar/2m = 1/2$ in our units. The wave function will be independent of $\nu$ in GSM. This is because the following two equations are equivalent, as shown for real-valued constant $z$ in [18]

$$\left(-\frac{\hbar^2}{2m}\triangle + V\right) exp(R + iS_Q) = i\hbar \frac{\partial}{\partial t} exp(R + iS_Q) \tag{19}$$

$$\left(-\frac{(z\hbar)^2}{2m}\triangle + \left(V + \frac{\hbar^2}{2m}\left(z^2 - 1\right)\frac{\triangle\sqrt{\rho}}{\sqrt{\rho}}\right)\right) exp(R + iS_Q/z) = i\left(z\hbar\right)\frac{\partial}{\partial t} exp(R + iS_Q/z) \tag{20}$$

and where $\rho = exp(2R)$. This equivalence is also true for all complex values of z by simple analytic continuation from the real values. The extra nonlinear term in the potential in (20) is just Bohm's quantum potential up to a multiplicative constant. The second equation looks like a modified Schrödinger equation with the replacements $\hbar \to z\hbar$ and $V \to \left(V + \frac{\hbar^2}{2m}\left(z^2-1\right)\frac{\triangle\sqrt{\rho}}{\sqrt{\rho}}\right)$, so we can immediately, model this with a stochastic process of the form (2) with

$$\nu = z\frac{\hbar}{2m} \tag{21}$$

$$b(x,t,\nu) = 2\nu\nabla\left(R(x,t) + S_Q(x,t)/z\right) = \nabla\left(2\nu R(x,t) + \frac{\hbar}{m}S_Q(x,t)\right) \tag{22}$$

This then defines the generalized stochastic process for any value of the diffusion constant. The generalized "equation of motion" is given in (31) along with (32).

## IV. AN ACTION PRINCIPAL COMPATIBLE WITH ARBITRARY DIFFUSION CONSTANT

Here we utilize the theorem and technique of Yasue [53], but with a modified Lagrangian. Given an action of the form

$$I = E\left[\int_a^b L(x(t), Dx(t), D_*x(t))dt\right] \tag{23}$$

where the operators $D$ and $D_*$ are defined in (A10) and (A11). He obtained the extended Euler-Lagrange equation

$$D\left(\frac{\partial L}{\partial D_*x(t)}\right) + D_*\left(\frac{\partial L}{\partial Dx(t)}\right) - \frac{\partial L}{\partial x(t)} = 0 \tag{24}$$

(see also Zambrini, equation 25 in [54]). The operators $D$ and $D_*$ are the forward and backward time derivatives of [41]. Yasue showed that the following time reversible Lagrangian

$$L_N = \frac{m}{2}\left(\frac{(Dx(t))^2 + (D_*x(t))^2}{2}\right) - V(x) \tag{25}$$

leads to the Euler-Lagrange equation if $Dx(t)$ and $D_*x(t)$ are varied independently

$$m\frac{DD_* + D_*D}{2}x(t) = -\frac{\partial V(x(t))}{\partial x(t)} \tag{26}$$

and this is the same equation of motion that Nelson found which gives the Schrödinger equation provided the diffusion constant is given by

$$\nu_N = \frac{\hbar}{2m} \tag{27}$$

and where we have chosen units so that $\hbar = 1$. The following acceleration is called the mean acceleration

$$a_N = \frac{DD_* + D_*D}{2}x(t) \tag{28}$$

Now consider modifying the Lagrangian to read

$$L_\nu = A\left(L_N + V(x)\right) + B\left(Dx(t)\right)\left(D_*x(t)\right) - V(x) \tag{29}$$

where $A$ and $B$ are constants. Note that this modified Lagrangian is still invariant under time reversal. The Euler-Lagrange equations for this Lagrangian are

$$A\left(m\frac{DD_* + D_*D}{2}x(t)\right) + B\left(m\frac{DD + D_*D_*}{2}x(t)\right) = -\frac{\partial V(x(t))}{\partial x(t)} \tag{30}$$

It was shown in [18] that the following "equation of motion" also leads to

Schrödinger's equation

$$\left(m\frac{DD_* + D_*D}{2}x(t)\right) + m\frac{\beta}{8}\left(D - D_*\right)^2 x(t) = -\frac{\partial V(x(t))}{\partial x(t)} \tag{31}$$

provided that

$$\nu = \frac{\hbar}{2m}\frac{1}{\sqrt{1-\beta/2}} \tag{32}$$

or equivalently

$$\beta = 2\left(1 - \left(\frac{\hbar}{2m\nu}\right)^2\right) \tag{33}$$

and therefore, we get Schrödinger's equation provided that

$$A = 1 - \frac{\beta}{4},\ and\ B = \frac{\beta}{4} \tag{34}$$

and therefore we can have an action principal for any value of the diffusion constant. Nelson has argued in [3] that although we cannot physically measure the diffusion constant according to existing quantum theory, maybe someday a means will be found to get around this, which would be a violation of our current understanding of what is possible. The action chosen in [3] for simplicity is divergent, but when the infinite part is subtracted off, one gets a remainder that gives a unique diffusion constant that is Nelson's original value of $\hbar/2m$. It is argued that the infinite part of the action depends only on the diffusion constant, and therefore does not enter into the variation. One might question the mathematical rigor of this argument because of the equality: $\infty + a = \infty + b$ for all finite values of $a$ and $b$. Since the diffusion constant cannot be measured, this argument is metaphysical, but it could nevertheless be the way that nature works. Jaekel has also given a rendition of this result [23]. For the purpose of tying together the Heisenberg operator approach to stochastic mechanics, we consider the variable diffusion constant general theory of GSM in this paper.

## V. DERIVING THE HEISENBERG OPERATOR FORMALISM FROM GENERALIZED STOCHASTIC MECHANICS

The simplest way to illustrate or derive the Heisenberg operator formalism in the framework of stochastic mechanics is to use diffusion in a complex space or space-time. Here we start with real $\nu$ and use results from [15, 16], but we alert the reader that the definition of $\nu$ in these two papers differ by a factor of 2. Here in (5) we use the definition in [15] which coincides with ([41], § 13). The basic idea of the derivation of non-commuting operators in stochastic mechanics is that the order of operations at equal times is simply the infinitesimal time-ordering of certain expectation values in the limit where the difference between the times tends to zero. Consider the random variable $x(t)$ in SM setting. The time derivative does not exist for these functions as is well known. However, if we consider the two-point function $E(x(t)x(s))$,then this function is typically differentiable with respect to either time if the forward velocity $b(x,t)$ is sufficiently well-behaved which we assume. Now to get an idea of where the noncommutativity comes from in this framework, consider the following limiting procedure with $t > s$

$$Commutator = \lim_{t \uparrow s} \left( \frac{\partial}{\partial t} - \frac{\partial}{\partial s} \right) E(x(t)x(s)) \tag{35}$$

The expectation $E()$ here is just the classical expectation for a stochastic process $x(t)$. It is <u>not</u> a quantum expectation. The sample paths for $x(t)$ are continuous but non-differentiable trajectories. They are fractal curves if you like. It is imperative that the expectation value be taken first, before the differentiation, because of this. The only exception is the Bohmian limit where $\nu = 0$ and the paths become differentiable. This microscopic time ordering is essentially the same technique employed by Feynman in his thesis [33], but for a true diffusion process here. Our treatment here is based entirely on stochastic mechanics, though as modified in GSM. If the sample paths were smooth, this commutator would vanish. It is a difference between the two time orderings of a velocity times a position. So It's like a commutator. It can be evaluated by means of the forward and backward derivatives in the present context

$$Commutator = E\left(\left(b_*(x,s) - b(x,s)\right)x(s)\right) \tag{36}$$

Now we use (A6) to obtain by using integration by parts

$$Commutator = -\int 2\nu \frac{grad\, \rho(x,t)}{\rho(x,t)} x\rho(x,t)dx = 2\nu \tag{37}$$

This simple result suggests that we can use the microscopic time ordering of expectations to define a non-commuting operator algebra. If we consider Nelson's value of $\nu = \hbar/2m$, and that the momentum is m times the velocity, it looks like the commutator of p and x is simply $\hbar$ up to a sign at least, since we could have chosen the opposite ordering. Thus the time-ordering of expectations of this type induce a non commutative structure as was described in [15, 16], but the commutators are real-valued if the diffusion constant is real. Note that the definition of $\nu$ in [16] differs by a factor of 2 compared with [15] and here. In order to get the usual Heisenberg commutation relations, we must analytically continue to imaginary values of $\nu$ and set

$$\nu = -i\frac{\hbar}{2m} \tag{38}$$

The interpretation of this analytical continuation is non-trivial, and the reader should not be surprised if he has trouble visualizing what this means. It's basically an area of active research. A few papers dedicated to this problem are discussed briefly below. We saw that Schrödinger's equation was invariant as a function of $\nu$ in GSM for positive $\nu$. But then, assuming that both sides of the equation are analytic in $\nu$, the equality extends automatically to all complex values of $\nu$. We assume this is the case. In this way we can achieve three desirable goals:

1. We describe quantum mechanics with a genuine diffusion process with positive definite probability densities.

2. We can describe these processes utilizing non-commuting operators as in quantum mechanics, but for real values of the diffusion constant, the commutator of $\hat{x}$ and $\hat{\dot{x}}$ is real. We can replicate the Heisenberg algebra with an imaginary commutator.

3. We can describe Bohmian mechanics, stochastic mechanics, and generalized stochastic mechanics in a single framework.

Given that the Schrödinger equation is always the same, independent of $\nu$ in GSM, then according the Nelson and Feynman's position mandate, it doesn't matter what value of $\nu$ we use. But, if we desire to replicate the Heisenberg algebra in this way, it requires that the particle diffuses in a complex space, and if we consider that space and time mix under Lorentz transformations, then this would ultimately require motion in a complex space-time. In order to develop the Heisenberg algebra further, we follow the approach in [15, 16]. In the Heisenberg picture the coordinates and momenta operators are time-dependent, but the wave function is fixed, as opposed to the Schrödinger picture where the wave function is time-dependent and the coordinates and momenta operators are time independent. Towards this end we introduce a base time $s$ which will be fixed, and we derive time-dependent operators from this point using the Markov transition functions. This base time is arbitrary. We introduce it so that we can immediately utilize the various equations of the stochastic process which all depend on time. It is assumed that all other choices of base time are equivalent and can be mapped from one to another using the equations of Markov processes. After the commutation rules for operators are established, we can transform them by similarity transformations that leave them unchanged, and in this way recover the usual quantum mechanical Hilbert space formulation. So first let's introduce a Hilbert space $\mathcal{H}_s$ with the following inner product:

$$(f,g) = E\left(f^*(x(s),s)g(x(s),s)\right) = \int \rho(x,s)f^*(x,s)g(x,s)dx \tag{39}$$

The functions $f$ and $g$ that we consider are assumed to be differentiable to all orders in both independent variables. Now let us define operators on these functions. The operator for $x$ we denote by simple multiplication, and since the base time $s$ is fixed, we drop it from the argument of $\hat{x}$ notational simplicity

$$\hat{x}f(x,s) \equiv xf(x,s) \tag{40}$$

Next we define an operator for $\dot{x}$ that acts on elements of the Hilbert space $\mathcal{H}_s$ and is defined as the following limit of a derivative of a conditional expectation

$$\hat{\dot{x}}f(x,s) \equiv \lim_{t\downarrow u}\lim_{u\downarrow s}\frac{\partial}{\partial u}E\left(x(u)f(x(t),t)|x(s)=x\right),\ t>u>s,\ f\in\mathcal{H}_s \tag{41}$$

where the notation $\lim_{t\downarrow u}$ means take the limit as $t$ approaches $u$ from above. Now it's convenient to write this expectation in terms of the Markov transition function densities because we can then use the forward and backward equations to simplify it. Using (A20) we can write the probability density for three times, where $t_1 < t_2 < t_3$, as

$$\rho(x_3,t_3;x_2,t_2;x_1,t_1) = P(x_3,t_3;x_2,t_2)P(x_2,t_2;x_1,t_1)\rho(x_1,t_1) \tag{42}$$

and this approach can be extended to any number of different times. To get the conditional expectation we simply drop the term $\rho(x_1,t_1)$. So we rewrite (41) as

$$\hat{\dot{x}}f(x,s) = \lim_{t\downarrow u}\lim_{u\downarrow s}\frac{\partial}{\partial u}\int f(x_t,t)x_uP(x_t,t;x_u,u)P(x_u,u;x,s)dx_tdx_u \tag{43}$$

We assume uniform convergence so that we can bring the $\dfrac{\partial}{\partial u}$ inside the integral.

$$\hat{\dot{x}}f(x,s) = \lim_{t\downarrow u}\lim_{u\downarrow s}\int f(x_t,t)\int x_u\frac{\partial}{\partial u}\left(P(x_t,t;x_u,u)P(x_u,u;x,s)\right)dx_u dx_t \tag{44}$$

$$\hat{\dot{x}}f(x,s) = \lim_{t\downarrow u}\lim_{u\downarrow s}\int f(x_t,t)\times$$
$$\times\int x_u\left(\frac{\partial P(x_t,t;x_u,u)}{\partial u}P(x_u,u;x,s) + P(x_t,t;x_u,u)\frac{P(x_u,u;x,s)}{\partial u}\right)dx_u dx_t \tag{45}$$

Use the backward equation A18 (with $\nabla$ and $\triangle$ symbols for derivatives in 1D):

$$\frac{\partial P(x_t,t;x_u,u)}{\partial u} = -\left(b(x_u,u)\cdot\nabla_{x_u} + \nu\triangle_{x_u}\right)P(x_t,t;x_u,u),\ t>u \tag{46}$$

and use the forward equation A17 to obtain:

$$\frac{P(x_u,u;x_s,s)}{\partial u} = -\left(\nabla_{x_u}\cdot b(x_u,u) + b(x_u,u)\cdot\nabla_{x_u} - \nu\triangle_{x_u}\right)P(x_u,u;x_s,s),\ u>s \tag{47}$$

and integrating this by parts, results in :

$$\int x_u\left(\frac{\partial P(x_t,t;x_u,u)}{\partial u}P(x_u,u;x,s) + P(x_t,t;x_u,u)\frac{P(x_u,u;x,s)}{\partial u}\right)dx_u$$
$$= \int P(x_t,t;x_u,u)\left(-x_u\left(\nabla_{x_u}\cdot b(x_u,u) + b(x_u,u)\cdot\nabla_{x_u} - \nu\triangle_{x_u}\right)\right)P(x_u,u;x,s)dx_u +$$
$$\int P(x_t,t;x_u,u)\left(\left(\nabla_{x_u}\cdot b(x_u,u) + b(x_u,u)\cdot\nabla_{x_u} - \nu\triangle_{x_u}\right)x_u\right)P(x_u,u;x,s)dx_u \tag{48}$$

This can be expressed more compactly in terms of a commutator

$$= \int P(x_t,t;x_u,u)\left[\left(\nabla_{x_u}\cdot b(x_u,u) + b(x_u,u)\cdot\hat{\nabla}_{x_u} - \nu\hat{\triangle}_{x_u}\right),x_u\right]P(x_u,u;x,s)dx_u \tag{49}$$

but $\left[\left(\nabla_{x_u}\cdot b(x_u,u)\right),x_u\right] = 0$ and so

$$= \int P(x_t,t;x_u,u)\left[\left(b(x_u,u)\cdot\hat{\nabla}_{x_u} - \nu\hat{\triangle}_{x_u}\right),x_u\right]P(x_u,u;x,s)dx_u \tag{50}$$

and we have $\left[\left(b(x_u,u)\cdot\hat{\nabla}_{x_u} - \nu\hat{\triangle}_{x_u}\right),x_u\right] = \left(b(x_u,u) - 2\nu\hat{\nabla}_{x_u}\right)$ (51)

$$= \int P(x_t, t; x_u, u) \left(b(x_u, u) - 2\nu\nabla_{x_u}\right) P(x_u, u; x, s) dx_u \tag{52}$$

Now we can take the limits

$$\lim_{t \downarrow u} P(x_t, t; x_u, u) = \delta(x_t - x_u) \tag{53}$$

$$\lim_{u \downarrow s} P(x_u, u; x_s, s) = \delta(x_u - x_s) \tag{54}$$

$$\hat{\dot{x}} f(x, s) = \lim_{t \downarrow u} \lim_{u \downarrow s} \int f(x_t, t)\delta(x_t - x_u) \left(b(x_u, u) - 2\nu\nabla_{x_u}\right) \delta(x_u - x_s) dx_t dx_u, \; t > u > \tag{55}$$

now we integrate over $x_t$

$$\hat{\dot{x}} f(x, s) = \lim_{t \downarrow u} \lim_{u \downarrow s} \int f(x_u, t) \left(b(x_u, u) - 2\nu\nabla_{x_u}\right) \delta(x_u - x_s) dx_u, \; t > u > s \tag{56}$$

now we integrate by parts once again

$$\hat{\dot{x}} f(x, s) = \lim_{t \downarrow u} \lim_{u \downarrow s} \int \delta(x_u - x_s) \left(b(x_u, u) + 2\nu\nabla_{x_u}\right) f(x_u, t) dx_u, \; t > u > s \tag{57}$$

$$\hat{\dot{x}} f(x, s) = \left(b(x, s) + 2\nu\nabla_x\right) f(x, s) \tag{58}$$

$$\hat{\dot{x}} = \left(b(x, s) + 2\nu\hat{\nabla}_x\right) \tag{59}$$

It follows then that a commutator relation exists

$$\left[\hat{\dot{x}}(s), \hat{x}(s)\right] = 2\upsilon \tag{60}$$

and this is very similar to the Heisenberg commutation rule, except that its value is real and not imaginary. Note that $\hat{\dot{x}}(s)$ is non-Hermitian in general, unless $\nu$ is imaginary. Since $\hat{x}$ is just a simple multiplication operator, we will drop the hat over this symbol for simplicity. To proceed further, we note that we can write

$$\hat{\dot{x}} = \left[(\nabla_x \cdot b(x,s)) + \left(b(x,s) + \nu\hat{\nabla}_x\right) \cdot \hat{\nabla}_x, x\right] \tag{61}$$

For a smooth function $\hat{j}(x,s)$ of $x$ and $s$ let us define an operator corresponding to a time derivative by a commutation rule derived from this as

$$\hat{\dot{j}}(x,s) = \left[(\nabla_x \cdot b(x,s)) + \left(b(x,s) + \nu\hat{\nabla}_x\right) \cdot \hat{\nabla}_x, \hat{j}(x,s)\right] + \frac{\partial \hat{j}(x,s)}{\partial s} \tag{62}$$

To justify this equation, replace the term $x_u$ in (43) with the function $j(x_u, u)$. Then assuming that this function is smooth, and that the integral is uniformly convergent, the result follows.
Next, we define an acceleration operator

$$\hat{\ddot{x}} f(x,s) \equiv \lim_{t\downarrow u} \lim_{u\downarrow s} \frac{\partial^2}{\partial u^2} E\left(x(u) f(x(t),t) | x(s) = x\right), \; t > u > s, \; f \in \mathcal{H}_s \tag{63}$$

This simplifies to

$$\hat{\ddot{x}} = \frac{\partial b(x,s)}{\partial s} + \nu\left(\triangle_x b(x,s)\right) + \frac{1}{2}\left(\nabla_x\left(b^2(x,s)\right)\right) \tag{64}$$

*Proof*

$$\hat{\ddot{x}} f(x,s) \equiv \lim_{t\downarrow u} \lim_{u\downarrow s} \frac{\partial^2}{\partial u^2} \int f(x_t,t) x_u P(x_t,t;x_u,u) P(x_u,u;x,s) dx_t dx_u, \; t > u > s \tag{65}$$

again we assume uniform convergence so that we can bring the $\frac{\partial^2}{\partial u^2}$ inside the integral and use the forward and backward equations to obtain

$$\hat{\ddot{x}} = \lim_{t\downarrow u} \lim_{u\downarrow s} \int P(x_t,t;x_u,u) \times$$
$$\left[\left(\nabla_{x_u} \cdot b(x_u,u) + b(x_u,u) \cdot \hat{\nabla}_{x_u} - \nu\hat{\triangle}_{x_u}\right),\right.$$

$$\left[\left(\nabla_{x_u}\cdot b(x_u,u)+b(x_u,u)\cdot\hat{\nabla}_{x_u}-\nu\hat{\triangle}_{x_u}\right),x_u\right]\,]$$
$$\times P(x_u,u;x,s)dx_u+\frac{\partial b}{\partial s} \tag{66}$$

$$\hat{\ddot{x}}=\lim_{t\downarrow u}\lim_{u\downarrow s}\int P(x_t,t;x_u,u)$$
$$\left[\left(\nabla_{x_u}\cdot b(x_u,u)+b(x_u,u)\cdot\hat{\nabla}_{x_u}-\nu\hat{\triangle}_{x_u}\right),\left(b(x_u,u)-2\nu\hat{\nabla}_{x_u}\right)\right]\times$$
$$\times P(x_u,u;x,s)dx_u+\frac{\partial b}{\partial s} \tag{67}$$

But we have

$$\begin{aligned}
&\left[\left(\hat{\nabla}_{x_u}\left(b(x_u,u)-2\nu\hat{\nabla}\right)+\nu\hat{\triangle}_{x_u}\right),\left(b(x_u,u)-2\nu\hat{\nabla}_{x_u}\right)\right]\\
&=\left[\left(\hat{\nabla}_{x_u}\left(b(x_u,u)-2\nu\hat{\nabla}\right)\right),\left(b(x_u,u)-2\nu\hat{\nabla}_{x_u}\right)\right]+\left[\nu\hat{\triangle}_{x_u},\left(b(x_u,u)-2\nu\hat{\nabla}_{x_u}\right)\right]\\
&=\left[\left(\hat{\nabla}_{x_u}\left(b(x_u,u)-2\nu\hat{\nabla}\right)\right),\left(b(x_u,u)-2\nu\hat{\nabla}_{x_u}\right)\right]+\left[\nu\hat{\triangle}_{x_u},b(x_u,u)\right]
\end{aligned} \tag{68}$$

and now we use the commutator identity $[AB,B]=[A,B]\,B$ to obtain

$$\begin{aligned}
&\left[\left(\hat{\nabla}\left(b-2\nu\hat{\nabla}\right)+\nu\hat{\triangle}\right),\left(b-2\nu\hat{\nabla}\right)\right]\\
&=\left[\hat{\nabla}\left(b-2\nu\hat{\nabla}\right),b-2\nu\hat{\nabla}\right]+\left[\nu\hat{\triangle},\left(b-\cancel{2\nu\hat{\nabla}}\right)\right]\\
&=\left[\hat{\nabla},\left(b-\cancel{2\nu\hat{\nabla}}\right)\right]\left(b-2\nu\hat{\nabla}\right)+\left[\nu\hat{\triangle},b\right]\\
&=\left[\hat{\nabla},b\right]\left(b-2\nu\hat{\nabla}\right)+\left[\nu\hat{\triangle},b\right]\\
&=(\nabla\cdot b)\left(b-2\nu\hat{\nabla}\right)+\nu\left[\hat{\triangle},b\right]\\
&=(\nabla\cdot b)\left(b-2\nu\hat{\nabla}\right)+\nu\left(\left[\hat{\nabla},b\right]\hat{\nabla}+\hat{\nabla}\left[\hat{\nabla},b\right]\right)\\
&=(\nabla\cdot b)\left(b-2\nu\hat{\nabla}\right)+\nu\left((\nabla\cdot b)\,\hat{\nabla}+\hat{\nabla}\,(\nabla\cdot b)\right)\\
&=\frac{1}{2}\nabla b^2+\nu\Delta b
\end{aligned} \tag{69}$$

Then adding in the term $\partial b/\partial s$ from 66, the result 64 is obtained. This is the same acceleration operator that was found in both [16] and [15] which use slightly different manipulations to arrive at the same result, up the the factor of 2 difference in the definition of $\nu$ mentioned above. This can be conveniently rewritten in the following form [15] as

$$\hat{\ddot{x}} = \nabla_x \left( exp(-R - S_N) \left( 2\nu \frac{\partial}{\partial t} + 2\nu^2 \triangle_x \right) exp(R + S_N) \right), b = 2\nu \nabla (R + S_N) \tag{70}$$

Let's compare (64) with the swapped version by interchanging $b$ and $b_*$ which is

$$\hat{\ddot{x}}_* = \frac{\partial b_*(x,s)}{\partial s} + \nu \triangle_x b_*(x,s) + \frac{1}{2} \nabla_x \left( b_*^2(x,s) \right) \tag{71}$$

Now consider the difference

$$\frac{\hat{\ddot{x}} - \hat{\ddot{x}}_*}{2} = \frac{\partial u(x,s)}{\partial s} + \nu \triangle_x u(x,s) + \nabla_x \left( u(x,s) v(x,s) \right) \tag{72}$$

Comparing this with the kinematic equation (13.5) in [41] which is

$$\frac{\partial u(x,s)}{\partial s} + \nu \triangle_x v(x,s) + \nabla_x \left( u(x,s) v(x,s) \right) = 0 \tag{73}$$

We see that

$$\frac{\hat{\ddot{x}} - \hat{\ddot{x}}_*}{2} = \nu \triangle_x u(x,s) - \nu \triangle_x v(x,s) \tag{74}$$

Kinematic here means that this result does not depend on the particular equation of motion chosen. Now we can write

$$\hat{\ddot{x}} = \frac{\left( \hat{\ddot{x}} + \hat{\ddot{x}}_* \right) + \left( \hat{\ddot{x}} - \hat{\ddot{x}}_* \right)}{2} \tag{75}$$

$$\hat{\ddot{x}} = \left(\frac{\partial v(x,s)}{\partial s} + \cancel{\nu \triangle_x v(x,s)} + \frac{1}{4}\nabla_x \left(b^2(x,s) + b_*^2(x,s)\right)\right) + \left(\nu \triangle_x u(x,s) - \cancel{\nu \triangle_x v(x,s)}\right) \tag{76}$$

$$\hat{\ddot{x}} = \frac{\partial v(x,s)}{\partial s} + \frac{1}{4}\nabla_x \left(b^2(x,s) + b_*^2(x,s)\right) + \nu \triangle_x u(x,s) \tag{77}$$

Let's compare this with Nelson's mean acceleration in equation ([2] eqn. 13.6) in

$$a_N = \frac{\partial v(x,s)}{\partial s} - u \cdot \nabla_x u + v \cdot \nabla_x v - \nu \triangle_x u \tag{78}$$

$$a_N = \frac{\partial v(x,s)}{\partial s} + \frac{1}{2}\nabla_x \left(v(x,s)^2 - u(x,s)^2\right) - \nu \triangle_x u(x,s) \tag{79}$$

Let's now find the difference

$$\hat{\ddot{x}} - a_N = \left(\frac{1}{4}\nabla_x \left(b^2 + b_*^2\right) + \nu \triangle_x u(x,s)\right) - \left(\frac{1}{2}\nabla_x \left(v(x,s)^2 - u(x,s)^2\right) - \nu \triangle_x u(x,s)\right) \tag{80}$$

$$\hat{\ddot{x}} - a_N = 2\nu \triangle_x u(x,s) + \frac{1}{4}\nabla_x \left(b(x,s)^2 + b_*(x,s)^2 - 2b(x,s)b_*(x,s)\right) \tag{81}$$

$$\hat{\ddot{x}} - a_N = 2\nu \triangle_x u(x,s) + \nabla_x \left(u(x,s)^2\right) \tag{82}$$

But $u = 2\nu \nabla_x R$, and therefore

$$\hat{\ddot{x}} - a_N = 2\nu \triangle_x \left(2\nu \nabla_x R\right) + \nabla_x \left(\left(2\nu \nabla_x R\right)^2\right) \tag{83}$$

$$\hat{\ddot{x}} - a_N = \left(2\nu\right)^2 \nabla_x \left(\triangle_x R + \left(\nabla_x R\right)^2\right) \tag{84}$$

$$\hat{\ddot{x}} - a_N = \left(2\nu\right)^2 \nabla_x \left(\frac{\triangle_x \sqrt{\rho}}{\sqrt{\rho}}\right) \tag{85}$$

It follows from the GSM equation of motion (31) that

$$m a_N = -\nabla_x V(x,t) - m\frac{\beta}{8}\left(D - D_*\right)^2 x(t) \tag{86}$$

$$m\frac{\beta}{8}\left(D-D_*\right)^2 x(t)=\nabla_x\left(m\beta\nu^2\frac{\triangle\sqrt{\rho(x,t)}}{\sqrt{\rho(x,t)}}\right) \tag{87}$$

$$ma_N=-\nabla_x\left(V(x,t)+m\beta\nu^2\frac{\triangle\sqrt{\rho(x,t)}}{\sqrt{\rho(x,t)}}\right) \tag{88}$$

$$ma_N=-\nabla_x\left(V(x,t)+m\left(2\left(\nu^2-\left(\frac{\hbar}{2m}\right)^2\right)\right)\frac{\triangle\sqrt{\rho(x,t)}}{\sqrt{\rho(x,t)}}\right) \tag{89}$$

Therefore

$$\hat{\ddot{x}}=a_N+(2\nu)^2\nabla_x\left(\frac{\triangle_x\sqrt{\rho}}{\sqrt{\rho}}\right) \tag{90}$$

$$m\hat{\ddot{x}}=-\nabla_x\left(V(x,t)+m\left(2\left(\nu^2-\left(\frac{\hbar}{2m}\right)^2\right)\right)\frac{\triangle\sqrt{\rho(x,t)}}{\sqrt{\rho(x,t)}}\right)+m\left(2\nu\right)^2\nabla_x\left(\frac{\triangle_x\sqrt{\rho}}{\sqrt{\rho}}\right) \tag{91}$$

$$m\hat{\ddot{x}}=-\nabla_x\left(V(x,t)+m\left(2\left(-\nu^2-\left(\frac{\hbar}{2m}\right)^2\right)\right)\frac{\triangle\sqrt{\rho(x,t)}}{\sqrt{\rho(x,t)}}\right) \tag{92}$$

$$m\hat{\ddot{x}}=-\nabla_x V(x,t)+\nabla_x\left(2m\nu^2+\frac{\hbar^2}{2m}\right)\frac{\triangle_x\sqrt{\rho(x,t)}}{\sqrt{\rho(x,t)}} \tag{93}$$

This is the same formula that was found in ([15], eqn. 35). Here we have presented the tedious derivation which was omitted there. We see that the simplest possibility occurs when $\nu=\pm i\hbar/2m$, for then we get the Heisenberg formula of quantum mechanics

$$m\hat{\ddot{x}}=-\nabla_x V(x,t),\ if\ \nu=\pm i\hbar/2m \tag{94}$$

In general we have

$$m\hat{\ddot{x}} = -\nabla_x U(x,t) \tag{95}$$

for some function $U(x,t)$, and in particular, for the GSM equations of motion, we further have

$$U(x,t) = V(x,t) - \left(2m\nu^2 + \frac{\hbar^2}{2m}\right)\frac{\triangle\sqrt{\rho(x,t)}}{\sqrt{\rho(x,t)}} \tag{96}$$

The zero diffusion limit of the stochastic process gives Bohmian mechanics [51]. This is a significant unification of two of the principal alternative interpretations of quantum mechanics, namely the Bohmian and the stochastic interpretations. In the Bohmian limit the velocity and position operators commute.

$$\lim_{\nu\to 0} m\hat{\ddot{x}} = -\frac{\partial\left(V - \frac{\hbar^2}{2m}\frac{\triangle\sqrt{\rho}}{\rho}\right)}{\partial x} \tag{97}$$

Here $V$ is the classical potential, and the extra term is called the quantum potential of Bohm [55]. The theoretical description of the Markov process is concisely summarized in the following two equations which were called "Markov wave equations" in [16].

$$\left[m\frac{(2\nu)^2}{2}\triangle_x + W\right]e^{R\pm S_N} = \mp m2\nu\frac{\partial}{\partial t}e^{R\pm S_N} \tag{98}$$

These are true for any given function $W(x,t)$, and not just for the GSM equations. These equations were also used in [3, 21], and were almost certainly found earlier than [16], but I don't know who found them first. Notice that $\hat{\dot{x}}$ in (59) can be put into canonical form by a similarity transformation which does not affect the commutation rules

$$\hat{\dot{x}}_{canonical} = e^{-R-S_N}\hat{\dot{x}}e^{R+S_N} = 2\nu\nabla_x \tag{99}$$

The operator $\hat{x}$ doesn't change under this transformation.

$$\hat{x}_{canonical} = \hat{x} \tag{100}$$

So we can have the exact Heisenberg momentum operator at an imaginary value for the diffusion constant

$$\hat{P}_{Heisenberg} = m\hat{\dot{x}}_{canonical} = -i\hbar\nabla_x, \; if \; \nu = -i\hbar/2m \tag{101}$$

$$\left[\hat{x}, \hat{P}_{Heisenberg}\right] = i\hbar \tag{102}$$

We can now define a Hamiltonian operator by

$$\hat{H} = \left[m\frac{(2\nu)^2}{2}\Delta + U\right] = \left[m\frac{(2\nu)^2}{2}\left(\hat{\dot{x}}_{canonical}\right)^2 + U\right] \tag{103}$$

$$\hat{\dot{x}}_{canonical} = \frac{1}{2m\nu}\left[H, \hat{x}\right] = 2\nu\nabla_x \tag{104}$$

$$\hat{\ddot{x}}_{canonical} = \frac{1}{2m\nu}\left[H, \hat{\dot{x}}\right] = -\frac{1}{m}\nabla_x U \tag{105}$$

Comparing this with (95) we find

$$\hat{\ddot{x}} = \hat{\ddot{x}}_{canonical} \tag{106}$$

We can construct higher derivatives of order $n$ using the recursive formula

$$\hat{x}_n = \frac{1}{2m\nu}\left[H, \hat{x}_{n-1}\right] + \frac{\partial \hat{x}_{n-1}}{\partial t}, \; \hat{x}_0 = \hat{x} \tag{107}$$

and in this way we can build up a Taylor's expansion for propagating in time, very similar to the time evolution operator in quantum mechanics. At the special imaginary value of $\nu = -i\hbar/2m$ this time evolution becomes exactly the usual time evolution operator. For other values, so long as the equation of motion is that of GSM, the probability density at different times propagated in this way is always the same as Schrödinger's equation gives, provided we calculate the density only on the real subspace of complex space-time using Born's rule there. If the probability density in position is all that can be measured, as discussed above, then even if the diffusion constant is imaginary but different from the Heisenberg value, the measurements will still be the same in GSM because the probability densities for positions will be the same. It's conceivable though that someday we might find a way to measure the diffusion constant, and this would open a door to a sub-quantum world that we can only imagine now. In X we discuss some speculations about several ideas about measuring the diffusion constant, thereby going beyond conventional quantum mechanics.

## VI. COMPARISON WITH THE FEYNMAN PATH INTEGRAL METHOD

It is quite a bit simpler to derive the Heisenberg operator algebra from Feynman path integrals than from stochastic mechanics [33]. The advantage of a stochastic process is that it can be understood as a proper diffusion process, unlike the Feynman path integral which uses effectively a complex measure to sum the different paths, and then requires one to square the absolute value of the resulting probability amplitude to obtain a probability. One might say that GSM is quantum mechanics made complicated, but from the point of view of emergent quantum mechanics, it's desirable to have a true diffusion process that can replicate quantum mechanics. Also, Born's rule is rigorously derivable from GSM, at least for real-valued diffusion constants. Then one can consider all sorts of possible diffusion models that might lead to quantum behavior, and the playground is enlarged.

To derive the commutation rules from Feynman path integrals, we can write simply

$$\langle \hat{x}(t)\hat{x}(s)\rangle = \int \mathcal{D}[x]x(t)x(s)e^{iS/\hbar} \tag{108}$$

and then we have

$$\lim_{s\uparrow t}\left(\frac{\partial}{\partial t}-\frac{\partial}{\partial s}\right)\langle \hat{x}(t)\hat{x}(s)\rangle = \lim_{s\uparrow t}\left(\frac{\partial}{\partial t}-\frac{\partial}{\partial s}\right)\int \mathcal{D}[x]x(t)x(s)e^{iS/\hbar} \tag{109}$$

$$\lim_{s\uparrow t}\left(\frac{\partial}{\partial t}-\frac{\partial}{\partial s}\right)\langle \hat{x}(t)\hat{x}(s)\rangle = \frac{1}{m}\langle [p(t),x(t)]\rangle = -i\hbar \tag{110}$$

and thus we have a time ordered commutation rule expressed as a limiting procedure applied to a Feynman path integral, and from this one can reconstruct the full Heisenberg algebra. Also, this commutator is imaginary, and so it does not require any analytic continuation to a complex space. Because the Feynman path integral does not represent a true diffusion theory, however, it does not give us a picture of reality, although it feels like it does when we draw Feynman diagrams on the blackboard. This is deceptive. GSM gives us a large variety of truly stochastic realistic pictures of quantum processes, but it's more complicated and it seems to favor a complex space-time embedding, although this is not absolutely necessary. The Feynman path integral approach has proven very useful, and it is seductive to imagine that quantum particles are somehow moving along Feynman paths in our imagination, but it is not a road-map to a stochastic interpretation of quantum mechanics. In a sense it's a deterrent because it's mathematically much simpler than stochastic mechanics or GSM, and it's tempting to not bother with a rigorous diffusion model.

The limiting case of Bohmian mechanics gives a deterministic picture of reality within the framework of GSM. It is interesting to speculate that the underlying diffusion parameter may depend on the local environment. So the diffusion parameter in a vacuum may be different than in a condensed matter setting. We speculate in X below on some experiments that might possibly allow us to determine the diffusion constant in some special cases.

## VII. THE CASE FOR COMPLEX SPACE-TIME

We have seen that generalized stochastic mechanics takes a particularly simple form for a diffusion constant which is purely imaginary with a value $\nu = \pm i\hbar/2m$. In this case the commutation relations also become exactly the Heisenberg rules. Such a diffusion constant would require the particle's motion to be in a complex space and because of space-time mixing in special relativity, in a complex space-time. If there are any values of the diffusion constant which are uniquely associated with quantum mechanics it is these two imaginary ones. But if the diffusion constant is imaginary, then the position of the particle will take on complex values. There is a history of suggestions that the space-time manifold might be embedded in a complex one. Perhaps the most interesting suggestion along these lines was made by Einstein. He believed that quantum theory might be derivable from classical field theory [9, 56, 57]. This is what he meant by unified field theory, and not the modern interpretation of the expression. He anticipated the complexification of space-time in this effort [58–60]. He used a Hermitian metric so that the metric length between points would be real-valued. Although he limited consideration to real space-time, many others have pursued this line of research in complex space-time[61–75] An entirely different application of complex space-time comes from a fascinating method of solving difficult problems in general relativity involving analytical continuation of electromagnetic charges to complex-valued coordinates [40, 75–77]. In quantum mechanics there have been a number of applications as well [40, 71, 78–87]

David Bohm introduced the concept of the Implicate Order to physics [88]. A mathematically valid example was proposed based on higher order algebras by Frescura and Hiley [89]. The simplest algebraic extension of space-time is to replace the algebra of real numbers with that of complex numbers for the coordinates. The idea of the implicate order is that behind our classical perception of reality lies a hidden realm which interconnects non-locally events in our world. This is a manifestation of a web of quantum entanglement that permeates our universe. Freshly minted physicists tend to find the concepts of wholeness and implicate order to be too metaphysical to be of interest, but as one matures in the study of quantum foundations they tend to increase in perceived importance, as the other alternatives like the Many Worlds or 't Hooft determinism seem equally strange. In a complex space-time interpretation there are many Riemann sheets for the electromagnetic fields and for the particle motions. This Riemann sheet web embedded in a complex space-time manifold is what we would compare to Bohm's implicate order. This comparison resonates with recent ideas in quantum gravity interpreting quantum entanglement in terms of Einstein-Rosen bridges [40, 90].

There is literature from stochastic mechanics extending it to complex space and space-time. These techniques give precise mathematical meaning to an imaginary value for the diffusion constant. See for example Wang [91, 92], Rosenbrock [93–96], and Kuipers [97]. These papers all provide a framework for analytically continuing the diffusion constant to complex space and time. Therefore it seems that the combination of Heisenberg algebra with stochastic mechanics strongly favors a universe which is embedded in a complex manifold. Complex time models have also been proposed [98]. This body of research suggests a possible connection between stochastic mechanics and the geometric Langlands program as applied to physics [99]. The paper by Kuipers also considers relativistic wave equations. Complex space-time models have also been utilized in condensed matter physics [100]. These topics are subjects of current and future research.

Wang frames the basis of his stochastic model of quantum mechanics in the language of weak measurement theory. He writes in [91]:

"In summary, we have proposed a new stochastic interpretation of quantum mechanics in which a quantum system is associated to a stochastic process in complex space. The real part of the trajectory of the stochastic process, which we call a weak trajectory, is interpreted as the trajectory of a particle in real configuration space. We showed that, on the one hand, the ensemble averaged moments of the weak trajectory are equal to the weak values of moments of the position operator of the quantum system, which are complex valued but nevertheless have well defined physical meanings according to the weak measurement theory [101, 102] and that, on the other hand, a single weak trajectory reduces to the correct classical trajectory in the classical limit."

The basis of his theory [91, 92] is exactly the same as our stochastic differential equation (2) with a diffusion constant given by $\nu = i\hbar/2m$. He shows that Schrödinger's equation can be derived with suitable choices of the backward derivative.

Wang uses a backward equation, but other than this, his stochastic differential equation is the same as GSM. The time reversal symmetry of GSM suggests that Wang's theory should have time reversal symmetry too. He finds in equation 36 of [92] an interesting result that the expectation value of a function of complex position z is equal to the usual quantum expectation value of the same function of a real position variable x. But this is only valid, he shows, if at some future time the full complex probability density $\rho(x, y, t_f)$ factorizes into a product $|\psi(x, t_f)|^2\,\delta(y)$, where $x$ and $y$ are the real and imaginary parts of z.

Rosenbrock uses the variational formalism of dynamic programming and control theory to arrive at a stochastic model of quantum mechanics. His model is equivalent to GSM at the value of $\nu = -i\hbar/2m$. As a step towards interpreting the results of complex diffusion, and towards obtaining Born's rule for probability densities, he makes the following postulate.

"The properties of the physical particle are related to those of the complex image by the following postulate: If the expected value of some property (such as momentum $\widetilde{p}$ or energy $\widetilde{H}$) of the complex images in an ensemble has the real value $\alpha$ on the real axis, then $\alpha$ is the value of the corresponding property of the physical particles. Whereas the usual formulation of quantum mechanics deals with complex functions on a real space, the above procedure embeds this theory in a theory of complex functions on a complex space. The properties of physical particles in a real space are then regained by the postulate."

He develops an elaborate theory which boils down to GSM evaluated at an imaginary value of the diffusion constant. So he has the same stochastic process as Wang, except that he has $\nu = -i\frac{\hbar}{2m}$ instead of $\nu = i\frac{\hbar}{2m}$

$$dz = b(z,t)dt + dw(t,\nu) \tag{111}$$

$$\nu = -i\frac{\hbar}{2m} \tag{112}$$

$$w(t,\nu) = \sqrt{-i\frac{\hbar}{2m}}w(t,1) = (1-i)\frac{1}{2}\sqrt{\frac{\hbar}{m}}w(t,1) \tag{113}$$

He writes the complex position $z$ in terms of real and imaginary parts

$$z = x + iy \tag{114}$$

He changes variables and chooses to work with two real variables $X$ and $Y$

$$X = x + y \tag{115}$$

$$Y = x - y \tag{116}$$

These are generated by the following stochastic differential equations

$$dX = (Re\, b(X,Y) + Im\, b(X,Y)dt \tag{117}$$

$$dY = (Re\, b(X,Y) - Im\, b(X,Y)dt + \sqrt{\hbar/m}dw(t) \tag{118}$$

where $Re$ and $Im$ denote real and imaginary parts. Note that the $X$ differential does not have a diffusion term. The following equation follows from the forward equation of Kolmogorov applied to the probability density $\rho(X,Y,t)$

$$\frac{\partial}{\partial t}\rho + \frac{\partial}{\partial X}\left((Re\, b + Im\, b)\rho\right) + \frac{\partial}{\partial Y}\left((Re\, b - Im\, b)\rho\right) - \frac{\hbar}{2m}\frac{\partial^2}{\partial Y^2}\rho = 0 \quad (119)$$

After transforming back to the original coordinates $x, y$ Rosenbrock shows that

$$\rho(x,y)|_{y=0} = \psi^*(x,t)\psi(x,t) \quad (120)$$

So $\rho(x,y)$ serves as a probability density function for the full complex space, and $\rho(x,0)$ serves as a probability density function on the real space, and moreover $\rho(x,0)$ is the usual quantum position density. This seems to be the best stand-in for Born's rule in the extant literature on complex space-time and stochastic mechanics. All of this is a special case of GSM.

Kuipers' mathematical analysis is sophisticated and powerful [97]. He considers non-relativistic and relativistic particle diffusions in complex space for the non relativistic case, and complex space and time in the relativistic case. He generalizes the usual stochastic mechanics to allow stochastic processes which are semimartingales.

From Wikipedia, we have the following definition [103]

> "In probability theory, a real valued stochastic process X is called a semimartingale if it can be decomposed as the sum of a local martingale and a càdlàg adapted finite-variation process. Semimartingales are "good integrators", forming the largest class of processes with respect to which the Itô integral and the Stratonovich integral can be defined.
>
> The class of semimartingales is quite large (including, for example, all continuously differentiable processes, Brownian motion and Poisson processes). Submartingales and supermartingales together represent a subset of the semimartingales."

In particular, he allows for the possibility that the real and imaginary parts of the coordinates may be driven by independent Wiener processes. He does not derive Born's rule however. His analysis includes as a special case the stochastic differential equation $dz(t) = b(z,t) + dw(t, \nu = -i\hbar/2m)$, which is the result of analytic continuation in GSM to the Heisenberg imaginary value for the diffusion constant, and is also the same as in Wang and Rosenbrock's papers.

He treats the relativistic case by showing the equivalence of his stochastic processes by introducing a "proper time" variable and deriving what has been termed the relativistic Schrödinger equation, or alternately the Stueckelberg-Horwitz-Piron (SHP) wave equation [104]. This yields the Klein-Gordon equation when the mass constraint is imposed on the solution. Other authors have proposed relativistic proper-time diffusion models for quantum mechanics in real space-time, For example [105–108]. The interpretation of these types of particle models is discussed in [109]. GSM has been extended to diffusions on Riemannian manifolds [110]. These results apply as well to pseudo-Riemannian manifolds like Minkowski space or the space-time in general relativity. So in particular, we could apply GSM to the relativistic diffusion models of Kuipers or those of Oron and Horwitz [105–108] and thereby obtain stochastic models of relativistic quantum mechanics which are embedded in real or complex space-time.

## VIII. A POSSIBLE CONNECTION TO ADLER'S TRACE DYNAMICS

A preliminary exploration of a possible connection between stochastic mechanics and Adler's theory of emergent quantum mechanics based on trace dynamics [12] was made in [111]. It was found that the non-commutative structure of stochastic mechanics might be used to provide an explanation for trace dynamics. However, it was also found that the result tended to lead to the thermal diffusion equation rather than the Schrödinger equation unless a nonlinear potential was added without any justification. However, only real-valued diffusion constants were considered in that work. If one revisits this mathematics with now an imaginary diffusion constant, it seems plausible that the Schrödinger equation can be understood as resulting from the trace-dynamics equilibrium for a suitably defined trace Lagrangian. This is still another reason to consider complex space-time to be an ontological reality.

## IX. VISUALIZATIONS

The mathematics of GSM is a bit dry, and so here we will try and give some pictorial insight into the subject. Visualizations of stochastic mechanics were given by Yasue and Zambrini in [112]. We have used an open source Python library called Qmsolve [113]. This convenient software generates solutions to Schrödinger's equation. We use this to calculate the drift terms in GSM, and then using standard Wiener process simulations we generate sample trajectories for GSM. We also utilized a Python stochastic differential equation package called sdeint [114].

In atomic units, with $\hbar = 1$ and $m_e = 1$, Schrödinger's equation takes a simple form convenient for numerical analysis

$$\left(-\frac{1}{2}\Delta + V(x,t)\right)\psi(x,t) = i\frac{\partial\psi(x,t)}{\partial t} \tag{121}$$

In these units, the diffusion constant of stochastic mechanics is just 1/2, and the unit of velocity is $\alpha c$ so that $c \approx 137$. The diffusion model in GSM depends on $\psi(x,t)$ and on $\nu$ by the formulas $\psi = e^{R+iS}$ and $b(x,t,\nu) = 2\nu\nabla R + \nabla S$. We generate the sample trajectories step-wise from the following stochastic equation

$$\mathbf{x}(t+\epsilon,\nu) - \mathbf{x}(t,\nu) = \mathbf{b}(x(t),t,\nu)\epsilon + \mathbf{W}(t+\epsilon,\nu) - \mathbf{W}(t,\nu) \tag{122}$$

### Case 1 A simple plane wave

We consider an electron with velocity in the y direction in the x-y plane. Though we've been considering the 1D case above, it's more interesting to see images in two space dimensions. Here's a simple plane wave in the y direction. Simulations are shown in figure 1. The velocity was 10 and the initial standard deviation was 1.

$$\psi(x, y, t) = exp(i(ky - (k^2/2)t)) \tag{123}$$

$$b(x, t, \nu) = \nabla S = k\hat{y} \tag{124}$$

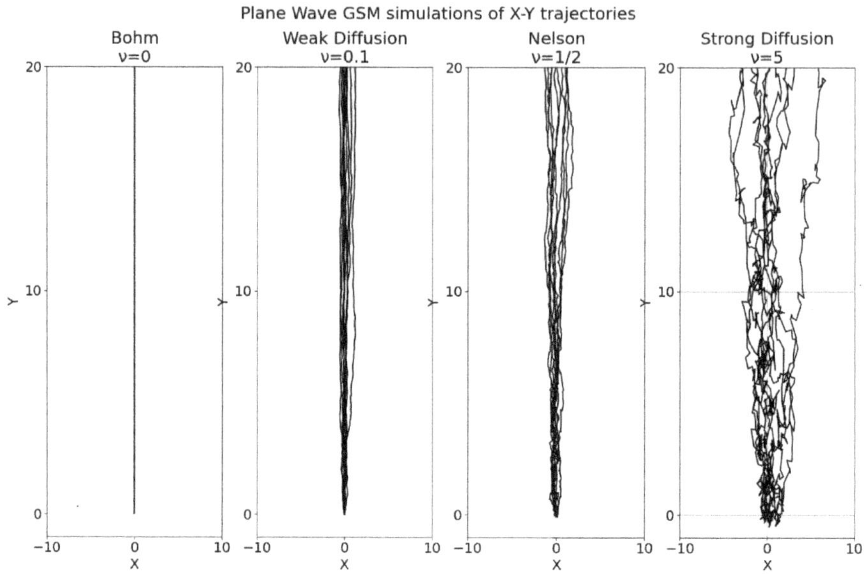


Figure 1. A plane wave as in Case 1 is plotted here. Only one starting point is shown since all others would look similar as the drift function is independent of starting x position for a plane wave. The diffusion constant is increasing from left to right, starting with a value of zero which gives the deterministic Bohmian trajectory. Next we have a weak diffusion case with $\nu < 1/2$, then we plot the standard Stochastic Mechanics case of Nelson with $\upsilon = 1/2$, and at the left we plot a strong diffusion case where $\nu > 1/2$

### Case 2 A Gaussian wave function

A Gaussian wave function is an analytic solution to Schrödinger's equation of the form (in 3 dimensions)

$$\psi(\mathbf{x}, t) = \left(2\pi s_t^2\right)^{-3/4} e^{i\mathbf{u}\cdot\left(\mathbf{x}-\frac{1}{2}\mathbf{u}t\right)-(\mathbf{x}-\mathbf{u}t)^2/4s_t\sigma_0} \tag{125}$$

$$s_t = \sigma_0 \left(1 + it/2\sigma_0^2\right) \tag{126}$$

A nice visualization of the Bohmian orbits for this and other wave functions together with an excellent elucidation of the theory is in Holland [115]. We find

$$S_Q(x,t) = -\frac{3}{2}\tan^{-1}(\frac{1}{2}t\sigma_0^2) + \mathbf{u}\cdot(\mathbf{x} - \frac{1}{2}\mathbf{u}t) + (\mathbf{x} - \mathbf{u}t)^2 t/8\sigma_0^2 \left|s_t\right|^2 \tag{127}$$

$$R(x,t) = -\frac{3}{4}\ln(2\pi\left|s_t\right|^2) - (x - ut)^2/4\left|s_t\right|^2 \tag{128}$$

We calculate the drift term from (22).

$$\mathbf{b}(x,t,\nu) = 2\nu\nabla R(x,t) + \nabla S_Q(x,t) \tag{129}$$

$$\mathbf{b}(x,t,\nu) = -2\nu\left(\mathbf{x} - \mathbf{u}t\right)/2\left|s_t\right|^2 + \mathbf{u} + (\mathbf{x} - \mathbf{u}t)t/4\sigma_0^2\left|s_t\right|^2 \tag{130}$$

Now with this drift term, we can calculate the orbits for various values of $\nu$ by integrating the stochastic differential equation $d\mathbf{x} = \mathbf{b}(x,t,\nu)dt + d\mathbf{w}(t,\nu)$. Simulations are shown in the figure 2. Note that these are x-y trajectory plots. The group velocity was 10 and the initial standard deviation was 1.

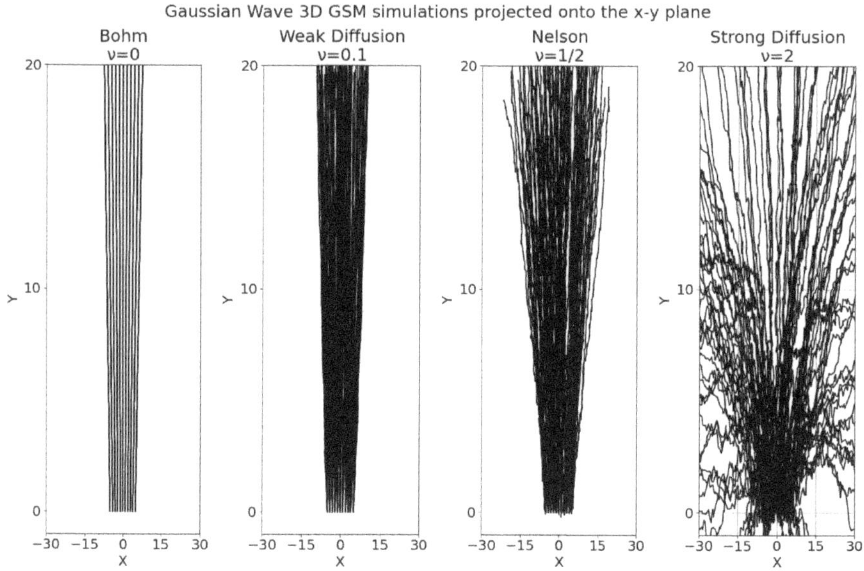


Figure 2. A Gaussian wave described in Case 2 is plotted here. We have plotted the trajectories projected onto the x-z plane. On the leftmost plot we see the Bohmian rays and there is no diffusion. The next three plots show increasing diffusion values. Although the trajectory plots look very different, when you do all the appropriate calculations to calculate the probability densities, they would all give the same result. We have only plotted results for real diffusion constants, but one can also have complex diffusion constants which still give the same probability densities.

### Case 3 Two slit diffraction of a Gaussian beam

This case cannot be solved analytically, and therefore we use a numerical Schrödinger equation solver [116]. We launch a Gaussian wave towards a two-slit diffraction potential barrier, and then from the solution to this wave equation we generate various trajectories of GSM for this problem. Qmsolve can generate

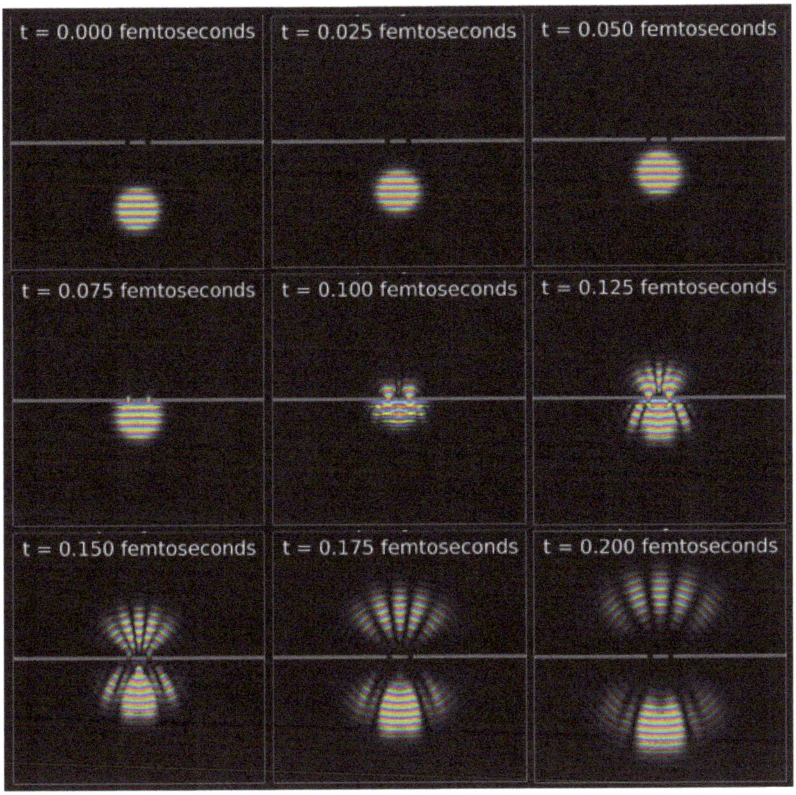


Figure 3. A Gaussian wave incident on a 2 Slit potential barrier as in Case 3

animations, but we can only show discrete times in the paper. The evolution of the wave function in time is illustrated in figure 3. The mass is the electron mass. The x and y limits are both $\pm 15\text{Å}$. The group velocity of the incident Gaussian wave is $80\text{Å}/\text{femtosecond}$ in the positive y direction. This is a velocity of 3.657 in atomic units. The standard deviation of the initial Gaussian is $1\text{Å}$. The center to center slit separation is $2.5\text{Å}$, and the width of each slit is $1/2\text{Å}$. The phase of wave function is encoded in the color hue.

Next we show ray tracing for rays starting at the two slits at about the time that the Gaussian wave is passing through them in Figure 4. We plot x vs y. This is a time-dependent simulation. You can think of the stochastic average density of the x-y trajectories as approximating the smeared probability distribution $\int_{t_1}^{t_2} \varrho(x, y, t)dt$ up to a normalization constant. In order to faithfully represent the probability density, we would need to increase the number of independent paths, and also to be precise, we would need to include rays starting from all possible initial positions at the start time, and weighted by their initial probability density. We added a color map to the Bohm orbits to make them more visible.

**Case 4 Geometrical phase quantum waveguide of Lévy-Leblond**

Quantum waveguide effects and associated phase delays, familiar from microwave and optical waveguides, have been a topic of interest in quantum physics [117, 118]. Here we present a GSM stochastic simulation for this process. Consider a particle with unit mass moving down a containing tube with square cross section such that $\psi$ vanishes on the boundary of the tube. The wave function in this cases is:

$$\psi(x, y, z, t) = sin(n_x \pi x/a) sin(n_y \pi y/a) e^{ipz} e^{-iEt} \tag{131}$$

Defining a parameter $sign$ by

$$sign \equiv sgn(sin(n_x \pi x/a) sin(n_y \pi y/a)) \tag{132}$$

Then we can calculate $R$ and $S_Q$ by (where $m = 1$)

$$\psi(x, y, z, t) = |sin(n_x \pi x/a) sin(n_y \pi y/a)|\, e^{i\pi(1-sign)/2} e^{ipz} e^{-iEt} \tag{133}$$

$$R(x, y, z, t) = ln(|sin(n_x \pi x/a) sin(n_y \pi y/a)|) \tag{134}$$

$$R(x, y, z, t) = ln(|sin(n_x \pi x/a)|) + ln(|sin(n_y \pi y/a)|) \tag{135}$$

$$S_Q(x, y, z, t) = pz - Et + \pi\left(1 - sign\right)/2 \tag{136}$$

$$E = p^2/2m + \frac{\left(n_x^2 + n_y^2\right)\pi^2}{2ma^2} \tag{137}$$

$$\mathbf{b}(\mathbf{x}, t, \nu) = 2\nu \nabla R(\mathbf{x}, t) + \nabla S_Q(\mathbf{x}, t) \tag{138}$$

$$\nabla R(\mathbf{x}, t) = \frac{\pi}{a}\left[n_x cotan\left(n_x \pi x/a\right)\hat{x} + n_y cotan\left(n_y \pi x/a\right)\hat{y}\right] \tag{139}$$

The results of a simulation are shown in 5. The velocity used for this was 10 in AU units.

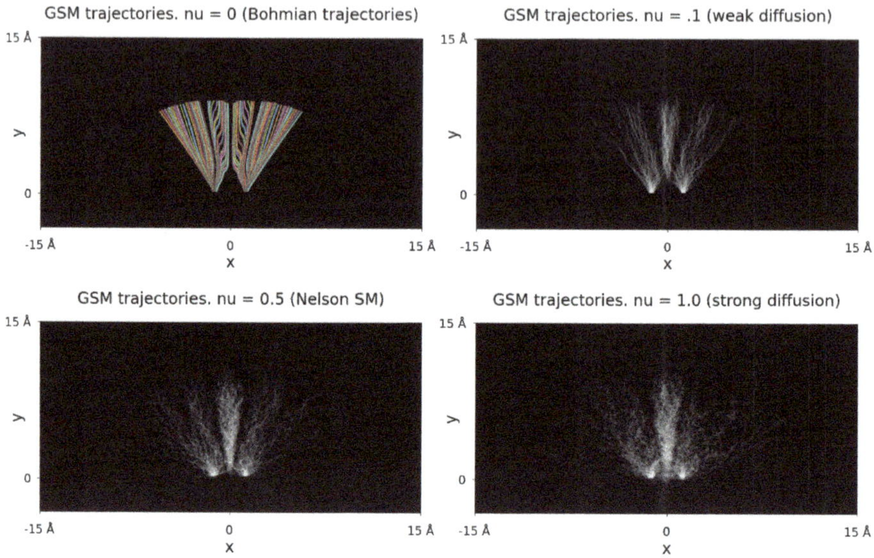


Figure 4. Ray tracing of GSM trajectories for 2 Slit diffraction as in Case 3

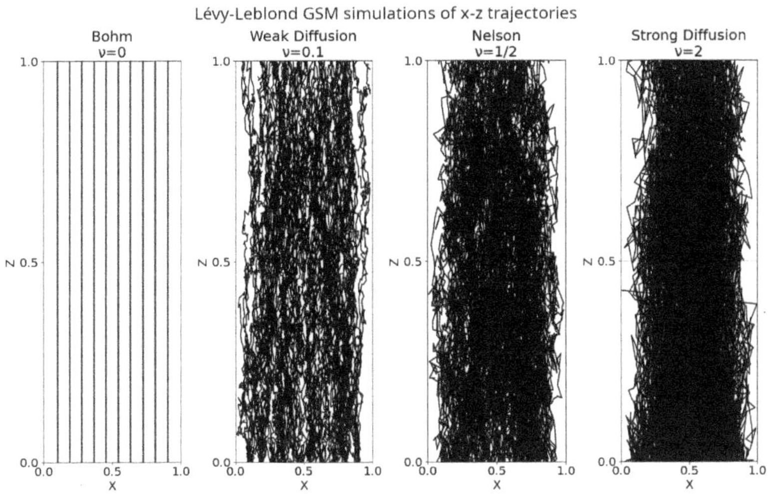


Figure 5. Lévy-Leblond GSM simulations. nx = ny = a =1, v=10, as in Case 4

## X. SOME SPECULATIONS ABOUT MEASURING THE DIFFUSION CONSTANT

If the assertion about measurement limitations discussed in (II) is correct, then it is impossible to measure the diffusion constant in GSM. However, it is after all an assertion, and not a proof. I can think of two phenomena that might be places to look where a different diffusion constant might lead to different experimental results in the context of GSM.

*1. Lattice confined fusion of deuterium* One possible experimental application might be as an explanation for anomalously large fusion rates that are observed in experiments with low energy deuterium beams incident on Palladium and other metals, and related experiments where they are irradiated by plasma discharge [119–124]. After accounting for electron screening which enhances the fusion rates significantly, the current theory still cannot explain the observed rates without unreasonably large screening energies. There are other possible mechanisms that might explain these results [125]. These experiments do produce the expected fusion products resulting from two deuterons fusing to a final state of either a neutron and $^3$He, or a proton and tritium. Therefore, they are not examples of cold fusion where the claimed fusion product is $^4$He, although they are occurring in a similar environment. Advocates of LENR or cold fusion do interpret these results as partial confirmation that something unusual is going on in deuterated Palladium experiments. The reader is urged not to be discouraged by this association, because these results can be taken as just an example where the theory of electron screening is having difficulty explaining repeatable experimental results, or in other words, squarely in the domain of conventional physics. They are results that physicists should be aware of. Also, the arguments of Huw Price should be kept in mind [126].

D-D fusion is a very local phenomenon. The two fusing nuclei need to be within a Compton wavelength or so of one another in order to fuse. This is an experiment where the relative position of two quantum particles matters critically. If we take a stochastic picture of this, then the two deuterons are both diffusing in GSM, and then the chance of two trajectories intersecting within a Compton wavelength of one another might be enhanced if the diffusion constant underlying the motion is large. If we assume that the diffusion constant in a vacuum is lower than that for deuterons moving in Palladium metal, this could explain the increased fusion rate. These ideas would require additional research but could be topics for future publications. Although the fusion rates in these experiments are not useful for energy production yet, further research may reveal ways to increase the yields. They are quite reproducible now, and NASA has claimed that fusion rocket engines based on these effects have been prototyped and have potential applications for long-distance space travel, where chemical rockets emit too much mass to be useful. [127, 128].

*2. Quantum behavior near event horizons and Hawking radiation* Hawking radiation might be affected by the value of the diffusion constant in GSM. A crude explanation of Hawking radiation is that when a particle-anti-particle pair is created as a vacuum fluctuation right at the event horizon of a black hole, the two particles might be on opposite sides of the event horizon so that one particle can escape the black hole while the other is trapped. Accepting this crude picture, let us consider that the two particles are each undergoing a stochastic motion described by GSM after the creation event. As we have seen, the sample paths of the particles are diffusing in GSM. If the diffusion constant $\nu$ is large, then the probability that the escaping particle will cross over into the interior of the black hole's event horizon will be increased, and then this particle that started off outside the event horizon has an increased chance of diffusing inside it and thus being presumably trapped by the black hole. So we would expect that the Hawking radiation might be reduced for larger diffusion constant, or at least affected by it. The same argument could be made for Unruh radiation [129]. In the limit of very large diffusion constant, we might expect that both Hawking radiation and Unruh radiation would tend to zero. So far neither radiation has been observed in nature. We might also expect that an isolated black hole would have a lower diffusion constant than a black hole with a significant accretion disk.

## XI. CONCLUSION

Stochastic mechanics provides a thorough derivation of the non-commutative algebra of Heisenberg. The fact that an imaginary diffusion constant is preferred adds theoretical evidence that complex space-time is related to a deeper understanding of quantum mechanics. It's amazing that Einstein not only declared the need for a deeper theory but also that some of his last papers dealt with complexification pointing the way. Not only that, but he also invented the theory of Brownian motion, which is the basis of stochastic mechanics, and he made critical insights into quantum theory. It's also remarkable that Heisenberg's non-commuting operators should point to the same conclusion, and that his postulate of wave function collapse along with his uncertainty principle plays a key role in it. Nelson was perhaps correct that there is a preferred value for the diffusion constant, but it might not be $\hbar/2m$, but rather it could well be $\pm i\hbar/2m$. This leads to the proposition that we are possibly living in a complex space-time, but our normal sense perception has evolved us to be aware of only the real subspace. But even if this is not the case, having GSM to provide a variety of diffusion models with which to model quantum mechanics has benefits in unifying several branches of quantum interpretations and stimulating efforts to someday-somehow perhaps measure the diffusion constant. Other models that are supported in GSM are Bohmian mechanics, diffusion parameters varying with position, complex time models, and weak and strong diffusion models, and complex diffusion models where the complex diffusion constant differs from the Heisenberg values of $\pm i\hbar/2m$.

## Appendix A: Compendium of useful formulae for stochastic mechanics

(We use the notation of [41], for the most part, in this appendix)

Fokker-Planck equation or Kolmogorov forward eqn.: $\frac{\partial \rho}{\partial t} = -div(b\rho) + \nu\triangle\rho$ (A1)

Schrödinger wave function: $\psi = e^{R+iS_Q}$ (A2)

Probabilty density: $\rho = \psi^*\psi = e^{2R}$ (A3)

Fokker-Planck equation time reversed or Kolmogorov backward eqn.: $\frac{\partial \rho}{\partial t} = -div(b_*\rho) - \nu$ (A4)

Backward velocity: $b_* = b - 2\nu(grad\,\rho)/\rho$ (A5)

Osmotic velocity: $u = \frac{b - b_*}{2} = \nu\frac{grad\,\rho}{\rho} = 2\nu grad(R)$ (A6)

$$\text{Current velocity: } \mathsf{v} = \frac{b + b_*}{2} = 2\nu grad(S_N) \tag{A7}$$

$$\text{Equation of continuity: } \frac{\partial \rho}{\partial t} = -div(\mathsf{v}\rho) \tag{A8}$$

$$\text{Time deriviative of osmotic velocity: } \frac{\partial \mathsf{u}}{\partial \mathsf{t}} = -\nu grad\, div\, \mathsf{v} - grad\, \mathsf{v} \cdot u \tag{A9}$$

$$\text{Forward time derivative formula: } Df(x(t), t) = \left(\frac{\partial}{\partial t} + b \cdot \nabla + \nu \triangle\right) f(x(t), t) \tag{A10}$$

$$\text{Backward time derivative formula: } D_* f(x(t), t) = \left(\frac{\partial}{\partial t} + b_* \cdot \nabla - \nu \triangle\right) f(x(t), t) \tag{A11}$$

$$\text{Mean acceleration of Nelson: } \mathsf{a}(t) = \frac{DD_* + D_* D}{2} x(t) \tag{A12}$$

Markov transition function:

$$P(x, t; y, s) = \lim_{d^3x \to 0} \frac{1}{d^3x} P(x(t) \in d^3x \mid x(s) = y), \quad t > s \tag{A13}$$

Chapman-Kolmogorov equation:

$$P(x, t; y, r) = \int P(x, t; z, s) P(z, s; y, r) d^3z, \ for\ times\ t > s > r \tag{A14}$$

$$\text{Continuity of paths: } \lim_{t \downarrow s} P(x, t; y, s) = \delta^3(x - y) \tag{A15}$$

$$\text{Time evolution of the density function: } \rho(x, t) = \int P(x, t; z, s) \rho(z, s) d^3z, \ t > s \tag{A16}$$

Forward equation for Markov transition (§ 3, [130]):

$$\left[\frac{\partial}{\partial t} + \nabla_x \cdot b(x, t) + b(x, t) \cdot \nabla_x - \nu \triangle_x\right] P(x, t; y, s) = 0, \ t > s \tag{A17}$$

Backward equation for Markov transition (§ 3, [130]):

$$\left[\frac{\partial}{\partial s} + b(y, s) \cdot \nabla_y + \nu \triangle_y\right] P(x, t; y, s) = 0, \ t > s \tag{A18}$$

Integration by parts when justified (§ 13, [2]):

$$\int_{-\infty}^{\infty} E(Df(x(t), t) g(x(t), t)) dt = -\int_{-\infty}^{\infty} E(f(x(t), t) D_* g(x(t), t)) dt \tag{A19}$$

Expectation value of a function of multiple positions at different times:

$$E(F(x(t_n),...,x(t_0))) =$$
$$\int F(x_n,...,x_0)P(x_n,t_n;x_{n-1},t_{n-1})\cdots P(x_1,t_1;x_0,t_0)\rho(x_0)\times$$
$$\times dx_0...dx_n,\ t_0 \le t_1 \le ... \le t_n \tag{A20}$$

## ACKNOWLEDGMENTS

I would like to thank the editor Tuck Choy for very helpful discussions and insights. I also thank the reviewers for their diligence and valuable comments.

---

[1] W. Heisenberg, Zeitschrift für Physik **33**, 879 (1925).

[2] E. Nelson, *Dynamical Theories of Brownian Motion*, second edition, 2001 ed. (Princeton University Press, Princeton, NJ, 1967).

[3] E. Nelson, *Quantum Fluctuations* (Princeton University Press, 1985).

[4] F. Guerra, Physics Reports **77**, 263 (1981).

[5] E. Nelson, Journal of Mathematical Physics **5**, 332 (1964), aDS Bibcode: 1964JMP.....5..332N.

[6] S. A. Albeverio, R. J. Høegh-Krohn, and S. Mazzucchi, *Mathematical Theory of Feynman Path Integrals*, edited by J. M. Morel, F. Takens, and B. Teissier, Lecture Notes in Mathematics, Vol. 523 (Springer, Berlin, Heidelberg, 2008).

[7] S. Albeverio and S. Mazzucchi, Scholarpedia **6**, 8832 (2011).

[8] Z. Haba, Journal of Mathematical Physics **35**, 6344 (1994).

[9] A. Einstein, Journal of the Franklin Institute **221**, 349 (1936).

[10] L. De La Peña, A. M. Cetto, and A. Valdés Hernández, *The Emerging Quantum: The Physics Behind Quantum Mechanics* (Springer International Publishing, Cham, 2015).

[11] G. Grössing, S. Fussy, J. M. Pascasio, and H. Schwabl, Journal of Physics: Conference Series **361**, 012008 (2012).

[12] S. L. Adler, *Quantum Theory as an Emergent Phenomenon: The Statistical Mechanics of Matrix Models as the Precursor of Quantum Field Theory* (Cambridge University Press, Cambridge, 2004).

[13] S. Wolfram, Finally We May Have a Path to the Fundamental Theory of Physics... and It's Beautiful—Stephen Wolfram Writings (2020).

[14] G. t. Hooft, *The Cellular Automaton Interpretation of Quantum Mechanics*, 1st ed. (Springer, 2016).

[15] M. Davidson, Letters in Mathematical Physics **3**, 367 (1979).

[16] M. Davidson, Physica A: Statistical Mechanics and its Applications **96**, 465 (1979).

[17] M. Davidson, Journal of Mathematical Physics **19**, 1975 (1978).

[18] M. Davidson, Letters in Mathematical Physics **3**, 271 (1979).

[19] M. T. Jaekel and D. Pignon, International Journal of Theoretical Physics **24**, 557 (1985).

[20] T. C. Choy, *Effective Medium Theory: Principles and Applications* (Oxford University Press, 2015).

[21] M. Nagasawa, *Schrodinger Equations and Diffusion Theory* (Basel ; Boston, 1993).

[22] G. Ghirardi and A. Bassi, in *The Stanford Encyclopedia of Philosophy*, edited by E. N. Zalta (Metaphysics Research Lab, Stanford University, 2020) summer 2020 ed.

[23] M. T. Jaekel, J. Phys. A **23**, 3497 (1990).

[24] W. Heisenberg, Zeitschrift für Physik **43**, 172 (1927).

[25] J. Von Neumann, *Mathematical Foundations of Quantum Mechanics* (Princeton University Press, 2018).

[26] P. A. M. Dirac, *The Principles of Quantum Mechanics* (Clarendon Press, 1981) google-Books-ID: XehUpGiM6FIC.

[27] S. L. Adler, Studies in History and Philosophy of Science Part B: Studies in History and Philosophy of Modern Physics **34**, 135 (2003).

[28] N. C. Petroni and L. M. Morato, Journal of Physics A: Mathematical and General **33**, 5833 (2000).

[29] S. Goldstein, Bohmian mechanics, in *The Stanford Encyclopedia of Philosophy*, edited by E. N. Zalta (Metaphysics Research Lab, Stanford University, 2021) fall 2021 ed.

[30] S. Goldstein, Journal of Statistical Physics **47**, 645 (1987).

[31] G. Bacciagaluppi, in *Endophysics, Time, Quantum and the Subjective* (WORLD SCIENTIFIC, 2005) pp. 367–388.

[32] D. Bohm and B. J. Hiley, Physics Reports **172**, 93 (1989), aDS Bibcode: 1989PhR...172...93B.

[33] B. L. M, *Feynman's Thesis - A New Approach To Quantum Theory* (World Scientific Publishing Company, 2005) google-Books-ID: z93ICgAAQBAJ.

[34] T. Maudlin, *Quantum Non-Locality and Relativity: Metaphysical Intimations of Modern Physics* (Blackwell, 2002).

[35] H. P. Stapp, Foundations of Physics **18**, 427 (1988).

[36] H. P. Stapp, American Journal of Physics **72**, 30 (2004).

[37] E. Nelson, in *Stochastic Processes in Classical and Quantum Systems*, Lecture Notes in Physics, edited by S. Albeverio, G. Casati, and D. Merlini (Springer, Berlin, Heidelberg, 1986) pp. 438–469.

[38] M. Derakhshani and G. Bacciagaluppi, On Multi-Time Correlations in Stochastic Mechanics (2022), arXiv:2208.14189 [quant-ph].

[39] M. Gardner, Scientific American **223**, 120 (1970).

[40] M. Davidson, Int. J. Theor. Phys. **62**, 154 (2023).

[41] E. Nelson, *Dynamical Theories of Brownian Motion*, second edition, 2001 ed. (Princeton University Press, Princeton, NJ, 1967).

[42] R. P. Feynman and A. R. Hibbs, *Quantum Mechanics and Path Integrals* (McGraw-Hill Companies, New York, 1965).

[43] H.-D. Doebner and G. A. Goldin, Physical Review A **54**, 3764 (1996), publisher: American Physical Society.

[44] L. Vaidman, Foundations of Physics **26**, 895 (1996).

[45] Y. Couder and E. Fort, Physical Review Letters **97**, 154101 (2006).

[46] R. Brady and R. Anderson, arXiv:1401.4356 [physics, physics:quant-ph] (2014), arXiv: 1401.4356.

[47] R. Brady and R. Anderson, arXiv:1305.6822 [physics, physics:quant-ph] (2013), arXiv: 1305.6822.

[48] D. Das, D. Home, H. Ulbricht, and S. Bose, Physical Review Letters **132**, 030202 (2024), publisher: American Physical Society.

[49] C. Emary, N. Lambert, and F. Nori, Reports on Progress in Physics **77**, 016001 (2013), publisher: IOP Publishing.

[50] G. C. Knee, K. Kakuyanagi, M.-C. Yeh, Y. Matsuzaki, H. Toida, H. Yamaguchi, S. Saito, A. J. Leggett, and W. J. Munro, Nature Communications **7**, 13253 (2016), publisher: Nature Publishing Group.

[51] D. S. Shucker, Letters in Mathematical Physics **4**, 61 (1980).

[52] E. A. Abbott and I. Stewart, *Flatland: a romance of many dimensions*, Dover thrift editions (Dover Publications, New York, 1992) oCLC: 26054205.

[53] K. Yasue, Journal of Functional Analysis **41**, 327 (1981).

[54] J. C. Zambrini, International Journal of Theoretical Physics **24**, 277 (1985).

[55] D. Bohm, Physical Review **85**, 166 (1952).

[56] C. Lehner, in *The Cambridge Companion to Einstein*, Cambridge Companions to Philosophy, edited by C. Lehner and M. Janssen (Cambridge University Press, Cambridge, 2014) pp. 306–353.

[57] T. Sauer, in *The Cambridge Companion to Einstein*, Cambridge Companions to Philosophy, edited by C. Lehner and M. Janssen (Cambridge University Press, Cambridge, 2014) pp. 281–305.

[58] A. Einstein and E. G. Straus, Annals of Mathematics **47**, 731 (1946), publisher: Annals of Mathematics.

[59] A. Einstein, Annals of Mathematics **46**, 578 (1945), publisher: Annals of Mathematics.

[60] A. Einstein, Reviews of Modern Physics **20**, 35 (1948), publisher: American Physical Society.

[61] E. H. Brown, Journal of Mathematical Physics **7**, 417 (1966), publisher: American Institute of Physics.

[62] A. Chamseddine, Communications in Mathematical Physics **264**, 291 (2006).

[63] A. Das, Journal of Mathematical Physics **7**, 45 (1966), publisher: American Institute of Physics.

[64] N. Debergh, G. D'Agostini, and J.-P. Petit, Journal of Modern Physics **12**, 218 (2021), number: 3 Publisher: Scientific Research Publishing.

[65] E. Goulart, EPL (Europhysics Letters) **115**, 10004 (2016), publisher: IOP Publishing.

[66] P. O. Hess and W. Greiner, International Journal of Modern Physics E **18**, 51 (2009), publisher: World Scientific Publishing Co.

[67] G. Kaiser, *Quantum Physics, Relativity, and Complex Spacetime: Towards a New Synthesis* (North-Holland, Amsterdam ; New York : New York, N.Y., U.S.A, 1990).

[68] M. K.-H. Kiessling and A. S. Tahvildar-Zadeh, Journal of Physics A: Mathematical and Theoretical **49**, 135301 (2016), publisher: IOP Publishing.

[69] I. V. B. a. B. L. Kogan, Progress In Electromagnetics Research B **69**, 103 (2016), publisher: EMW Publishing.

[70] C. Mantz and T. Prokopec, *Hermitian Gravity and Cosmology*, Tech. Rep. arXiv:0804.0213 (arXiv, 2008) arXiv:0804.0213 [astro-ph, physics:gr-qc, physics:hep-th] type: article.

[71] E. A. Rauscher, J. J. Hurtak, and D. E. Hurtak, Journal of Physics: Conference Series **1251**, 012042 (2019), publisher: IOP Publishing.

[72] E. Witten, Physical Review Letters **61**, 670 (1988), publisher: American Physical Society.

[73] E. Witten, *A Note On Complex Spacetime Metrics*, Tech. Rep. arXiv:2111.06514 (arXiv, 2022) arXiv:2111.06514 [gr-qc, physics:hep-th] type: article.

[74] G. Esposito, *Complex General Relativity* (Springer Science & Business Media, 2006) google-Books-ID: pp0SBwAAQBAJ.

[75] A. Burinskii, Gravitation and Cosmology **14**, 109 (2008).

[76] T. Adamo and E. T. Newman, *The Kerr-Newman metric: A Review*, Tech. Rep. arXiv:1410.6626 (arXiv, 2016) arXiv:1410.6626 [gr-qc, physics:hep-th] type: article.

[77] D. Lynden-Bell, in *Stellar Astrophysical Fluid Dynamics* (2003) pp. 369–375.

[78] M. Davidson, Foundations of Physics **48**, 1590 (2018).

[79] M. Davidson, General Relativity and Gravitation **44**, 2939 (2012), arXiv: 1109.4923.

[80] C.-D. Yang, Annals of Physics **319**, 444 (2005).

[81] C.-D. Yang, Annals of Physics **321**, 2876 (2006).

[82] C.-D. Yang, Chaos, Solitons & Fractals **32**, 312 (2007).

[83] C.-D. Yang, Chaos, Solitons & Fractals **33**, 1073 (2007).

[84] C.-D. Yang, Physics Letters A **372**, 6240 (2008).

[85] C.-D. Yang, Chaos Solitons & Fractals - CHAOS SOLITON FRACTAL **41**, 317 (2009).

[86] C.-D. Yang and S.-Y. Han (2020).

[87] C.-D. Yang and S.-Y. Han, Entropy **23**, 210 (2021), number: 2 Publisher: Multidisciplinary Digital Publishing Institute.

[88] D. Bohm, Wholeness and the Implicate Order.

[89] F. A. M. Frescura and B. J. Hiley, Foundations of Physics **10**, 7 (1980).

[90] J. Maldacena and L. Susskind, Fortschritte der Physik **61**, 781 (2013).

[91] M. S. Wang, Physical Review Letters **79**, 3319 (1997), publisher: American Physical Society.

[92] M. Wang, Chinese Journal of Physics **38** (2000).

[93] H. H. Rosenbrock, Physics Letters A **110**, 343 (1985).

[94] H. H. Rosenbrock, Physics Letters A **254**, 307 (1999).

[95] H. Rosenbrock, IEEE Transactions on Automatic Control **45**, 73 (2000).

[96] H. H. Rosenbrock, Proceedings of the Royal Society of London. Series A: Mathematical and Physical Sciences **450**, 417 (1997), publisher: Royal Society.

[97] F. Kuipers, J. Math. Phys. **63**, 042301 (2022), _eprint: 2109.10710.

[98] J. C. Zambrini, Physical Review A **33**, 1532 (1986), publisher: American Physical Society.

[99] A. Kapustin and E. Witten, Electric-Magnetic Duality And The Geometric Langlands Program (2007), arXiv:hep-th/0604151.

[100] R. Tsekov, in *New Advances in Physics*, Vol. 8 (2014) p. 111, journal Abbreviation: New Advances in Physics.

[101] Y. Aharonov, D. Z. Albert, and L. Vaidman, Physical Review Letters **60**, 1351 (1988), publisher: American Physical Society.

[102] Y. Aharonov and L. Vaidman, Physical Review A **41**, 11 (1990), publisher: American Physical Society.

[103] Semimartingale (2024), page Version ID: 1223154876.

[104] L. P. Horwitz, *Relativistic Quantum Mechanics* (Springer Netherlands, Dordrecht, 2015).

[105] O. Oron and L. P. Horwitz, Rel. Grav. Cosmol. **1**, 109 (2004), _eprint: physics/0212036.

[106] O. Oron and L. P. Horwitz, Foundations of Physics **35**, 1181 (2005), publisher: Kluwer Academic Publishers-Plenum Publishers.

[107] O. Oron and L. P. Horwitz, Relativistic Brownian Motion in 3+1 Dimensions (2003), arXiv:math-ph/0312003.

[108] M. Pavon, Journal of Mathematical Physics **42**, 4846 (2001).

[109] M. Land, Entropy **19**, 234 (2017), number: 5 Publisher: Multidisciplinary Digital Publishing Institute.

[110] M. Davidson, Letters in Mathematical Physics **4**, 475 (1980).

[111] M. Davidson, Stochastic mechanics, trace dynamics, and differential space - a synthesis (2006), arXiv:quant-ph/0602211.

[112] K. Yasue and J.-C. Zambrini, Annals of Physics **159**, 99 (1985).

[113] R. De la Fuente and M. Lamorena, QMsolve: A module for solving and visualizing the Schrödinger equation (2024).

[114] M. J. Aburn, sdeint: Numerical integration of stochastic differential equations (SDE).

[115] P. R. Holland, *The Quantum Theory of Motion: An Account of the de Broglie-Bohm Causal Interpretation of Quantum Mechanics*, revised ed. edition ed. (Cambridge University Press, Cambridge, 1995).

[116] R. d. l. Fuente and M. Lamorena, qmsolve: A module for solving and visualizing the schrÃ¶dinger equation.

[117] J.-M. Lévy-Leblond, Physics Letters A **125**, 441 (1987).

[118] R. E. Kastner, American Journal of Physics **61**, 852 (1993).

[119] K. Czerski, P. Heide, A. Huke, L. Martin, and G. Ruprecht, in *International Symposium on Nuclear Astrophysics – Nuclei in the Cosmos – IX* (CERN, Geneva, Switzerland, 2006).

[120] K. Czerski, P. Heide, A. Huke, A. I. Kilic, I. Kulesza, and N. Targosz-Sleczka, ACTA PHYSICA POLONICA B **40**, 903 (2009).

[121] K. Czerski, A. Huke, P. Heide, and G. Ruprecht, in *The 2nd International Conference on Nuclear Physics in Astrophysics*, edited by Z. Fülöp, G. Gyürky, and E. Somorjai (Springer Berlin Heidelberg, Berlin, Heidelberg, 2006) pp. 83–88.

[122] T. Schenkel, A. Persaud, H. Wang, P. A. Seidl, R. MacFadyen, C. Nelson, W. L. Waldron, J.-L. Vay, G. Deblonde, B. Wen, Y.-M. Chiang, B. P. MacLeod, and Q. Ji, Journal of Applied Physics **126**, 203302 (2019).

[123] J. Vesić, Particles **7**, 834 (2024), number: 3 Publisher: Multidisciplinary Digital Publishing Institute.

[124] H. Yuki, J. Kasagi, A. G. Lipson, T. Ohtsuki, T. Baba, T. Noda, B. F. Lyakhov, and N. Asami, Journal of Experimental and Theoretical Physics Letters **68**, 823 (1998).

[125] F. Metzler, C. Hunt, and N. Galvanetto, Known mechanisms that increase nuclear fusion rates in the solid state (2022), arXiv:2208.07245.

[126] H. Price, Risk and Scientific Reputation: Lessons from Cold Fusion (2022), arXiv:2201.03776.

[127] V. Pines, M. Pines, A. Chait, B. M. Steinetz, L. P. Forsley, R. C. Hendricks, G. C. Fralick, T. L. Benyo, B. Baramsai, P. B. Ugorowski, M. D. Becks, R. E. Martin, N. Penney, and C. E. Sandifer, Physical Review C **101**, 044609 (2020), publisher: American Physical Society.

[128] B. M. Steinetz, T. L. Benyo, A. Chait, R. C. Hendricks, L. P. Forsley, B. Baramsai, P. B. Ugorowski, M. D. Becks, V. Pines, M. Pines, R. E. Martin, N. Penney, G. C. Fralick, and C. E. Sandifer, Physical Review C **101**, 044610 (2020), publisher: American Physical Society.

[129] S. A. Fulling and G. E. A. Matsas, Scholarpedia **9**, 31789 (2014).

[130] J. L. Doob, *Stochastic Processes P*, reprint edition ed. (WILEY, New York, NY, 1990).

# Chapter 6

Uncertainty over terms such as 'apparatus' is still rife in serious discussions of quantum mechanics, over 60 years after its conception

## Against 'measurement'

JOHN BELL

SURELY, after 62 years, we should have an exact formulation of some serious part of quantum mechanics? By 'exact' I do not of course mean 'exactly true'. I mean only that the theory should be fully formulated in mathematical terms, with nothing left to the discretion of the theoretical physicist . . . until workable approximations are needed in applications. By 'serious' I mean that some substantial fragment of physics should be covered. Nonrelativistic 'particle' quantum mechanics, perhaps with the inclusion of the electromagnetic field and a cut-off interaction, is serious enough. For it covers 'a large part of physics and the whole of chemistry' (P A M Dirac 1929 *Proc. R. Soc.* A **123** 714). I mean too, by 'serious', that 'apparatus' should not be separated off from the rest of the world into black boxes, as if it were not made of atoms and not ruled by quantum mechanics.

The question, '. . . should we not have an exact formulation . . . ?', is often answered by one or both of two others. I will try to reply to them: *Why bother? Why not look it up in a good book?*

### Why bother?

Perhaps the most distinguished of 'why bother?'ers has been Dirac (1963 *Sci. American* **208** May 45). He divided the difficulties of quantum mechanics into two classes, those of the first class and those of the second. The second-class difficulties were essentially the infinities of relativistic quantum field theory. Dirac was very disturbed by these, and was not impressed by the 'renormalisation' procedures by which they are circumvented. Dirac tried hard to eliminate these second-class difficulties, and urged others to do likewise. The first-class difficulties concerned the role of the 'observer', 'measurement', and so on. Dirac thought that these problems were not ripe for solution, and should be left for later. He expected developments in the theory which would make these problems look quite different. It would be a waste of effort to worry overmuch about them now, especially since we get along very well in practice without solving them.

Dirac gives at least this much comfort to those who are troubled by these questions: he sees that they exist and are difficult. Many other distinguished physicists do not. It seems to me that it is among the most sure-footed of quantum physicists, those who have it *in their bones*, that one finds the greatest impatience with the idea that the 'foundations of quantum mechanics' might need some attention. Knowing what is right by instinct, they can become a little impatient with nitpicking distinctions between theorems and assumptions. When they do admit some ambiguity in the usual formulations, they are likely to insist that ordinary quantum mechanics is just fine 'for all practical purposes'. I agree with them about that: ORDINARY QUANTUM MECHANICS (as far as I know) IS JUST FINE FOR ALL PRACTICAL PURPOSES.

Even when I begin by insisting on this myself, and in capital letters, it is likely to be insisted on repeatedly in the course of the discussion. So it is convenient to have an abbreviation for the last phrase: FOR ALL PRACTICAL PURPOSES = FAPP.

I can imagine a practical geometer, say an architect, being impatient with Euclid's fifth postulate, or Playfair's axiom: of *course* in a plane, through a given point, you can draw only one straight line parallel to a given straight line, at least FAPP. The reasoning of such a natural geometer might not aim at pedantic precision, and new assertions, known in the bones to be right, even if neither among the originally stated assumptions nor derived from them as theorems, might come in at any stage. Perhaps these particular lines in the argument should, in a systematic presentation, be distinguished by this label – FAPP – and the conclusions likewise: QED FAPP.

I expect that mathematicians have classified such fuzzy logics. Certainly they have been much used by physicists.

But is there not something to be said for the approach of Euclid? Even now that we know that Euclidean geometry is (in some sense) not quite true? Is it not good to know what follows from what, even if it is not really necessary FAPP? Suppose for example that quantum mechanics were found to *resist* precise formulation. Suppose that when formulation beyond FAPP is attempted, we find an unmovable finger obstinately pointing outside the subject, to the mind of the observer, to the Hindu scriptures, to God, or even only Gravitation? Would not that be very, very interesting?

But I must say at once that it is not mathematical precision, but physical, with which I will be concerned here. I am not squeamish about delta functions. From the present point of view, the approach of von Neumann's book is not preferable to that of Dirac's.

### Why not look it up in a good book?

But *which* good book? In fact it is seldom that a 'no problem' person is, on reflection, willing to endorse a treatment already in the literature. Usually the good unproblematic formulation is still in the head of the person

This article is published with the permission of Plenum Publishing, New York; it will appear in the proceedings of *62 Years of Uncertainty* (Erice, 5–14 August 1989).

in question, who has been too busy with practical things to put it on paper. I think that this reserve, as regards the formulations already in the good books, is well founded. For the good books known to me are not much concerned with physical precision. This is clear already from their vocabulary.

Here are some words which, however legitimate and necessary in application, have no place in a *formulation* with any pretension to physical precision: *system, apparatus, environment, microscopic, macroscopic, reversible, irreversible, observable, information, measurement*.

The concepts 'system', 'apparatus', 'environment', immediately imply an artificial division of the world, and an intention to neglect, or take only schematic account of, the interaction across the split. The notions of 'microscopic' and 'macroscopic' defy precise definition. So also do the notions of 'reversible' and 'irreversible'. Einstein said that it is theory which decides what is 'observable'. I think he was right – 'observation' is a complicated and theory-laden business. Then that notion should not appear in the *formulation* of fundamental theory. *Information? Whose* information? Information about *what*?

On this list of bad words from good books, the worst of all is 'measurement'. It must have a section to itself.

## Against 'measurement'

When I say that the word 'measurement' is even worse than the others, I do not have in mind the use of the word in phrases like 'measure the mass and width of the Z boson'. I do have in mind its use in the fundamental interpretive rules of quantum mechanics. For example, here they are as given by Dirac (*Quantum Mechanics* Oxford University Press 1930):

> '. . . any result of a measurement of a real dynamical variable is one of its eigenvalues . . .'

> '. . . if the measurement of the observable . . . is made a large number of times the average of all the results obtained will be . . .'

> '. . . a measurement always causes the system to jump into an eigenstate of the dynamical variable that is being measured . . .'

It would seem that the theory is exclusively concerned about 'results of measurement', and has nothing to say about anything else. What exactly qualifies some physical systems to play the role of 'measurer'? Was the wavefunction of the world waiting to jump for thousands of millions of years until a single-celled living creature appeared? Or did it have to wait a little longer, for some better qualified system . . . with a PhD? If the theory is to apply to anything but highly idealised laboratory operations, are we not obliged to admit that more or less 'measurement-like' processes are going on more or less all the time, more or less everywhere? Do we not have jumping then all the time?

The first charge against 'measurement', in the fundamental axioms of quantum mechanics, is that it anchors there the shifty split of the world into 'system' and 'apparatus'. A second charge is that the word comes loaded with meaning from everyday life, meaning which is entirely inappropriate in the quantum context. When it is said that something is 'measured' it is difficult not to think of the result as referring to some pre-existing property of the object in question. This is to disregard Bohr's insistence that in quantum phenomena the apparatus as well as the system is essentially involved. If it were not so, how could we understand, for example, that 'measurement' of a component of 'angular momentum' – in an arbitrarily chosen direction – yields one of a discrete set of values? When one forgets the role of the apparatus, as the word 'measurement' makes all too likely, one despairs of ordinary logic – hence 'quantum logic'. When one remembers the role of the apparatus, ordinary logic is just fine.

In other contexts, physicists have been able to take words from everyday language and use them as technical terms with no great harm done. Take for example, the 'strangeness', 'charm', and 'beauty' of elementary particle physics. No one is taken in by this 'baby talk', as Bruno Touschek called it. Would that it were so with 'measurement'. But in fact the word has had such a damaging effect on the discussion, that I think it should now be banned altogether in quantum mechanics.

## The role of experiment

Even in a low-brow practical account, I think it would be good to replace the word 'measurement', in the formulation, by the word 'experiment'. For the latter word is altogether less misleading. However, the idea that quantum mechanics, our most fundamental physical theory, is exclusively even about the results of experiments would remain disappointing.

In the beginning natural philosophers tried to understand the world around them. Trying to do that they hit upon the great idea of contriving artificially simple situations in which the number of factors involved is reduced to a minimum. Divide and conquer. Experimental science was born. But experiment is a tool. The aim remains: to understand the world. To restrict quantum mechanics to be exclusively about piddling laboratory operations is to betray the great enterprise. A serious formulation will not exclude the big world outside the laboratory.

## The quantum mechanics of Landau and Lifshitz

Let us have a look at the good book *Quantum Mechanics* by L D Landau and E M Lifshitz. I can offer three reasons for this choice:

(i) It is indeed a good book.

(ii) It has a very good pedigree. Landau sat at the feet of Bohr. Bohr himself never wrote a systematic account of the theory. Perhaps that of Landau and Lifshitz is the nearest to Bohr that we have.

(iii) It is the only book on the subject in which I have

read every word.

This last came about because my friend John Sykes enlisted me as technical assistant when he did the English translation. My recommendation of this book has nothing to do with the fact that one per cent of what you pay for it comes to me.

LL emphasise, following Bohr, that quantum mechanics requires for its formulation 'classical concepts' – a classical world which intervenes on the quantum system, and in which experimental results occur (brackets after quotes refer to page numbers):

'. . . It is in principle impossible . . . to formulate the basic concepts of quantum mechanics without using classical mechanics.' (LL2)

'. . . The possibility of a quantitative description of the motion of an electron requires the presence also of physical objects which obey classical mechanics to a sufficient degree of accuracy.' (LL2)

'. . . the 'classical object' is usually called *apparatus* and its interaction with the electron is spoken of as *measurement*. However, it must be emphasised that we are here not discussing a process . . . in which the physicist-observer takes part. By *measurement*, in quantum mechanics, we understand any process of interaction between classical and quantum objects, occurring apart from and independently of any observer. The importance of the concept of measurement in quantum mechanics was elucidated by N Bohr.' (LL2)

And with Bohr they insist again on the inhumanity of it all:

'. . . Once again we emphasise that, in speaking of 'performing a measurement', we refer to the interaction of an electron with a classical 'apparatus', which in no way presupposes the presence of an external observer.' (LL3)

'. . . Thus quantum mechanics occupies a very unusual place among physical theories: it contains classical mechanics as a limiting case, yet at the same time it requires this limiting case for its own formulation . . .' (LL3)

'. . . consider a system consisting of two parts: a classical apparatus and an electron . . . The states of the apparatus are described by quasiclassical wavefunctions $\Phi_n(\xi)$, where the suffix $n$ corresponds to the 'reading' $g_n$ of the apparatus, and $\xi$ denotes the set of its coordinates. The classical nature of the apparatus appears in the fact that, at any given instant, we can say with certainty that it is in one of the known states $\Phi_n$ with some definite value of the quantity $g$; for a quantum system such an assertion would of course be unjustified.' (LL21)

'. . . Let $\Phi_0(\xi)$ be the wavefunction of the initial state of the apparatus . . . and $\Psi(q)$ of the electron . . . the initial wavefunction of the whole system is the product $\Psi(q)\Phi_0(\xi)$. After the measuring process we obtain a sum of the form

$$\sum_n A_n(q)\Phi_n(\xi)$$

where the $A_n(q)$ are some functions of $q$.' (LL22)

'The classical nature of the apparatus, and the double role of classical mechanics as both the limiting case and the foundation of quantum mechanics, now make their appearance. As has been said above, the classical nature of the apparatus means that, at any instant, the quantity $g$ (the 'reading of the apparatus') has some definite value. This enables us to say that the state of the system apparatus + electron after the measurement will in actual fact be described, not by the entire sum, but by only the one term which corresponds to the 'reading' $g_n$ of the apparatus $A_n(q)\Phi_n(\xi)$. It follows from this that $A_n(q)$ is proportional to the wavefunction of the electron after the measurement . . .' (LL22)

This last is (a generalisation of) the Dirac jump, not an assumption here but a theorem. Note, however, that it has become a theorem only by virtue of another jump being assumed – that of a 'classical' apparatus into an eigenstate of its 'reading'. It will be convenient later to refer to this last, the *spontaneous* jump of a macroscopic system into a definite macroscopic configuration, as the LL jump. And the *forced* jump of a quantum system as a result of 'measurement' – *an external intervention* – as the Dirac jump. I am not implying that these men were the inventors of these concepts. They used them in references that I can give.

According to LL (LL24), measurement (I think they mean the LL jump) '. . . brings about a new state . . . Thus the very nature of the process of measurement involves a far-reaching principle of irreversibility . . . causes the two directions of time to be physically non-equivalent, i.e. creates a difference between the future and the past.'

The LL formulation, with vaguely defined wavefunction collapse, when used with good taste and discretion, is adequate FAPP. It remains that the theory is ambiguous in principle, about exactly when and exactly how the collapse occurs, about what is microscopic and what is macroscopic, what quantum and what classical. We are allowed to ask: is such ambiguity dictated by experimental facts? Or could theoretical physicists do better if they tried harder?

## The quantum mechanics of K Gottfried

The second good book that we will look at here is that of Kurt Gottfried (*Quantum Mechanics* Benjamin 1966). Again I can give three reasons for this choice:

(i) It is indeed a good book. The CERN library had four copies. Two have been stolen – already a good sign. The two that remain are falling apart from much use.

(ii) It has a very good pedigree. Kurt Gottfried was inspired by the treatments of Dirac and Pauli. His personal teachers were J D Jackson, J Schwinger, V F Weisskopf and J Goldstone. As consultants he had P Martin, C Schwartz, W Furry and D Yennie.

(iii) I have read some of it more than once.

This last came about as follows. I have often had the pleasure of discussing these things with Viki Weisskopf. Always he would end up with 'you should read Kurt Gottfried'. Always I would say 'I *have* read Kurt Gottfried'. But Viki would always say again next time 'you should read Kurt Gottfried'. So finally I read again some parts of KG, and again, and again, and again.

At the beginning of the book there is a declaration of priorities (KG1): '. . . The creation of quantum mechanics in the period 1924–28 restored logical consistency to its rightful place in theoretical physics. Of even greater importance, it provided us with a theory that appears to be in complete accord with our empirical knowledge of all nonrelativistic phenomena . . .'

The first of these two propositions, admittedly the less important, is actually given rather little attention in the book. One can regret this a bit, in the rather narrow context of the particular present enquiry – into the possibility of precision. More generally, KG's priorities are those of all right-thinking people.

The book itself is above all pedagogical. The student is taken gently by the hand, and soon finds herself or himself *doing* quantum mechanics, without pain – and almost without thought. The essential division of KG's world into system and apparatus, quantum and classical, a notion that might disturb the student, is gently implicit rather than brutally explicit. No explicit guidance is then given as to how in practice this shifty division is to be made. The student is simply left to pick up good habits by being exposed to good examples.

KG declares that the task of the theory is (KG16) '. . . to predict the results of measurements on the system . . .' The basic structure of KG's world is then $W = S + R$ where $S$ is the quantum system, and $R$ is the rest of the world – from which measurements on $S$ are made. When our *only* interpretive axioms are about measurement results (or findings (KG11)) we absolutely *need* such a base $R$ from which measurements can be made. There can be no question then of identifying the quantum system $S$ with the whole world $W$. There can be no question – without changing the axioms – of getting rid of the shifty split. Sometimes some authors of 'quantum measurement' theories seem to be trying to do just that. It is like a snake trying to swallow itself by the tail. It can be done – up to a point. But it becomes embarrassing for the spectators even before it becomes uncomfortable for the snake.

But there is something which can and must be done – to analyse theoretically not *removing* the split, which cannot be done with the usual axioms, but *shifting* it. This is taken up in KG's chapter 4: 'The Measurement Process . . .' Surely 'apparatus' can be seen as made of atoms? And it often happens that we do not know, or not well enough, either *a priori* or by experience, the functioning of some system that we would regard as 'apparatus'. The theory can help us with this only if we take this 'apparatus' $A$ out of the rest of the world $R$ and treat it together with $S$ as part of an enlarged quantum system $S'$: $R = A + R'$; $S + A = S'$; $W = S' + R'$. The original axioms about 'measurement' (whatever they were exactly) are then applied not at the $S/A$ interface, but at the $A/R'$ interface – where for some reason it is regarded as more safe to do so. In real life it would not be possible to find any *such* point of division which would be *exactly* safe. For example, strictly speaking it would not be exactly safe to take it between the counters, say, and the computer – slicing neatly through some of the atoms of the wires. But with some idealisation, which might '. . . be highly stylised and not do justice to the enormous complexity of an actual laboratory experiment . . .' (KG165), it might be possible to find more than one not too implausible way of dividing the world up. Clearly it is necessary to check that different choices give consistent results (FAPP). A disclaimer towards the end of KG's chapter 4 suggests that that, and only that, is the modest aim of that chapter (KG189): '. . . we emphasise that our discussion has merely consisted of several demonstrations of internal consistency . . .' But reading reveals other ambitions.

Neglecting the interaction of $A$ with $R'$, the joint system $S' = S + A$ is found to end, in virtue of the Schrödinger equation, after the 'measurement' on $S$ by $A$, in a state

$$\Psi = \sum_n c_n \Psi_n$$

where the states $\Psi_n$ are supposed each to have a definite apparatus pointer reading $g_n$. The corresponding density matrix is

$$\rho = \sum_n \sum_m c_n c_m^* \Psi_n \Psi_m^*$$

At this point KG insists very much on the fact that $A$, and so $S'$, is a macroscopic system. For macroscopic systems, he says, (KG186) '. . . $trA\hat{\rho} = trA\rho$ for all observables $A$ known to occur in nature . . .' where

$$\hat{\rho} = \sum_n |c_n|^2 \Psi_n \Psi_n^*$$

i.e. $\hat{\rho}$ is obtained from $\rho$ by dropping interference terms involving pairs of macroscopically different states. Then (KG188) '. . . we are free to replace $\rho$ by $\hat{\rho}$ after the measurement, safe in the knowledge that the error will never be found . . .'

Now, while quite uncomfortable with the concept 'all known observables', I am fully convinced of the practical elusiveness, even the absence FAPP, of interference between macroscopically different states (J S Bell and M Nauenberg 1966 'The moral aspects of quantum mechanics' in *Preludes in Theoretical Physics* North-Holland). So let us go along with KG on this and see where it leads: '. . . If we take advantage of the indistinguishability of $\rho$ and $\hat{\rho}$ to say that $\hat{\rho}$ is the state of the system subsequent to measurement, the intuitive interpretation of $c_m$ as a probability amplitude emerges without further ado. This is because $c_m$ enters $\hat{\rho}$ only via $|c_m|^2$, and the latter quantity appears in $\hat{\rho}$ in precisely the same manner as probabilities do in classical statistical physics . . .'

I am quite puzzled by this. If one were not actually on the lookout for probabilities, I think the obvious interpretation of even $\hat{\rho}$ would be that the system is in a state in which the various $\Psi$s somehow coexist: $\Psi_1\Psi_1^*$ *and* $\Psi_2\Psi_2^*$ *and* . . .

This is not at all a probability interpretation, in which the different terms are seen not as coexisting, but as alternatives: $\Psi_1\Psi_1^*$ *or* $\Psi_2\Psi_2^*$ *or* . . .

The idea that elimination of coherence, in one way or another, implies the replacement of 'and' by 'or', is a very common one among solvers of the 'measurement problem'. It has always puzzled me.

It would be difficult to exaggerate the importance attached by KG to the replacement of $\rho$ by $\hat{\rho}$: '. . . To the extent that nonclassical interference terms (such as $c_m c_m^*$) are present in the mathematical expression for $\rho$ . . . the numbers $c_m$ are intuitively uninterpretable, and the theory is an empty mathematical formalism . . .' (KG187)

But this suggests that the original theory, 'an empty mathematical formalism', is not just being approximated – but discarded and replaced. And yet elsewhere KG seems clear that it is in the business of approximation that he is engaged, approximation of the sort that introduces irreversibility in the passage from classical mechanics to thermodynamics: '. . . In this connection one should note that in approximating $\rho$ by $\hat{\rho}$ one introduces irreversibility, because the time-reversed Schrödinger equation cannot retrieve $\rho$ from $\hat{\rho}$.' (KG188)

New light is thrown on KG's ideas by a recent recapitulation, referred to in the following as KGR (K Gottfried 'Does quantum mechanics describe the collapse of the wavefunction?' Presented at *62 Years of Uncertainty*, Erice, 5–14 August 1989). This is dedicated to the proposition that (KGR1) '. . . the laws of quantum mechanics yield the results of measurements . . .' These laws are taken to be (KGR1): '(1) a pure state is described by some vector in Hilbert space from which expectation values of observables are computed in the standard way; and (2) the time evolution is a unitary transformation on that vector' (KGR1). Not included in the laws is (KGR1) von Neumann's '. . . infamous postulate: the measurement act 'collapses' the state into one in which there are no interference terms between different states of the measurement apparatus . . .' Indeed, (KGR1) 'the reduction postulate is an ugly scar on what would be a beautiful theory if it could be removed . . .'

Perhaps it is useful to recall here just how the infamous postulate is formulated by von Neumann (J von Neumann 1955 *Mathematical Foundations of Quantum Mechanics* Princeton University Press). If we look back we find that what vN actually *postulates* (vN347, 418) is that 'measurement' – an external intervention by $R$ on $S$ – causes the state

$$\phi = \sum_n c_n \phi_n$$

to jump, with various probabilities into $\phi_1$ *or* $\phi_2$ *or* . . .

From the 'or' here, replacing the 'and', as a result of external intervention, vN infers that the resulting density matrix, averaged over the several possibilities, has no interference terms between states of the system which correspond to different measurement results (vN347). I would emphasise several points here.

(i) von Neumann presents the disappearance of coherence in the density matrix, not as a postulate, but as a consequence of a postulate. The postulate is made at the wavefunction level, and is just that already made by Dirac for example.

(ii) I cannot imagine von Neumann arguing in the opposite direction, that lack of interference in the density matrix implies, without further ado, 'or' replacing 'and' at the wavefunction level. A special postulate to that effect would be required.

(iii) von Neumann is concerned here with what happens to the state of the system that has suffered the measurement – an external intervention. In application to the extended system $S'(= S + A)$ von Neumann's collapse would not occur before external intervention from $R'$. It would be surprising if this consequence of external intervention on $S'$ could be inferred from the purely internal Schrödinger equation for $S'$. Now KG's collapse, although justified by reference to 'all known observables' at the $S'/R'$ interface, occurs after 'measurement' by $A$ on $S$, but before interaction across $S'/R'$. Thus the collapse which KG discusses is not that which von Neumann infamously postulates. It is the LL collapse rather than that of von Neumann and Dirac.

The explicit assumption that expectation values are to be calculated in the usual way throws light on the subsequent falling out of the usual probability interpretation 'without further ado'. For the rules for calculating expectation values, applied to projection operators for example, yield the Born probabilities for eigenvalues. The mystery is then: what has the author actually derived rather than assumed? And why does he insist that probabilities appear only after the butchering of $\rho$ into $\hat{\rho}$, the theory remaining an 'empty mathematical formalism' so long as $\rho$ is retained? Dirac, von Neumann, and the others, nonchalantly assumed the usual rules for expectation values, and so probabilities, in the context of the unbutchered theory. Reference to the usual rules for expectation values also makes clear what KG's probabilities are probabilities *of. They are probabilities of 'measurement' results*, of external results of external interventions, from $R'$ on $S'$ in the application. We must not drift into thinking of them as probabilities of intrinsic properties of $S'$ independent of, or before, 'measurement'. Concepts like that have no place in the orthodox theory.

Having tried hard to understand what KG has written, I will finally permit myself some guesses about what he may have in mind. I think that from the beginning KG tacitly assumes the Dirac rules at $S'/R'$ – including the Dirac–von Neumann jump, required to get the correlations between results of successive (moral) measurements. Then, for 'all known observables', he sees that the 'measurement' results at $S'/R'$ are AS IF (FAPP) the LL jump had occurred in $S'$. This is important, for it shows how, FAPP, we can get away with attributing definite classical properties to 'apparatus' while believing it to be governed by quantum mechanics. But a jump assumption remains. LL derived the Dirac jump from the assumed LL jump. KG derives, FAPP, the LL jump from assumptions at the shifted split $R'/S'$ which include the Dirac jump there.

It seems to me that there is then some conceptual drift in the argument. The qualification 'as if (FAPP)' is dropped, and it is supposed that the LL jump really takes place. The drift is away from the 'measurement' ( . . . external intervention . . .) orientation of orthodox quantum mechanics towards the idea that systems, such as $S'$ above, have intrinsic properties – independently of and before observation. In particular the readings of experimental apparatus are supposed to be really there before they are read. This would explain KG's reluctance to interpret the unbutchered density matrix $\rho$, for the interference terms there could seem to imply the simultaneous existence of different readings. It would explain his need to collapse $\rho$ into $\hat{\rho}$, in contrast with von Neumann and the others, without external intervention across the last split $S'/R'$. It would explain why he is anxious to obtain this reduction from the internal Schrödinger equation of $S'$. (It would not explain the reference to 'all known observables' – at the $S'/R'$ split.) The resulting theory would be one in which some 'macroscopic' 'physical attributes' *have* values at all times, with a dynamics that is related somehow to the butchering of $\rho$ into $\hat{\rho}$ – which is seen as somehow not incompatible with the internal Schrödinger equation of the system. Such a theory, assuming intrinsic properties, would not need external intervention, would not need the shifty split. But the retention of the vague word 'macroscopic' would reveal limited ambition as regards precision. To avoid the vague 'microscopic' 'macroscopic' distinction – again a shifty split – I think one would be led to introduce variables which *have* values even on the smallest scale. If the exactness of the Schrödinger equation is maintained, I see this leading towards the picture of de Broglie and Bohm.

## The quantum mechanics of N G van Kampen

Let us look at one more good book, namely *Physica* A **153** (1988), and more specifically at the contribution: 'Ten theorems about quantum mechanical measurements', by N G van Kampen. This paper is distinguished especially by its robust common sense. The author has no patience with '. . . such mind-boggling fantasies as the many world interpretation . . .' (vK98). He dismisses out of hand the notion of von Neumann, Pauli, Wigner – that 'measurement' might be complete only in the mind of the observer: '. . . I find it hard to understand that someone who arrives at such a conclusion does not seek the error in his argument' (vK101). For vK '. . . the mind of the observer is irrelevant . . . the quantum mechanical measurement is terminated when the outcome has been macroscopically recorded . . .' (vK101). Moreover, for vK, no special dynamics comes into play at 'measurement': '. . . The measuring act is fully described by the Schrödinger equation for object system and apparatus together. The collapse of the wavefunction is a consequence rather than an additional postulate . . .' (vK97).

After the measurement the measuring instrument, according to the Schrödinger equation, will admittedly be in a superposition of different readings. For example, Schrödinger's cat will be in a superposition $|cat\rangle = a|life\rangle + b|death\rangle$. And it might seem that we do have to deal with 'and' rather than 'or' here, because of interference: '. . . for instance the temperature of the cat . . . the expectation value of such a quantity $G$ . . . is not a statistical average of the values $G_{ll}$ and $G_{dd}$ with probabilities $|a|^2$ and $|b|^2$, but contains cross terms between life and death . . .' (vK103).

But vK is not impressed: 'The answer to this paradox is again that the cat is macroscopic. Life and death are macrostates containing an enormous number of eigenstates $|l\rangle$ and $|d\rangle$ . . .

$$|cat\rangle = \sum_l a_l|l\rangle + \sum_d b_d|d\rangle$$

. . . the cross terms in the expression for $\langle G\rangle$ . . . as there is such a wealth of terms, all with different phases and magnitudes, they mutually cancel and their sum practically vanishes. This is the way in which the typical quantum mechanical interference becomes inoperative between macrostates . . .' (vK103).

This argument for no interference is not, it seems to me, by itself immediately convincing. Surely it would be possible to find a sum of very many terms, with different amplitudes and phases, which is not zero? However, I am convinced anyway that interference between macroscopically different states is very, very elusive. Granting this, let me try to say what I think the argument to be, for the collapse as a 'consequence' rather than an additional postulate.

The world is again divided into 'system', 'apparatus', and the rest: $W = S + A + R' = S' + R'$. At first, the usual rules for quantum 'measurements' are assumed at the $S'/R'$ interface – including the collapse postulate, which dictates correlations between results of 'measurements' made at different times. But the 'measurements' at $S'/R'$ which can actually be done, FAPP, do not show interference between macroscopically different states of $S'$. It is *as if* the 'and' in the superposition had already, *before* any such measurements, been replaced by 'or'. So the 'and' *has* already been replaced by 'or'. It is *as if* it were so . . . so it *is* so.

This may be good FAPP logic. If we are more pedantic, it seems to me that we do not have here the proof of a theorem, but a *change of the theory* – at a strategically well chosen point. The change is from a theory which speaks *only* of the results of external interventions on the quantum system, $S'$ in this discussion, to one in which that system is attributed *intrinsic properties* – deadness or aliveness in the case of cats. The point is strategically well chosen in that the predictions for results of 'measurements' across $S'/R'$ will still be the same . . . FAPP.

Whether by theorem or by assumption, we end up with a theory like that of LL, in which superpositions of macroscopically different states decay somehow into one of the members. We can ask as before just how and how often it happens. If we really had a theorem, the answers to these questions would be calculable. But the only possibility of calculation in schemes like those of KG and vK involves shifting further the shifty split – and the questions with it.

For most of the paper, vK's world seems to be the petty world of the laboratory, even one that is not treated very realistically: '. . . in this connection the measurement is always taken to be instantaneous . . .' (vK100)

But almost at the last moment a startling new vista opens up – an altogether more vast one:

'*Theorem IX: The total system is described throughout by the wave vector* $\Psi$ *and has therefore zero entropy at all times* . . .

This ought to put an end to speculations about measurements being responsible for increasing the entropy of the universe. (It won't of course.)' (vK111)

So vK, unlike many other very practical physicists, seems willing to consider the universe as a whole. His universe, or at any rate some 'total system', has a wavefunction, and that wavefunction satisfies a linear Schrödinger equation. It is clear, however, that this wavefunction cannot be the whole story of vK's totality. For it is clear that he expects the experiments in his laboratories to give definite results, and his cats to be dead or alive. He believes then in variables $X$ which identify the realities, in a way which the wavefunction, without collapse, can not. His complete kinematics is then of the de Broglie–Bohm 'hidden variable' dual type: $(\Psi(t,q), X(t))$.

For the dynamics, he has exactly the Schrödinger equation for $\Psi$, but I do not know exactly what he has in mind for the $X$, which for him would be restricted to some 'macroscopic' level. Perhaps indeed he would prefer to remain somewhat vague about this, for

'*Theorem IV: Whoever endows* $\Psi$ *with more meaning than is needed for computing observable phenomena is responsible for the consequences* . . .' (vK99)

## Towards a precise quantum mechanics

In the beginning, Schrödinger tried to interpret his wavefunction as giving somehow the density of the stuff of which the world is made. He tried to think of an electron as represented by a wavepacket – a wavefunction appreciably different from zero only over a small region in space. The extension of that region he thought of as the actual size of the electron – his electron was a bit fuzzy. At first he thought that small wavepackets, evolving according to the Schrödinger equation, would remain small. But that was wrong. Wavepackets diffuse, and with the passage of time become indefinitely extended, according to the Schrödinger equation. But however far the wavefunction has extended, the reaction of a detector to an electron remains spotty. So Schrödinger's 'realistic' interpretation of his wavefunction did not survive.

Then came the Born interpretation. The wavefunction gives not the density of *stuff*, but gives rather (on squaring its modulus) the density of probability. Probability of *what*, exactly? Not of the electron *being* there, but of the electron being *found* there, if its position is 'measured'.

Why this aversion to 'being' and insistence on 'finding'? The founding fathers were unable to form a clear picture of things on the remote atomic scale. They became very aware of the intervening apparatus, and of the need for a 'classical' base from which to intervene on the quantum system. And so the shifty split.

The kinematics of the world, in this orthodox picture, is given by a wavefunction (maybe more than one?) for the quantum part, and classical variables – variables which *have* values – for the classical part: $(\Psi(t,q\ \ldots), X(t)\ \ldots\ldots\ldots)$. The $X$s are somehow macroscopic. This is not spelled out very explicitly. The dynamics is not very precisely formulated either. It includes a Schrödinger equation for the quantum part, and some sort of classical mechanics for the classical part, and 'collapse' recipes for their interaction.

It seems to me that the only hope of precision with the dual $(\Psi, x)$ kinematics is to omit completely the shifty split, and let both $\Psi$ and $x$ refer to the world as a whole. Then the $x$s must not be confined to some vague macroscopic scale, but must extend to all scales. In the picture of de Broglie and Bohm, every particle is attributed a position $x(t)$. Then instrument pointers – assemblies of particles *have* positions, and experiments *have* results. The dynamics is given by the world Schrödinger equation plus precise 'guiding' equations prescribing how the $x(t)$s move under the influence of $\Psi$. Particles are *not* attributed angular momenta, energies, etc, but *only* positions as functions of time. Peculiar 'measurement' results for angular momenta, energies, and so on, emerge as pointer positions in appropriate experimental setups. Considerations of the KG and vK type, on the absence (FAPP) of macroscopic interference, take their place here, and an important one, in showing how usually we do not have (FAPP) to pay attention to the whole world, but only to some subsystem and can simplify the wavefunction . . . FAPP.

The Born-type kinematics $(\Psi, X)$ has a duality that the original 'density of stuff' picture of Schrödinger did not. The position of the particle there was just a feature of the wavepacket, not something in addition. The Landau–Lifshitz approach can be seen as maintaining this simple nondual kinematics, but with the wavefunction compact on a macroscopic rather than microscopic scale. We know, they seem to say, that macroscopic pointers *have* definite positions. *And* we think there is nothing *but* the wavefunction. So the wavefunction must be narrow as regards macroscopic variables. The Schrödinger equation does not preserve such narrowness (as Schrödinger himself dramatised with his cat). So there must be some kind of 'collapse' going on in addition, to enforce macroscopic narrowness. In the same way, if we had modified Schrödinger's evolution somehow we might have prevented the spreading of his wavepacket electrons. But actually the idea that an electron in a ground-state hydrogen atom is as big as the atom (which is then perfectly spherical) is perfectly tolerable – and maybe even attractive. The idea that a macroscopic pointer can point simultaneously in different directions, or that a cat can have several of its nine lives at the same time, is harder to swallow. And if we have no extra variables $X$ to express macroscopic definiteness, the wavefunction itself must be narrow in macroscopic directions in the configuration space. This the Landau–Lifshitz collapse brings about. It does so in a rather vague way, at rather vaguely specified times.

In the Ghiradi–Rimini–Weber scheme (see the box and the contributions of Ghiradi, Rimini, Weber, Pearle, Gisin

### The Ghiradi–Rimini–Weber scheme

The GRW scheme represents a proposal aimed to overcome the difficulties of quantum mechanics discussed by John Bell in this article. The GRW model is based on the acceptance of the fact that the Schrödinger dynamics, governing the evolution of the wavefunction, has to be modified by the inclusion of stochastic and nonlinear effects. Obviously these modifications must leave practically unaltered all standard quantum predictions about microsystems.

To be more specific, the GRW theory admits that the wavefunction, besides evolving through the standard Hamiltonian dynamics, is subjected, at random times, to spontaneous processes corresponding to localisations in space of the microconstituents of any physical system. The mean frequency of the localisations is extremely small, and the localisation width is large on an atomic scale. As a consequence no prediction of standard quantum formalism for microsystems is changed in any appreciable way.

The merit of the model is in the fact that the localisation mechanism is such that its frequency increases as the number of constituents of a composite system increases. In the case of a macroscopic object (containing an Avogadro number of constituents) linear superpositions of states describing pointers 'pointing simultaneously in different directions' are dynamically suppressed in extremely short times. As stated by John Bell, in GRW 'Schrödinger's cat is not both dead and alive for more than a split second'.

- The original and technically detailed presentation of GRW can be found in 1986 *Phys. Rev D* 34 470; a brilliant and simple presentation has been given by John Bell in *Schrödinger: Centenary Celebration of a Polymath* C W Kilmister (ed) 1987 Cambridge University Press p41
- A general discussion of the conceptual implications of the scheme can be found in 1988 *Foundation of Physics* 81 1
- The GRW model has been the object of many recent papers and a lively debate on its implications is going on. Recently the model has been generalised to cover the case of systems of identical particles and to meet the requirements of relativistic invariance.

**G C Ghiradi, A Rimini and T Weber**

and Diosi presented at *62 Years of Uncertainty*, Erice, 5–14 August 1989) this vagueness is replaced by mathematical precision. The Schrödinger wavefunction even for a single particle, is supposed to be unstable, with a prescribed mean life per particle, against spontaneous collapse of a prescribed form. The lifetime and collapsed extension are such that departures of the Schrödinger equation show up very rarely and very weakly in few-particle systems. But in macroscopic systems, *as a consequence of the prescribed equations*, pointers very rapidly point, and cats are very quickly killed *or* spared.

The orthodox approaches, whether the authors think they have made derivations or assumptions, are just fine FAPP – when used with the good taste and discretion picked up from exposure to good examples. At least two roads are open from there towards a precise theory, it seems to me. Both eliminate the shifty split. The de Broglie–Bohm-type theories retain, exactly, the linear wave equation, and so necessarily add complementary variables to express the non-waviness of the world on the macroscopic scale. The GRW-type theories have nothing in their kinematics but the wavefunction. It gives the density (in a multidimensional configuration space!) of *stuff*. To account for the narrowness of that stuff in macroscopic dimensions, the linear Schrödinger equation has to be modified, in the GRW picture by a mathematically prescribed spontaneous collapse mechanism.

The big question, in my opinion, is which, if either, of these two precise pictures can be redeveloped in a Lorentz invariant way.

'. . . All historical experience confirms that men might not achieve the possible if they had not, time and time again, reached out for the impossible.' **Max Weber**

'. . . we do not know where we are stupid until we stick our necks out.' **R P Feynman**

## Further reading

J S Bell and M Nauenberg 1966 The moral aspect of quantum mechanics in *Preludes in theoretical physics* (in honour of V F Weisskopf) (North-Holland) 278–286

P A M Dirac 1929 *Proc. R. Soc.* A **123** 714

P A M Dirac 1948 *Quantum mechanics* third edn (Oxford University Press)

P A M Dirac 1963 *Sci. American* **208** May 45

K Gottfried 1966 *Quantum mechanics* (Benjamin)

K Gottfried Does quantum mechanics describe the collapse of the wavefunction? Presented at *62 Years of Uncertainty*, Erice, 5–14 August 1989

L D Landau and E M Lifshitz 1977 *Quantum mechanics* third edn (Pergamon)

N G van Kampen 1988 Ten theorems about quantum mechanical measurements *Physica* A **153** 97–113

J von Neumann 1955 *Mathematical foundations of quantum mechanics* (Princeton University Press)

**John Bell** is in the theory division, CERN, CH-1211, Geneva, Switzerland

# Chapter 7

# In defence of "measurement"

Rudolf Peierls

In a stimulating article (*Physics World* August p33) the late John Bell professed dissatisfaction with the foundations of quantum mechanics as usually presented, particularly in connection with the so-called "collapse of the wavefunction" as a result of a measurement. He agreed that *for all practical purposes* the use of quantum mechanics by qualified practitioners leads to well defined answers which, where they can be checked, agree with experiment.

However, he regarded it as necessary to have a clearly formulated presentation of the physical significance of the theory without relying on ill-defined concepts. I agree with him that this is desirable, and, like him, I do not know of any textbook which explains these matters to my satisfaction. I agree in particular that the books he quoted do not give satisfactory answers (I assume that they are fairly quoted; I have not re-read them).

But I do not agree with John Bell that these problems are very difficult. I think it is easy to give an acceptable account, and in this article I shall try to do so. I shall not aim at a rigorous axiomatic, but only at the level of the logic of the working physicist.

In my view the most fundamental statement of quantum mechanics is that the wavefunction, or more generally the density matrix, represents our *knowledge* of the system we are trying to describe. I shall return later to the question "whose knowledge?". It is well known that we have to use a wavefunction if we have a "pure state" i.e. if our knowledge of the system is complete, in the sense that any further knowledge is barred by the uncertainty principle. Failing such complete knowledge we must use a density matrix, which therefore contains both quantum and classical ignorance. The wavefunction is a special case of a density matrix, and I shall here talk about "density matrix" when I mean "wavefunction or density matrix".

More precisely, while the time variation of the density matrix is given by Schrödinger's equation, the initial values represent knowledge usually obtained from observations. (There are not always measurements; for example, if an atom has been for a reasonable time in free space we know it must be in its ground state.)

Our knowledge is not fixed, but may increase or decrease. It increases if further observations are made; it decreases if the system is disturbed by external factors which we cannot control. There is nothing new in this. In classical physics our knowledge may increase and decrease in the same way. The only difference is that in quantum mechanics we have to be specific about what we know, because our possible knowledge is confined by the uncertainty principle. In classical physics there is no reason in principle why we cannot know everything about the system and we usually argue as if we did. But in a practical situation our knowledge may increase or decrease as indicated.

In quantum mechanics any increase in our knowledge is usually accompanied by a decrease in some other respect, because of the uncertainty principle. This applies particularly when we are concerned with a "pure state". Then we can gain no new information (other than confirming what we know already) without losing some of the existing information.

Once this significance of the density matrix is understood, it is clear that upon a change in our knowlege the density matrix must change. This is not a physical process, and we certainly cannot expect it to follow from the Schrödinger equation. It is just the fact that our knowledge has changed, and thus must be represented by a new density matrix.

When I refer to "observation", this term has its common-sense meaning. The observation usually (but not necessarily) involves an apparatus which interacts with the system in question, and which produces a signal (visible, audible, or other) which we can recognise, and which is correlated with the variables of the system. Bell quoted the view of Landau and Lifshitz (and therefore of Bohr) that the apparatus must necessarily obey classical physics. In my view this is not correct. It is of course true that our senses are macroscopic, and that the instruments we find convenient are also macroscopic and in practice classical. But this is a practical point, not one of principle. The sensitivity of the human eye is almost sufficient to detect a single photon. If some experimentalist has sufficient vision to see one photon, the observation of that photon might perfectly well serve as a measurement.

The apparatus usually consists of a chain of correlated events. I have elsewhere (Peierls 1979, 1985) discussed as an example the observation of a spin component of a spin−½ atom by a Stern–Gerlach magnet. The first step, the passage through the inhomogeneous magnetic field, sets up a correlation between the spin component and the position of the atom. It is not yet a measurement; we have not yet gathered any information. This requires determining the position of the particle, i.e. in which part of the split beam it travels. To find this out, we may use a counter, but again this conveys no information – and nothing collapses! – until we find out whether the counter has been activated. We can obviously pursue this chain: the counter will be part of an electrical circuit, the circuit will operate a digital recorder, we may read this recorder by means of the light it reflects into our eye, etc. Each step is correlated with the preceding ones and therefore with the spin component of the particle. Each step keeps both options open until we "see" the result, and then we revise our density matrix.

Because of the uncertainty principle we cannot acquire knowledge of, say, the $z$ component of the spin without losing what information we had previously about, say, the $x$ component. Is this happening in the first step, the passage through the magnet? At first sight this looks likely, because information about $s_x$ is contained in the phase relation between the components of the wavefunction belonging to $s_z = +½$ and $s_z = -½$. Since the beams corresponding to the two $s_z$ values are now split, they do not overlap and do not interfere, so their phase relationship is not observable. However, the information is not irretrievably lost. By arranging a further magnet we could recombine the two beams and observe their phase relation (thereby foregoing the possibility of observing $s_z$). We do finally lose the "forbidden" information when we "see" the atom in one of the beams. We then have to replace our density matrix by one containing only the one $s_z$ value, so there is no interference.

As long as we do not "see" the atom in the beam, the reconstruction of the seemingly lost information is troublesome, but easy to visualise. At the next stage, i.e. after the counter, it becomes much more involved. Since the density matrix now contains the variables of the counter, interference requires not only that the two atomic beams be made to overlap, but in addition that there be an overlap between the density matrices for the activated and unactivated states of the counter. The observation of the phase relation therefore requires an operator capable of deactivating the counter coherently. This is possible in principle, but in practice prohibitively difficult. As we go further down the chain of connections involved in our "measurement", this difficulty gets worse.

> "I must confess that the scheme, with both hidden variables and probability rules, seems to me exceedingly ugly, but of course one cannot argue about this"

This is the origin of the belief that the apparatus makes the off-diagonal matrix elements of the density matrix disappear. In most cases that is true "for all practical purposes", but not in principle. The off-diagonal matrix elements disappear only when we *know* the result of the measurement.

The "system" to which we apply our description can be as large as we like, including the whole world if we want. However, if we make the system too large, the amount of information we can obtain is relatively small, so that the density matrix is made up mostly of parts proportional to the unit matrix (which denotes complete ignorance) and it becomes hard to do any useful physics. In any case the "system" cannot include the mind of the observer and his knowledge, because present physics is not able to describe mind and knowledge (and it is not obvious that this is a proper subject for physics).

The objection is sometimes made: "How can one apply quantum mechanics to the early Universe, when there were no observers around?" The answer is that the observer does not have to be contemporaneous with the event. We can, from present evidence, draw conclusions about the early Universe, the classical example being the cosmic microwave background. In this sense we are observers. If there is a part of the Universe, or a period in its history, which is not capable of influencing present-day events directly or indirectly, then indeed there would be no sense in applying quantum mechanics to it.

That leaves the question: whose knowledge should be represented in the density matrix? In general there will be many who may have some information about the state of a physical system. Each of them has to use his or her density matrix. These may differ, as the nature and amount of knowledge may differ. People may have observed the system by different methods, with more or less accuracy; they may have seen part of the results of another physicist. However, there are limitations to the extent to which their knowledge may differ. This is imposed by the uncertainty principle. For example if one observer has knowledge of $s_z$ of our Stern–Gerlach atom, another may not know $s_x$, since the measurement of $s_x$ would have destroyed the other person's knowledge of $s_z$, and vice versa. This limitation can be compactly and conveniently expressed by the condition that the density matrices used by the two observers must commute with each other.

John Bell referred to two alternative interpretations of quantum mechanics, that of de Broglie–Bohm (BB), and that of Ghiradi–Rimini–Weber (GRW). As far as I know the BB scheme reproduces all predictions of quantum mechanics. A decision can therefore be made only on aesthetic grounds. I must confess that the scheme, with both hidden variables and probability rules, seems to me exceedingly ugly, but of course one cannot argue about this. I have not studied the implications of the GRW scheme in detail, but I believe that there must be cases where it makes predictions differing from those of quantum mechanics, which would be observable in principle.

## Further reading

J S Bell 1990 Against "measurement" *Physics World* August 33–40

R Peierls 1979 *Surprises in Theoretical Physics* Princeton section 1.6 (Some of the points made in this article will also be discussed in a forthcoming volume, *More Surprises in Theoretical Physics*, Princeton)

R Peierls 1985 Observations in Quantum Mechanics and the "Collapse of the Wave Function" in *Symposium on the Foundations of Modern Physics* World Scientific

# Commentary on Peierls's "In defence of "measurement"" [1].

Tuck C. Choy*

*Center for Space, Time and the Quantum, Parc Maraveyre Bât. 1,*

*13260, Cassis, Bouches du Rhône, France and*

*Departmento de Fisica-CIOyN, Universidad de Murcia, Murcia 30071, Spain.*

"I do not know what's reality, that is a question for metaphysics, not physics."

Rudolph Peierls in private discussions with the author 1992.

This year's centenary celebrations for the founding of quantum mechanics by Werner Heisenberg in 1925 also coincides with the thirtieth anniversary of the death of Rudolph Peierls in 1995. Peierls was Heisenberg's student, perhaps his most illustrious one, well known to be the co-author of the Frisch-Peierls Memoranda (1940) [2] that initiated the UK development of atomic weapons. In writing this commentary on Peierls's paper in Physics World (1991) [1] I thought it might be useful to recall some memories and discussions I had with Rudy, as we called him in Oxford, pertaining in particular to fundamental questions in quantum mechanics, as a tribute to both men and as a prelude to my views on his response to Bell [3]. Heisenberg was of course famous for his discovery of matrix mechanics in 1925 and as a student Peierls had learnt a lot from him and his way of doing physics and also from Pauli who had many anecdotes which Peierls recalled to me on occasions over the dinner table. Concerning quantum mechanics, I was always interested in discussions with him on the Landau and Peierls (1930) [4] paper on the uncertainty in the measurement of fundamental quantities imposed by relativity (finite speed of light $c$), which is the Landau-Peierls expression for the relativistic Heisenberg uncertainty principle: $\Delta p \Delta t \sim \hbar / c$. This would imply amongst other things that a precise measurement for the momentum say of the electromagnetic field i.e. the photon (which by default is a relativistic object) would require an infinite amount of time. The result came as such a shock to Niels Bohr that Peierls and Landau, both young men at the time, were detained in Copenhagen in 1930, unable to leave until Bohr was satisfied that a resolution has been found. In fact, it was not until 1950 that Bohr and Rosenfeld finally published their paper [5], which was a masterpiece on measurement analysis, which Bohr insisted must be conducted with classical apparatus. The psychological effects on both Peierls and Landau must have been deep and also on Heisenberg who later became involved [5]. Heisenberg later changed his own earlier views on quantum mechanics and adopted the S matrix theory for particle interaction processes [6], as the only description that has physical meaning. As for Peierls, my feeling is that he was completely broken down by Bohr (a very persistent and persuasive man) while Landau may not have been so. Landau never accepted quantum field theory as a logically consistent

* tuckvk3cca@gmail.com

theory because of the problem of infinities, and partly because of the Landau-Peierls results of 1930 [4, 7]. Prior to this they had also together penned their paper on quantum electrodynamics in configuration space, proposing for the first time a Schrödinger equation for light quanta [8]. The paper inspired Enrico Fermi to use their methodology for the theory of beta decay.

I only met John Bell once, in 1987 at the Schrödinger centenary at Imperial College London; his paper on the Ghiradi-Rimini-Weber (GRW) model, which preceded CN Yang's did not leave a mark in my memory. However later conversations with Peierls taught me a few things about Bell, who was a student at Birmingham in 1956 while Peierls was the Head of Department. He was a clever student, difficult at times as they all are, but had deep insight in all aspects of physics. I recall also my ex-boss and mentor Peter Schofield (while I was a research scientist at the Atomic Energy Authority (AERE) at Harwell in 1988) telling me about Walter Marshall who was also a brilliant ex-student of Peierls's and a contemporary of Bell's. Walter later became director of the AERE in 1968 and then Chairman of the UK Atomic Energy Authority (UKAEA) in 1981. Peter told me that Marshall came to him one day in tears exasperated saying: "How come whenever I get into an argument with Bell, he is always right and I am wrong?". As to quantum mechanics Peierls, must have had a tough time teaching Bell anything concerning the status quo. Peierls told me that Bell had actually thought that experiments would show that he was right i.e. they would not violate his inequalities, but Aspect's experiments, and later, those of others had left him rather disappointed. Perhaps Bell's later devotion to the GRW model with its hidden variables and unconventional probability rules was his desperate effort to see the quantum world his way. The above quote from Peierls came out of one of my discussions with him on these issues. So now we can go on to the paper.

For a start, Bell's 1990 article actually disappointed me in later years. I have simply no idea why in his critique of "Against Measurement" [3] he chose three '*good*' books: 1) Landau and Lifshitz's Quantum Mechanics [9], 2) Kurt Gottfried's Quantum Mechanics [10], and 3) Nicole van Kampen's Ten theorems about quantum mechanical measurements [11]. I would have thought he should have gone for the jugular and tear to pieces (1) Dirac's : Principles of Quantum Mechanics [12], (2) Heisenberg's : Physical Principles of the Quantum theory [13] and (3) Herman Weyl's : The theory of Groups and Quantum Mechanics [14]. Those would be my three '*good*' books if I were daring enough to challenge the principles of quantum mechanics. Dirac never got involved in wavefunction collapse or the shifty split. The closest he got to that would be section 10 in his book Principles of Quantum Mechanics in which he assumed that if a measurement is made, "the disturbance causes a jump in the state of the dynamical system" . I do not however buy Bell's charge that Dirac was the most distinguished of the 'why bother?'ers. For others yes, but for Dirac, it was not the case. Bell himself had mitigated his own charge by saying that Dirac acknowledged that his class I type problems exist but the time was not ripe to be able to go into them. Again in his 1963 Scientific American paper [15] for example he considers that a consistent picture for quantum mechanics (of the type Bell was looking for) is a too hard Class I type problem. He had in fact worried a lot about quantum mechanics foundations and proposed that $\hbar$ may not stand as firm as $e$ and $c$ in the fine structure constant $\alpha = e^2/\hbar c$ and could change. He stubbornly refused to accept the Renormalization program as the final word, but in this even Richard Feynman was the same. Modern particle physics vindicated Dirac's concerns to some extent as, at sufficiently high energies, it was found that the value of $\alpha$, now known to 8 or 9 decimal places does indeed change.

In Peierls's article two things stand out. In such a short two-page article Peierls emphasized the uncertainty principle four times and this is for him the very basis for the foundation and the formalism of quantum mechanics, not the other way round. In this he was true to the spirit of Heisenberg. Secondly he adopts the Heisenberg stand that the wavefunction or density matrix represents our knowledge of the system [16]. It has no other physical meaning beyond that. However because our knowledge can increase or decrease as a result of the act of measurement, the density matrix or wavefunction which represents this knowledge must change accordingly, subject to the uncertainty principle, but not at all following from a Schrödinger equation which might include the apparatus and an observer. Peierls disagrees with Bell over whether the measurement issues he mentioned are difficult ones. One gets the impression that these two men may have been arguing ever since Bell was a student at Birmingham where Peierls was Professor in 1956 and they have never come to any resolution. Peierls also challenges the Landau-Lifshitz and thus Bohr's view that the apparatus which performs the act of measurement must necessarily be classical i.e. obey classical physics. He cites the example that an experimentalist with good vision can detect down to one photon which can thus constitute a measurement by a non-classical system. What is classical according to him is that measurements generate a chain of correlated events.

Most of these beyond the first are essentially classical i.e. contain diagonal or near diagonal density matrices. He was not bothered by Bell's shifty split. His reasoning in his article [1] is a summary of his analysis in his Surprises in Theoretical Physics [17] section 1.6. To him the wavefunction or density matrix represents our knowledge of the system as in the Stern-Gerlach apparatus, since at any stage of a measurement it is in principle possible even after disturbing a system to unscramble it and recover all information until the final change when the observer has registered the result of the measurement. In this case there is new knowledge and therefore an update of the density matrix becomes necessary. The concept of any physical system state collapse has no meaning for him. In fact he was in the process in 1991 of finishing his More Surprises in Theoretical Physics [18] when I visited him from Australia. On that occasion he complaint he could not find time to finish it due to many distractions, and the publishers were getting impatient for the manuscript. In section 1.2 of More Surprises [18] he had elaborated a bit more than what was in his 1991 article [1] and it was during our conversations about it in reference to what 'reality' is that my quote from him above got stuck in my memory. I direct the reader to the opening in section 1.2, "I have discussed the interpretation of quantum mechanics and emphasized that there is *no* alternative to the standard view that the wavefunction (or more generally the density matrix) represents *our* knowledge of a physical system" (my italics). One point Peierls did not address was Bell's "and/or" arguments in his article, which was Bell's strong point ( I agree with Ian Aitchison who raised this with me in private correspondence). However upon closer examination, this dissection of the shifty split using Boolean logic does not worry me too much now. I used to worry about it because of my knowledge of the Temple paradox [19] to which Peierls' wrote a rebuttal in Nature soon after [20], not entirely satisfactory in my opinion. However it is interesting that I had noticed Peierls had quite a few books by Betrand Russell on his shelf in his flat during my visits. They must have both shared pacifism, atheism and perhaps even politics. Recently I came across the quote generally attributed to Russell: If you want to study Logic, don't read Aristotle [21]. Then it clicked. One just cannot justify applying Boolean logic to a superposition principle of complex valued wavefunctions or density matrices. So Bell's "and/or" arguments are no longer such a strong point after all in my humble opinion. In fact his interpretation of the quantum superposition of the system near the system-world S'/R' interface [3] as equivalent to an "and" Boolean statement, which he complained of as being 'strategically" switched at R' into an "or' by the likes of van Kampen and Landau and Lifzhitz is somewhat dubious. We now know that classical Boolean "and" and "or" logic gates do not exist naturally as basic constructs in quantum computing. To validate his 'and/or' interpretation would require a prior massaging of the system, or equivalently to construct interfaces between quantum gates and classical gates that would generally lead to irretrievable information loss [22].

Furthermore it is worth noting that quantum mechanics just does not deal with individual events, see for example Ballentine [23]. In any individual experiment the outcome is that one will observe a cat to be either alive or dead. The proposition that Schrödinger's superposition describes some kind of pseudo reality with the cat being half dead or half alive with complex phases in Peierls's view is incorrect (see again my above quote). After all we all know that the Heitler-London model of molecular bonding from the start involves a superposition of this very kind. The very structure of a lot of chemistry is explained this way. To question what is really happening at a deeper level is of course interesting, as we now have eta second pulse lasers and can do experiments that may provide a better picture if indeed quantum mechanics has a deeper structure than we have so far been able to find. I think this is what Kurt Gottfried [24] meant by telling Bell that a future theory may well involve abstractions and mathematics that make him yearn for good old fashion quantum theory.

---

[1] Rudolph Peierls, "In defence of "measurement"", Physics World, 19-20, January (1991).

[2] Frank Close, "TRINITY - The Treachery and Pursuit of the Most Dangerous Spy in History", Allen Lane, Penguin Random House UK 2019.

[3] John Bell, "Against measurement", Physics World, 33-40, August (1990)

[4] L. Landau and R. Peierls, "Extension of the Uncertainty Principle to Relativistic Quantum Theory", Zeitschrift für Physik **69**, 56-69 (1930)

[5] N. Bohr and L. Rosenfeld, "Field and Charge Measurements in Quantum Electrodynamics", Physical Review **78**(6), 794-98, (1950). As early as 1933, Bohr and Rosenfeld and later others, including Heisenberg in 1934, were seriously trying to resolve the issues raised by Peierls and Landau, see references cited in this paper.

[6] Werner Heisenberg first published his S matrix theory in 1943 in "Die beobachtbaren Größin der Theorie der Elementarteilchen", Zeitschrift für Physik **120** 7-10 (1943), though it must

have been known as early as 1938, see Landau and Lifshitz [7].

[7] V.B. Berestetskii and E.M. Lifshitz, Quantum Electrodynamics, Course of Theoretical Physics Vol 4, second edition Pergamon Press, Oxford, UK, 1979.

[8] L. Landau and R. Peierls, "Quantum Electrodynamics in Configuration Space", Zeitschrift für Physik **62**, 188-200 (1930).

[9] L.D.Landau and E.M. Lifshitz , Quantum Mechanics, Course of Theoretical Physics Vol 3, fourth edition, Pergamon Press, Oxford, UK, 1988.

[10] Kurt Gottfried, Quantum Mechanics, W.A. Benjamin, New York, 1966.

[11] N.G. van Kampen, "Ten Theorems about Quantum Mechanical Measurements", Physica **A153**, 97-113 (1988).

[12] P.A.M. Dirac, "Principles of Quantum Mechanics", the international series of monographs in physics Oxford University Press, fourth edition (revised), Oxford 84-89 1958.

[13] Werner Heisenberg, "The Physical Principles of the Quantum Theory" by Wener Heisenberg, Dover Publications Inc. New York 1949.

[14] Herman Weyl, "The theory of Groups and Quantum Mechanics", translated from German by H.P. Robertson, 272-276 Dover Inc., New York 1931.

[15] P. Dirac, "The Evolution of the Physicist's Picture of Nature", Scientific American **208**(5), 45-53 (1963).

[16] Heisenberg first alluded to the wavefunction as representing our knowledge of the system in his 1930 Chicago lecture [13]. I thank Mike Burt for pointing this out to me in an earlier preprint version of this manuscript. Rudolph Peierls developed this further in his first book [17] and Nicole van Kampen also adopted and developed this further in his paper [11]. In all cases I would prefer changing the word 'knowledge' to 'information'. The word 'knowledge' opens up all sorts of philosophical conundrums dating back to Heraclitus and Protagoras (500-420 BC), that in my opinion are not relevant to Peierls's thesis, see for example Russell [21].

[17] R. Peierls, "Surprises in Theoretical Physics", Princeton University Press, New Jersey 1979.

[18] R. Peierls, "More Surprises in Theoretical Physics", Princeton University Press, New Jersey 1991.

[19] G. Temple, "The Fundamental Paradox of the Quantum Theory", Nature June 8, 957 (1935).

[20] R. Peierls, "The Fundamental Paradox of the Quantum Theory", Nature Sep 7, 395 (1935).

[21] Betrand Russell, "History of Western Philosophy", Routledge London 1946. See also "The Art

of Drawing Inferences" in The Art of Philosophizing and other essays, New York: Philosophical Library (1968).

[22] Nielsen, Michael A.; Chuang, Isaac L. Quantum Computation and Quantum Information. Cambridge University Press. Camridge 2010.

[23] L.E. Ballentine, "The Statistical Interpretation of Quantum Mechanics ", Rev. Mod. Physics **42**,(4) 358-381 (1970).

[24] Kurt Gottfried, "Does quantum mechanics carry the seeds of its own destruction", Physics World, 34-40, August (1991).

# Chapter 8

Last year in *Physics World*, the late John Bell attacked several authors for their treatment of "measurement" in discussions of quantum mechanics. Here one of those authors replies

# Does quantum mechanics carry the seeds of its own destruction?

KURT GOTTFRIED

*. . . the quantum mechanical description will be superceded. In this it is like all theories made by man. But to an unusual extent its ultimate fate is apparent in its internal structure. It carries within itself the seeds of its own destruction.* (Bell and Nauenberg 1966)

QUANTUM mechanics is the most powerful theory of physics. It was John Bell's view, however, that no formulation of orthodox quantum mechanics is free of fatal flaws. While fully recognising the theory's indisputable utilitarian success, Bell was convinced that its failings meant that "something is rotten in the state of Denmark". Indeed, as many know from personal experience, he was in the habit of proclaiming that quantum mechanics is "rotten". What I do not know is whether he chose this particular adjective as a veiled condemnation of the House of Bohr.

What are the seeds of destruction that he saw so clearly?

To approach this question it is best to first examine what Bell meant by a satisfactory theory. I was fortunate to gain an insight into this through several long discussions we had during a bucolic June week in 1990 at a wonderful workshop on the foundations of quantum mechanics held at Amherst College, Massachusetts, US. Bell held out classical electrodynamics as the paradigm of a properly formulated theory of a "serious piece of physics". Classical electrodynamics is, however, beset by serious internal difficulties when one tries to incorporate a microscopic model of the sources in a Lorentz-invariant manner, and so I kept arguing that, at bottom, his "ideal" was not superior to quantum mechanics. Bell baffled me by rejecting these (valid) objections as irrelevant. It is only recently that I came to see clearly the point he sought to make. I would now put it this way.

Imagine someone handing Ampère a sheet of paper bearing Maxwell's equations and the Lorentz force law, and telling Ampère that these describe all the phenomena that he is investigating in his laboratory, and much beyond that, and further, that the crucial new concept in these equations are the symbols $\mathbf{E}(\mathbf{r},t)$ and $\mathbf{B}(\mathbf{r},t)$. Armed only with this cryptic message, and his knowledge of Newtonian dynamics, Ampère could then have deciphered precisely what is meant by the electromagnetic field. The essential point is that the new theory, Maxwell–Lorentz electrodynamics, defined its revolutionary basic concepts, *by its equations and without any scholastic crutches*, in terms of more primitive *pre-existing* concepts.

Quantum mechanics does not do this. If one were to hand the Schrödinger equation to Maxwell, and tell him that it describes the structure of matter in terms of various point particles whose masses and charges are to be seen in the equation, this knowledge would not, by itself, enable Maxwell to figure out what is meant by the wavefunction. Eventually he would need help: "Oh, I forgot to tell you that according to Rabbi Born, a great thinker in the yeshiva that flourished in Göttingen in the early part of the 20th century, $|\Psi(\mathbf{r}_1 \ldots \mathbf{r}_N,t)|^2$ is the joint probability for finding these charges at the indicated positions at time $t$, and its evolution is determined by the differential equation I revealed to you. But before you rush to the conclusion that Rabbi Born's interpretation answers all your

A slightly longer version of this article was presented at the symposium on quantum physics in memory of John Stewart Bell, CERN 2-3 May 1991 (to be published).

questions, you should reflect on a very profound observation by his great contemporary, Rabbi Dirac of Cambridge, to wit, that if $|\Psi_1(t_0)|^2, \ldots, |\Psi_K(t_0)|^2$ are probabilities of physical interest, then any linear superposition $\Xi(t_0) = \Sigma c_n \Psi_n$ is, evidently, a valid initial condition for the differential equation, and hence $|\Xi(t)|^2$ must again be a physically realisable probability distribution." Quite a Talmudic mouthful.

This contrast between the manner in which quantum mechanics is formulated, when compared to its great revolutionary predecessors, was an essential motive for Bell's conviction that the theory is fatally flawed. And this contrast stimulated his last publication, *Against "measurement"* (*Physics World* August 1990 pp33–40), which will be my principal focus. (All quotations that carry no citation are from this article.) A large part of this article is a biting critique of the treatment of the measurement problem in my 1966 textbook (and in a recent elaboration of it, Gottfried 1989). That he saw fit to devote so much space to decrying my efforts is the highest compliment I have ever been paid.

*Against "measurement"* displays all the subtlety and verve that characterised Bell's spoken and written word. Recall the opening passage:

> "Surely, after 62 years, we should have an exact formulation of some serious part of quantum mechanics. By 'exact' I do not of course mean 'exactly true'. I only mean that the theory should be fully formulated in mathematical terms, with *nothing left to the discretion of the theoretical physicist* [emphasis added] . . . By 'serious' I mean that some substantial fragment of physics should be covered. Non-relativistic 'particle' quantum mechanics, perhaps with the inclusion of the electromagnetic field . . . I mean too by 'serious', that 'apparatus' should not be separated off from the rest of the world into black boxes, as if it were not made of atoms and not ruled by quantum mechanics."

This is not a naive plea for a theory that would provide a perfect depiction of some "serious fragment" of the physical world, let alone a "theory of everything", the holy grail that so many physicists have chased in recent years, but a call for a theory *formulated in a logically precise manner*. And true to form, in the same breath as he set down his own preconditions for a satisfactory theory he excoriated the "the most sure-footed of quantum physicists, those who have it *in their bones* . . . [and] become a little impatient with nitpicking distinctions between theorems and assumptions . . . and are likely to insist that ordinary quantum mechanics is just fine 'for all practical purposes'. I agree with them about that: *ORDINARY QUANTUM MECHANICS (as far as I know) IS JUST FINE FOR ALL PRACTICAL PURPOSES*". For this last phrase Bell coined a marvellously sardonic acronym: *FOR ALL PRACTICAL PURPOSES = FAPP*, which he embellished into a new species of "theorem": *PROVEN TO BE CORRECT FAPP = QED FAPP*. FAPP and QED FAPP seem destined to become permanent additions to the lexicon of theoretical physics.

Bell held that if quantum theory is to be precise, in the sense just described, then not only must FAPPish arguments be eradicated from its formulation, but the language in which the theory is formulated must be cleansed:

> ". . . however legitimate and necessary in application, [the following words] have no place in a *formulation* with any pretention to physical precision: *system, apparatus, environment, microscopic, macroscopic, reversible, irreversible, observable, information, measurement* . . . on this list of bad words . . . the worst of all is 'measurement'."

And from that point on he always pinned stigmata – quotation marks – onto *measurement*. As he put it to me in Amherst, when we teach people Maxwell's theory, we don't drown the presentation of the theory *per se* in lengthy discussions of arcane minutiae about the use of antennas to measure properties of the ionosphere, etc.; the theory speaks for itself.

On further reflection, I now think that in arguing his case Bell exaggerated, oversimplified the historical development of both classical electrodynamics and quantum mechanics, and as a result arrived at a definition of "understanding" that may be too rigid in that it reflects a presumption – a prejudice, if you will – as to what constitutes physical reality. Maxwell and his contemporaries did not present the theory only in terms of the equations so familiar to us, but also employed intricate and highly sophisticated models of the vacuum – the aether – that were crucial to their mode of thought. Long ago, when I was an undergraduate at McGill University, Professor Norman Shaw lectured from notes he had taken while listening to J J Thomson lecture from notes Thomson had taken as a student of Maxwell! I can assure you that one cannot teach Maxwell's theory in this "authentic" way to 20th century students because they do not have the background needed to cope with wave propagation in conducting fluids permeated by vortices and other exotic objects. More importantly, a modern (post-Einstein) presentation, in which the vacuum is *really empty*, but is, nevertheless, purported to support the transport of energy and momentum, would anger Maxwell and his contemporaries – they would say that we have embraced an empty mathematical formalism that provides no *understanding* of what is "*actually*" going on. Their indictment of Jackson's *Classical Electrodynamics* would sound quite similar to Bell's against such vassals of the House of Bohr as Landau and Lifshitz.

Furthermore, it is not so that the Schrödinger equation (or its matrix mechanics equivalent) must be supplemented by at least a passable interpretation of the wavefunction (or the matrix elements) before the theory is of any use. Heisenberg, Pauli, Schrödinger, Dirac and their colleagues had firmly established the power of the theory by doing a host of intricate and highly successful calculations before the Born interpretation and the Uncertainty Principle had been put forward. This is not just a debating point. Under an enormously broad set of situations the physical world marches to a quantum mechanical drummer, and was doing so for eons before any conscious being had worried about whether some wave packet had just collapsed. The spectra of distant stars are identical to those of atoms that we humans torture through terrestial Stern–Gerlachesque mazes; the pace at which thermonuclear reactions proceed within stellar interiors is set by barrier penetration probabilities that bear a clear relation to those measured in our laboratories; and should a star collapse into a compact object its constitutents would do so in faithful obedience to the Pauli Principle even though not even one theoretical physicist was watching. Indeed, what are perhaps the two most important triumphs of quantum mechanics – the demonstration that a finite amount of energy is required to excite any bound system, and that every reasonably well isolated atom of some particular species in a particular quantum state has precisely the same properties wherever it happens to be in the Universe – do not hinge on a precise understanding of $\Psi$. Schrödinger, with his totally wrong conception of what his wavefunction meant, knew that his equation explained these most basic properties of matter. Bell was, therefore, resorting to rhetorical overkill when he wrote that

> "It would seem that the theory is exclusively concerned about 'results of measurement', and has nothing to say about anything else. What exactly qualifies some physical system to play the role of 'measurer'? Was the wavefunction of the world waiting to jump for [billions] of years until a single-celled living creature appeared? Or did it have to wait a little longer, for some better qualified system . . . with a PhD? If the theory is to apply to anything but highly idealised laboratory operations, are we not obliged to admit that more or less 'measurement-like' processes are going on all the time, more or less everywhere? Do we have jumping all the time?"

By taking issue with Bell on this score, I am not saying that I am satisfied with *my* understanding of quantum mechanics. I share Feynman's belief that no one really understands quantum mechanics. But we should not exaggerate our embarrassment; it is ample as it stands. In any event, I agree with Bell that the foundations continue to merit the closest scrutiny.

## Orthodoxy defended

Let me, then, examine Bell's critique of quantum mechanical orthodoxy. I do so as a fairly typical defender of established theology, though as with all venerable creeds that Bell and other heretics call "defenders of orthodoxy" is not one species of dinosaur. My "orthodoxy" is not identical to that of Bohr, nor to that of Peierls (*Physics World* January 1991 pp19–21), to mention two especially eminent examples. Hence I must state *my* definition of "orthodoxy".

I. The interpretation of the scalar product as a probability amplitude cannot be derived from the Schrödinger equation.

Various strategies have been advocated for coping with this conundrum. Those that modify the Schrödinger equation and/or the content of the wavefunction I call "heretical". My strategy is to seek a demonstration that, under certain appropriate and well specified circumstances, the statistical interpretation is *compatible* with the unitary evolution of state vectors. (In *Against "measurement"* (p36), Bell asserts that "reading [my book] reveals other ambitions". That is not so. The opening page (p165) of the chapter on measurement has the following statement (in italics): "Here, as nowhere else, it is crucial to keep one's eye firmly focused on the ultimate objective: *To verify the consistency of the mathematical formalism with the interpretation of the scalar product as a probability amplitude*.") What could be called "canonical orthodoxy" is to add von Neumann's wave packet reduction postulate to the axiomatic basis, and it is my principal objective to avoid this. Indeed, a central theme in the debate is whether the reduction postulate is, or is not, indispensible to a demonstration of compatibility between the interpretation and the superposition principle.

II. A *pure* quantum state, represented by a ray in the Hilbert space, describes an ensemble of identically prepared systems.

This also requires some elaboration. First, I see no need to apologise for this definition. Physics, in contrast to difficult pursuits, is the study of *reproducible* phenomena. In the microscopic realm it is an *empirical fact*, learned without *any* help from theory, that only the behaviour of ensembles is, in general, reproducible, whereas that of individual systems is not. At one time it was possible to entertain the conjecture that there are hidden variables which, when discovered, will remove the need for statistics at a fundamental level, but the experiments inspired by the Bell inequality have shut that escape hatch as they rule out all but nonlocal hidden variable theories. Hence a statistical theory of the microcosmos is all that theoretical physics *should* seek. Second, some critics of "orthodoxy" argue that by assuming that the state vector does not describe an individual system, but only ensembles, one has, by definition, done away with the measurement problem. I do not agree. Important problems remain even if one accepts the irreducibly statistical character of quantum mechanics. Furthermore, the treatment of these remaining problems that I shall sketch is not entirely satisfactory. In short, as I see it, statement II does not offer peace with honour between heresy and "orthodoxy".

III. I agree with Bell that the ". . . 'apparatus' should not be [treated] as if it were not made of atoms and ruled by quantum mechanics". Here my "orthodoxy" differs from Bohr's, but not from that of Peierls.

This definition of "orthodoxy" then requires the confirmation of the following statements:

A. Whenever an ensemble $\{S\}$ of systems $S$ is prepared in the same manner, upon being sent through a measurement apparatus $A$ one finds a joint probability distribution that correlates each state of $\{S\}$ measured by $A$ with one macroscopically distinguishable state of $A$, so that a macroscopic observation of the latter suffices to determine the former.

B. If a *single* system $S$ is sent through $A$, one cannot predict which of the several outcomes in the probability distribution will actually occur, but in each and every trial (assuming perfect counting efficiency) *one* definite outcome will be observed.

For the sake of definiteness, let $S$ be a neutral atom and $A$ a Stern–Gerlach set-up furnished with detectors that register whether or not an atom is in the neighbourhood of the classical trajectory $T_m$ as determined by the $z$-projection of the atoms' magnetic moment, $\mu_B g m$, where $\mu_B$ is the Bohr magneton.[†] After entry into $A$ the atoms' wavefunction is $\eta_m \varphi_m(\mathbf{R})$, where $\eta_m$ is the internal state and $\varphi_m(\mathbf{R})$ the centre-of-mass wavefunction which goes to zero rapidly as $\mathbf{R}$ recedes from $T_m$. Each detector $D_m$ is a system – such as a surface on which the atoms are deposited for subsequent observation as in the original experiment, or a counter that changes its state when traversed by a magnetic moment. $D_m$ is to be incorporated into the Schrödinger equation; it has variables designated by $x_m$, and a ground state $\chi_0(x_m)$. The essential property of the detectors is that they only have a short-range interaction with the atoms, so that when atoms in the state $\eta_m$ pass through $A$ they can *only* influence $D_m$.

When $\eta_m \varphi_{inc}(\mathbf{R})$ is incident, the state that emerges is

$$\Psi_m = \eta_m \varphi_m(\mathbf{R}) \chi_E(x_m) \prod_{n \neq m} \chi_0(x_n) \tag{1}$$

where $\chi_E$ describes the excited state of $D_m$. For a detector of high efficiency, the overlap between $\chi_0$ and $\chi_E$ is small, and for brevity's sake we shall treat it as if it vanished. If the initial state of the atoms is the arbitrary *pure* state $\Sigma c_m \eta_m$, the superposition principle tells us that the combined state of $S$ and $A$ that emerges is

$$\Psi = \sum_m c_m \Psi_m. \tag{2}$$

**"Under an enormously broad set of situations the physical world marches to a quantum mechanical drummer, and was doing so for eons before any conscious being had worried about whether some wave packet had just collapsed"**

The central objective of the exercise is then to demonstrate that the $|c_m|^2$ are probabilities for finding atoms with the eigenvalues $m$ of $\mathcal{J}_z$.

If $A$ is properly designed, it is an amplifier whose output is a superposition of states of the detecting systems, each of which has no overlap in configuration space (in this example, the space of the $x_m$) with any of the others. To be more precise, if $\boldsymbol{O}$ is any observable related to the atomic centre-of-mass and/or the detectors, and if $m \neq m'$, the matrix element $\langle \mathbf{R}, x_m | \boldsymbol{O} | \mathbf{R}', x'_{m'} \rangle$ is assumed to tend rapidly to zero when the separation between the coordinates in the bra and ket grows to macroscopic proportions. We call such observables *macroscopically local*. For any such $\boldsymbol{O}$ this apparatus has established a one-to-one correlation between the states of the micro-system $\boldsymbol{S}$ and macroscopically distinguishable states of the apparatus. If $\boldsymbol{A}$ cannot produce such an outcome it is a contraption, however useful or intriguing, that does not accomplish the measurement task, but such a failure would put no onus on quantum mechanics.

Obviously (2) is a pure state. But if all the observables $\boldsymbol{O}$ at our disposal are macroscopically local, this pure state $\Psi$ cannot be distinguished from the mixture obtained by deleting all interference terms between states of differing $m$ from the density matrix:

$$\hat{\rho} = \sum_m |c_m|^2 \varphi_m(\mathbf{R}) \varphi_m^*(\mathbf{R}') \chi_E(x_m) \chi_E^*(x'_m) \prod_{n \neq m} \chi_0(x_n) \chi_0^*(x_n). \qquad (3)$$

This "butchered" density matrix has the structure of a *classical* joint probability distribution, without mysterious interference terms, and displays the one-to-one system–apparatus correlations demanded by requirements A. Hence it identifies the $|c_m|^2$ as the probabilities that $\boldsymbol{S}$ will be found to "be in" the $m$th state $\eta_m$ with the indicated eigenvalue. Furthermore, in (3) each term has one, and only one, detector excited, so that a sufficiently dilute beam will only cause a single detector to trigger at a time. (Note that this conclusion does not follow from the most stripped-down description of the Stern–Gerlach experiment, in which $\boldsymbol{A}$ is represented only by the centre-of-mass $\mathbf{R}$.) In short, both the requirements A and B have been met. Indeed, as Bell (1976) put it:

> "In quantum measurement theory such elimination of coherence is the philosopher's stone. For with an incoherent mixture specialisation to one of its components can be regarded as a purely mental act, the innocent selection of a particular subensemble, from some total statistical ensemble . . ."

I must emphasise that while I agree with this statement, it turns out that Bell was in his sarcastic mode when he wrote it in 1976, and did not mean it to be taken at face value. (I learned this when I cited his statement in publicly presenting my position at Amherst, at which point Bell impatiently interrupted and said that I had failed to grasp his point! That was indeed so, and remains so, despite Abner Shimony's passionate and patient efforts to instruct me.) In *Against "measurement"* Bell called this the LL (for Landau & Lifshitz) jump – ". . . [the] jump . . . of a 'classical' apparatus into an eigenstate of one of its 'readings' . . . the *spontaneous* jump of a macroscopic system into a definite macroscopic configuration." Here the emphasis on "spontaneous" and the repetitive use of "jump" is a mystery to me: when the roulette wheel stops in Monte Carlo, and assuming no chicanery, is that not also a spontaneous jump into a definite state from what was a probability distribution before the wheel was spun? Why is the inference I have drawn from $\hat{\rho}$ different from what one would infer from ruinous nights spent in casinos? Admittedly, there is a residual coherence in the quantum case that has no Monte Carlo counterpart, but this is not observable FAPP, though the validity of this last statement does not hold forever, strictly speaking, a point to which I now turn.

The time $T_C$ over which the imposter $\hat{\rho}$ can masquerade successfully as the true density matrix with its interference terms is very sensitive to the extent to which a realistic description of the apparatus $\boldsymbol{A}$ is included in the Schrödinger equation. At the most naive extreme, when only the atoms' centre-of-mass $\mathbf{R}$ is considered to be $\boldsymbol{A}$, with the differing trajectories $\boldsymbol{T}_m$ playing the role of pointer, it is relatively straightforward to add fields that combine the separated beams and to re-establish coherence – and by that token, to simultaneously destroy the ability of $\boldsymbol{A}$ to effect any measurement. In this case $T_C$ is short.

But $T_C$ expands enormously as soon as a more realistic description of $\boldsymbol{A}$ is included in the Schrödinger equation. Consider, for example, the inclusion of the surface on which the atoms are deposited after traversal of the magnet, and assume, for simplicity's sake, that it is legitimate to treat this surface as being in its ground state before impact. (In this connection, see A Danieri, A Loinger and G M Prosperi *Nuclear Physics* (1962) **33** 297.) The initial state of surface plus atom is in the continuum, and a transition must then occur into a state in which the atom is localised about a point $X_m$ close to the intersection of the classical trajectory $\boldsymbol{T}_m$ with the surface. In general, the transition will be accompanied by non-localised excitations of the surface (e.g. phonons), and result in capture into a very narrow level out of which the atom can only tunnel to a neighbouring position after a very long time. That $\hat{\rho}$ is an imposter can only become known when the state just described attains detectable coherence with one in which the atom was deposited originally at a different location $X_{m'}$ that is macroscopically separated from $X_m$. But quantum mechanical coherence requires overlap in *configuration* space, and not just in 3-space. The wavefunctions for the two situations, localisation about $X_{m'}$ and $X_m$, involve a very large number of coordinates describing the local environments of the surface as well as excitations that may have propagated afar. Even if all but a small portion of these variables eventually come to overlap appreciably (e.g. through diffusion of the localised atoms by successive tunnellings) there will still be negligible coherence if the other variables (describing, in this example, non-localised excitations) fail to overlap. As we see, time scales of enormous extent characterise the interval over which coherence is undetectable, and it seems impossible to devise some external intervention (akin to the recombining fields in the primitive Stern–Gerlach illustration) that could enforce coherence.

**To summarise, I claim that**

- the desired statistical distribution – without interference terms between distinct outcomes – is indistinguishable from the density matrix that evolves in accordance with the *unitary* Schrödinger evolution once a reasonably realistic description of the measurement process is incorporated into the dynamics,

†The Stern-Gerlach paradigm has become antiquated. It should be replaced by examples drawn from quantum optics. See, for example, R J Glauber Amplifiers, attenuators, and Schrödinger's Cat *Ann. N.Y. Acad. Sci.* **480** 336; Y Yamamoto and H A Haus 1986 *Rev. Mod. Phys.* **58** 1001.

provided all the operators pertaining to $\boldsymbol{S}$ and $\boldsymbol{A}$ are macroscopically local;

- these interference terms can only reappear, if at all, in some exceedingly entangled fashion if we wait for a time that is enormous compared with macroscopic (and not atomic) time scales;
- QED FAPP.

The time $T_C$ over which coherence is undetectable therefore plays a role similar to that of the Poincaré recurrence time in classical statistical mechanics. Admittedly, it is somewhat unsettling that the argument has a flavour reminiscent of that which concludes that classical statistical mechanics can be used with confidence for FAPPish endeavours, for we know that statistical mechanics is only a superstructure on a fundamental theory, whereas quantum mechanics is supposed to be fundamental.

The restriction to macroscopically local apparatus observables also requires a comment. In contrast to Dirac's usage, I restrict the term "observable" to objects that actually occur in nature, or for that matter, that one might imagine to occur in nature. In Dirac's terminology, all Hermitian operators are observables. Apart from a set of measure zero (which includes those of physical interest), Dirac's "observables" are macroscopically *non*-local. The theory should not be expected to have an interpretation in terms of everyday (i.e. macroscopic) concepts unless there are restrictions that reflect the real world on the Hamiltonians admitted into Schrödinger's equation – there is no reason why the theory should produce outcomes of which we can make sense if we insert unphysical interactions between $\boldsymbol{S}$ and $\boldsymbol{A}$ into the equation of motion. Of course, superfluids possess macroscopically nonlocal observables, and can display interference over macroscopic separations. But if such a device were added to $\boldsymbol{A}$, it would not invalidate the interpretation of quantum mechanics, but only render the contraption in question useless for the measurement that was the original goal.

## Bell's critique

Let me now try to render Bell's criticisms of the above argument. First, it was not focused primarily on the replacement of the true density matrix by the butchered imposter $\hat{\rho}$: "I am fully convinced of the practical elusiveness, even the absence FAPP, of interference between macroscopically different states." What did vex him was (1) that I did not seem to understand the difference between *and* and *or*; (2) that my argument showed "some conceptual drift" which had, perhaps unwittingly, led to tacit use of the reduction postulate even though my proclaimed purpose was to shun it; and (3), that like all those unwilling to abandon "orthodoxy" (of whatever strain), I resorted to what Bell viewed as an unacceptable "shifty split" of the world into a part that quantum mechanics purports to describe and a remainder that falls outside the theory's realm.

First, then, the *and/or* issue. Bell argued that if the various terms in (3) correspond to *distinct alternatives*, what should be demonstrated (at least at the QED FAPP level) is not that the outcome is the *sum* of these terms, as in $\hat{\rho}$, but rather:

$$
\begin{aligned}
\Psi\Psi^* &\rightarrow |c_1|^2\eta_1\eta_1^*\varphi_1\varphi_1^* \ldots \\
&\rightarrow \text{or } |c_2|^2\eta_2\eta_2^*\varphi_2\varphi_2^* \ldots \\
&\rightarrow \text{or} \ldots
\end{aligned}
$$

This seems to be a devastating blow. But in $\hat{\rho}$ the different terms have no overlap in the space of the apparatus variables (i.e. $\mathbf{R}$ and the $x_m$); moreover, if a probability distribution (e.g. of a population's incomes) breaks up into two or more non-overlapping pieces, these can always be added together, normalised to unity, and the sum plotted as a function of the variables that label the individual members of the ensemble. After this is done, it is still legitimate to say that this sum of terms represents the probability that a member will fall into one *or* another of the subensembles. As this is obvious, Bell's criticism could not be so superficial. To my understanding, his point was that if *only* $\hat{\rho}$ is provided, that does not imply that $\hat{\rho}$ represents the probability distribution for a set of distinct events – as in the case of a beam sufficiently dilute so that only one atom at a time passes through $\boldsymbol{A}$. Put another way, the function $\hat{\rho}$ could equally well represent an indivisible continuum, as in Schrödinger's original misconception that $|\Psi|^2$ is a continuous charge distribution. Indeed, I seem to recall that Bell made this point to me in a 1964 correspondence while I was writing my book, and that this stimulated me to end my chapter on measurement with the disclaimer "we have not attempted to reconcile the fact that the theory only makes statistical predictions with our observation of individual, solitary, events".

Recently I have realised that this concedes too much: that an elementary exercise (see Appendix) shows that when nothing but the Schrödinger equation is used to describe a sufficiently dilute beam interacting with, say, a Stern–Gerlach set-up, the detecting system will record isolated hits – that the number of atoms trapped on the detecting surface climbs in very brief and well separated intervals by unit steps. That is to say, by using a more realistic description of the state of the system $\boldsymbol{S}$ before measurement (as compared to a more realistic description of $\boldsymbol{A}$, which heretofore has been my defensive tactic), one can infer that the density matrices under discussion are probability distributions for isolated events. What I do not know is whether this pedestrian argument suffices to change any minds, or whether it misses a point that is too subtle for me to have grasped thus far.

Next comes the second point – whether the reduction postulate slipped in through the back door, or as Bell put it: "The mystery is then: what has the author [KG] actually derived rather than assumed?" As my objective is only to establish consistency between the interpretation and the unitary dynamics, my argument is, by definition, circular. The question is what is on the circle, and what is not, and how close does the circle come to being a tautology.

The argument leading to $\hat{\rho}$ does not rely on an interpretation of the atoms' internal wavefunctions $\eta_m$ beyond what can be inferred directly from the Schrödinger equation, namely, that these functions contain all the information that the theory can provide regarding the internal structure and dynamics of the atoms. Concerning the centre-of-mass functions $\varphi(\mathbf{R})$, by solving the Schrödinger equation one can demonstrate that $\varphi$ vanishes unless $\mathbf{R}$ is in close proximity to the classical trajectory $T_m$ appropriate to the classical magnetic moment $\mu_B gm$. *Therefore the only assumption concerning the interpretation of wavefunctions that is needed is that the atom is never in places where $\varphi(\mathbf{R})$ vanishes*, for that suffices to tell us where, *on a macroscopic length scale*, atoms in the subensembles labelled by $m$ strike the detector surface, and from that the (eigen)values of the magnetic moment and the corresponding probabilities $|c_m|^2$. The statement in italics is thus the suspect on the circle; but it is a much weaker assumption than the fully fledged Born edict, which governs all length scales dealt with by non-relativistic quantum mechanics. To some extent, Bell acknowledges that my argument does accomplish something: "This is important, for it shows how, FAPP, we can get away with attributing definite classical properties to 'apparatus' while believing it to be governed by quantum mechanics. But a jump (i.e. reduction) assumption remains . . . KG derives, FAPP, the LL jump from assumptions about the shifted split [between $A$ and the rest of the world], which includes the [von Neumann] jump there." Unfortunately, I do not understand the last phrase – where I used the von Neumann jump, and must leave the matter there.

Finally, then, there is the shifty split, which Bell puts as follows in his indictment:

> "The basic structure of KG's world is then $W = S + R$ . . . where $S$ is the quantum system, and $R$ is the rest of the world – from which measurements on $S$ are made. When our *only* interpretive axioms are about measurement results we absolutely *need* such a base $R$ . . . there can be no question then of identifying the quantum system $S$ with the whole world $W$. There can be no question – without changing the axioms – of getting rid of the shifty split . . . But there is something which can and must be done – to analyse theoretically not *removing* the split, which cannot be done with the usual axioms, but of *shifting* it. This is taken up in KG's chapter 4 . . . there is then some conceptual drift in the argument . . . it is supposed that the LL jump really takes place. The drift is away from the 'measurement' ( . . . external intervention . . . ) orientation of orthodox quantum mechanics towards the idea that systems, such as $S + A$ [with $A$ sufficiently realistic], have intrinsic properties – independent of and before observation. In particular, the readings of experimental apparatus are supposed to be really there before they are read . . . The resulting theory would be one in which some 'macroscopic' 'physical attributes' *have* values at all times, with a dynamics that is related somehow to the butchering of $\rho$ into $\hat{\rho}$ . . . But the retention [by KG] of the vague word 'macroscopic' reveals limited ambition as regards precision . . . To avoid the vague 'microscopic' – 'macroscopic' distinction – again a shifty split – I think one would be led to introduce variables which *have* values [at all times] even on the smallest scale."

I plead guilty to some of these charges:

- I think it has been, and continues to be, a blessing that we need not take the whole world $W$ into account if we are to understand the microcosm. Nothing I have been able to understand has swayed me to accept Bell's assertion that "a serious formulation [of quantum mechanics] will not exclude the big world outside the laboratory".
- Nor has anything I have been able to understand led me to lose sleep over the "shifty split", or over demonstrations that only attain the QED FAPP level of rigour – to paraphrase Bell, I am not squeamish about QED FAPP. The theory itself tells us under what circumstances the classical approximation is, FAPP, valid for some particular system. Once the description of the apparatus crosses a reasonably well defined threshold of authenticity – a threshold that is not determined by the whim of some theoretical physicist but by the theory itself, the statistical interpretation becomes compatible with the mathematical formalism; further embellishments do not, FAPP, alter the situation: the "shifty split" becomes, so to say, "translation" invariant FAPP.
- I believe that the argument I have given shows that once the above-mentioned threshold is crossed the system $A + S$ does have properties "independent of and before observation" by still more complex systems, including, for example, systems that draw a CERN Grade 14 salary. (In this connection, see the argument on p183 of my text which concludes that it does not matter whether we "look" at the detectors *provided* there is an interaction between them and the atoms.)
- Given the enormous range of distances and times that separate everyday life from the microscopic world, I find it most impressive that quantum mechanics can tell us when, FAPP, the "shifty split" is permissible, and thus, when our gross, human-scale intuition will, by and large, not lead us astray.

On the other hand, I do, and with shame, plead guilty to having presented an argument that is slipshod. There is, indeed, a conceptual drift which reflects an inadequate understanding of what I "have actually derived rather than assumed". That is, I agree with Bell that I have a too "limited ambition as regards precision", but I do not agree that the "smoking gun" is a "vague" distinction between macroscopic and microscopic. The smoking gun is that I do not understand properly how much of the mathematical formalism – operators, expectation values, etc. – it is legitimate to use in a demonstration of consistency between the theory's mathematical formalism and its interpretation.

## Crisis? What crisis?

Finally, I return to the question of whether quantum mechanics "carries within itself the seeds of its own destruction". If this verdict is correct, it would mean that theoretical physics is in deep crisis, a condition that usually leads to a profound breakthrough. But we must then ask what, precisely, is the crisis we are faced with?

In addressing this question, it is useful to recall two very different crises that beset physics at the beginning of this century. The first was Einstein's realisation that the equivalence principle implied that one could not construct a theory of gravitation compatible with the principle of relativity by restricting the space–time transformations to those that form the Lorentz group. This, then, was a deep conceptual flaw in the fabric of existing theory formulated in a mathematically well defined fashion, and the evident need to remove this flaw led directly (at least in hindsight) to general relativity. The second example was the inability of classical physics to reconcile the compelling hypothesis that matter is composed of atoms with crucial *qualitative* properties of matter: that all atoms and molecules of any particular species are identical and do not alter their properties under a wide range of physical and chemical processes; and, in virtue of the equipartition theorem, that atoms and molecules behave as if they had a fixed and quite small number of degrees of freedom. Classical concepts imply that any species of molecules should display a continuum of structures, and that a fixed number of molecular degrees of freedom is incompatible with any internal structure whatsoever for atoms, which contradicts the existence of atomic spectra. As we now know, quantum theory removes both of these conundrums, but at the time of the theory's invention the crisis was only understood in another guise – that classical theory could not handle the interaction between matter and the radiation field, a dilemma that had been formulated in a precise mathematical form by Wien, Rayleigh and Jeans. That classical theory had catastrophic implications for the constitution of matter was barely appreciated by Planck and others at the turn of the century, and played no substantial role in the development of quantum theory until Bohr's work 13 years later.

The lesson that could be drawn from this, and also from the development of general relativity, is that a crisis will only become creative if it is formulated in a mathematically precise manner. This conclusion has an echo in Bell's insistence on having quantum mechanics "*fully* formulated in mathematical terms, with nothing left to the discretion of the theoretical physicist", but what it really calls for is a critique of the "orthodox" theory fully formulated in mathematical terms with nothing left to the discretion of the critic.

That such a critique is not yet in hand was acknowledged by Bell in the same breath as he held out the hope that the failings we saw in the "orthodox" theory held out great prospects for the future of physics:

> "Suppose for example that quantum mechanics was found to *resist* precise formulation. Suppose that when formulation beyond FAPP is attempted, we find an unmovable finger pointing outside the subject, to the mind of the observer, to the Hindu scriptures, to God, or even only to Gravitation? Would not that be very, very interesting?"

But the vision of the future that he conveyed was a mixed signal. For example, in the 1960s he had

> "look[ed] forward to a new theory which can refer meaningfully to events in a given system without requiring 'observation' by another system. The critical test cases requiring this conclusion are systems containing consciousness and the Universe as a whole . . . [we] share with most physicists a degree of embarrassment at consciousness being dragged into physics, and share the usual feeling that to consider the Universe as a whole is at least immodest, if not blasphemous . . . It seems likely . . . that physics will again adopt a more objective description of nature long before it begins to understand consciousness, and the Universe as a whole may well play no central role in this development." (Bell and Nauenberg p284)

On the other hand, he also thought that there are just two options for completing the theory:

> "Either the wavefunction, as given by the Schrödinger equation, is not everything or it is not right . . . [thus] two roads are open towards a precise theory . . . Both eliminate the shifty split. The de Broglie–Bohm ['pilot wave'] theories retain, exactly, the linear wave equation, and so necessarily add complementary variables to express the non-waviness of the world on the macroscopic scale. [In] the Ghirardi–Rimini–Weber type theories . . . the linear Schrödinger equation has to be modified . . . by a mathematically prescribed spontaneous collapse mechanism." (Bell 1987)

But neither of these theories include "the big world outside laboratory", nor give any discernable hint of why such an inclusion might be necessary, unless the "spontaneous collapse mechanism" is, in some not yet understood sense, actually brought about by the existence of "the big world".

I, for one, remain unpersuaded that a crisis of major proportions has been identified, but at the same time I am open to the possiblity that such a crisis has been glimpsed, though perhaps only subliminally, by Bell and his great predecessors: Einstein, Schrödinger, de Broglie, Bohm and Wigner. This is a roster which should give us all pause. To my understanding, however, *this crisis has not yet been formulated with sufficient precision to facilitate the birth of a great offspring*. Furthermore, with Bell I would wager that "what seems . . . likely is that new ways of seeing things will involve an imaginative leap that will astonish us". And as I told Bell in our last conversation, I would also wager that the new theory will be even further removed from our philosophical and conceptual prejudices than is quantum mechanics, and that he would then pine for the time when he had "good old" quantum mechanics to kick around.

## Appendix

The language of second quantisation is best suited to analysing whether the "orthodox" theory can account for the observation of isolated events. Let $\psi_m(\mathbf{R})$ create an atom in the state $\eta_m$ with its centre-of-mass at $\mathbf{R}$, and

$$\psi(\mathbf{R}) = \sum_m c_m\, \psi_m(\mathbf{R}).$$

Next, introduce a set of wave packets $u_n(\mathbf{R})$ which move with the mean velocity of the incident beam, take the time $T_{tr}$ to traverse the apparatus $A$, and perform the traversal in the interval $n\tau < t < (n + 1)\tau$, where $\tau \gg T_{tr}$. Finally, define the one-atom creation operators

$$\Phi_n^\dagger(t) = \int d^3R\; u_n(\mathbf{R})\psi^\dagger(\mathbf{R}). \tag{4}$$

We can now construct states of the system plus apparatus that describe very dilute atomic beams. For $t<0$, such a state is

$$|S + A\rangle = \prod_{n=0}^{\infty} \Phi_n^\dagger(t)|0\rangle \tag{5}$$

where $|0\rangle$ is a product of the detectors' ground states and the state with no atoms. For $t > 0$ this state will evolve in accordance with the interactions between $S$ and $A$. At any time, no more than one atom is on its way through $A$ towards impact on the detecting surface – the state always has either the eigenvalue 0 or 1 of the operator

$$N_A = \sum_m \int d^3R\, \psi_m^\dagger(\mathbf{R})\psi_m(\mathbf{R}), \tag{6}$$

where the integration is over the volume of $A$ *excluding* the surface of the detector.

The interaction between the atoms and the detecting surface is

$$V = \sum_m \int d^3R\, d^3x\, v(\mathbf{R} - \mathbf{x})\psi_m^\dagger(\mathbf{R})\psi_m(\mathbf{R})\Theta(\mathbf{x}), \tag{7}$$

where $\Theta(\mathbf{x})$ is an operator that causes a change of state of the detector emanating from a point $\mathbf{x}$ "on" its surface, and $v$ is a very short range, highly attractive effective interaction that traps atoms onto the surface. If we define $N_{surf}$ by the same expression as $N_A$, but with the integration running only over the detecting surface, the eigenvalue of $N_{surf}$ increases by 1 in every interval of duration $\tau$. Furthermore, if sufficiently short packets $u_n$ are used, the time over which this eigenvalue changes will be far briefer than $T_{tr}$.

## Further reading

J S Bell and M Nauenberg 1966 The moral aspects of quantum mechanics in *Preludes in Theoretical Physics* North Holland. With the exception of the *Physics World* reference, this and all other articles by Bell on the foundations of quantum mechanics are reprinted in J S Bell 1987 *Speakable and Unspeakable in Quantum Mechanics* Cambridge University Press

J S Bell 1975 *Helv. Phys. Acta* **48** 93

J S Bell 1987 Are there quantum jumps? in *Schrödinger, Centenary of a Polymath* Cambridge University Press

J S Bell 1990 *Physics World* August 33–40

K Gottfried 1966 *Quantum Mechanics* W A Benjamin (re-issued by Addison-Wesley 1989)

K Gottfried 1989 Does quantum mechanics describe the collapse of the wave function? in *62 Years of Uncertainty* unpublished

**Kurt Gottfried** is at the Newman Laboratory of Nuclear Studies, Cornell University, Ithaca, NY 14853, USA

# Commentary on Gottfried "Does quantum mechanics carry the seeds of its own destruction?" PHYSICS WORLD, 35-40, Oct 1991.

"I, for one, remain unpersuaded that a crisis of major proportions has been identified, but at the same time I am open to the possibility that such a crisis has been glimpsed, though perhaps only subliminally, by Bell and his great predecessors: Einstein, Schrödinger, de Broglie, Bohm and Wigner." Kurt Gottfried ([1], pg. 40)

## Introduction

Quantum mechanics poses perhaps the most profound conceptual challenge in the history of science. Here we have a theory that has completely transformed the world, that threatens its destruction or perhaps its salvation, and yet at its core we are confronted with astounding and unrelenting apparent paradoxes. It is impossible to do justice to the volumes of research devoted to its many interpretations here. The first waves of discontent occurred during the 1920s at the famous Solvay conferences where Bohr and Einstein debated its physical interpretation[2]. The eminent philosopher Karl Popper engaged with the early quantum theorists and criticized the Copenhagen interpretation for the first time in 1934 [3], and he remained engaged and critical for much of the rest of his life. Einstein too was of course never satisfied with the Copenhagen interpretation [4], and neither was Schrödinger [5]. There followed decades of what has been termed the "shut-up and calculate" epoch of physics (roughly from 1935 to 1980), where fundamental questions about quantum theory were discouraged as heresy. A second period of intense debate occurred in the wake of Bell's theorem, and the subsequent experiments by John Clauser [6, 7, 8] , Alain Aspect [9], and others confirming that quantum theory correctly describes the experimental results of spin correlated pairs of entangled particles. This cast severe doubt on the possibility that hidden-variable theories could be local. So this greatly tarnished the acceptability of models such as that proposed by David Bohm [10, 11], stochastic models [12, 13, 14, 15], hydrodyanmic models [16], or other de Broglie pilot wave theories [17]. These experiments were inspired by the seminal papers of Einstein, Podolsky, and Rosen [18], and of Bohm [19]. I think that even after Clauser's experiment Bell remained very dissatisfied with the Copenhagen picture. In particular he criticized the division by Bohr of the universe into a quantum world and a classical one which enabled or more accurately empowered an experimenter to perform a measurement on a quantum system and thereby seemingly cause a collapse of the wave function by the mere act of observing the result [20]. This collapse was cast in mathematical granite by Von Neumann in his treatise on quantum mechanics as a projection operator acting on a quantum state vector in a Hilbert space [21]. Bohr's Copenhagen interpretation was defended by many devotees over the decades. For them the Clauser experiment along with a subsequent one by Aspect that more rigorously tested locality [9] were a vindication of Bohr. There was a debate between John Bell and Kurt Gottfried which succinctly illustrates the zeitgeist in the early 1990s [20, 1] This debate has continued and grown more intense since then.

Today there is a cornucopia of ideas and interpretations of quantum mechanics vying for attention. It's clear that Einstein's and Popper's concerns about quantum theory are now recognized as relevant. Einstein's views on quantum mechanics were complicated and nuanced [22, 23]. He is commonly believed to be in favor of a local and realistic description underlying quantum mechanics. Most today believe that the Bell experiments rule out such a description. In any event, he was not wrong in his desire to find a deeper explanation for quantum mechanics than the Copenhagen interpretation offered. To paraphrase a famous quote by Feynman, anybody who says that they understand quantum mechanics is fooling themselves, and nobody understands it. The mystery of quantum mechanics reminds one of the monolith in Kubrick's 2001 which is taken from Arthur C. Clark's Space Oddyssey novels. We see its shape, but we don't see its core workings, structure, or origin.

## Gottfried's defense of orthodoxy

Bell attacked Gottfried in his paper "Against Measurement" [20]. The disagreement was over the measurement problem. Bell argued that the concept of measurement implies, according to Bohr, a strict division between a "system" being measured and an "apparatus" doing the measuring. Bohr treated the measuring apparatus as classical, at least in part, so that it can be observed by an observer. Bell argued that this distinction is artificial and leads to conceptual difficulties. This division is sometimes called the "Heisenberg cut". Bell termed it the shifty-split. He thought that measurement should have no place in a physical theory. Gottfried explained that Bell wants a theory like classical electrodynamics that needs no measurement postulate and that the word 'measurement' should be banned. He illustrated his argument by reference to Gottfiried's treatment of the measurement problem in [24].

Gottfried defended "the orthodoxy" in [1], which he described by these properties:
I. The interpretation of the scalar product as a probability amplitude cannot be derived from the Schrödinger equation.
II. A pure quantum state, represented by a ray in the Hilbert space, describes an ensemble of identically prepared systems.
III. He agrees with Bell that the "apparatus" should not be treated as if it were not made of up of atoms and ruled by quantum mechanics. In this he differs from Bohr and denies the existence of the Heisenberg cut

The second property is critical. Bohr postulated that a pure quantum state for a single particle was a complete description of that particle. Gottfried rather assumes that the wave function or Hilbert space ray describes an ensemble of similarly prepared particles, and not just a single particle. This ensemble assumption is called the statistical interpretation of quantum mechanics. Einstein was an advocate of this interpretation. Although this difference might seem rather minor, it has historically been a topic of great debate. If the wave function describes only an ensemble, it raises the question of what else is required to completely define a single particle within the ensemble. Because of this, Einstein argued that quantum mechanics was incomplete. This is discussed in Ballentine [25]. Here is a qoutation from this book (§6):

"The Statistical Interpretation, which regards quantum states as being descriptive of ensembles of similarily prepared systems, is completely open with respect to hidden variables. It does not demand them, but it makes the search for them entirely reasonable ... On the other hand, the Copenhagen Interpretation, which regards a state vector as an exhaustive description of an individual system, is antipathetic to the idea of hidden variables, since a more complete description than that provided by a state vector would contradict that interpretation". Einstein is reputed to have written in a letter to Max Born that Bohm's hidden variable theory was "too cheap", but Bohm's model is not the only possible hidden variable theory.

The Copenhaganists realized that the statistical interpretation of quantum mechanics greatly improved, if not fully resolved the measurement problem. Therefore they incorporated it into a modified Copenhagen interpretation, but accompanied this grafting with a fierce opposition to any suggestion that a deeper understanding of qauntum mechanics was desirable or possible. This took its most vigorous form in their opposition to hidden variable theories. A good example of this movement is Léon Rosenfeld [26]. The interplay of Marxist philosophy, cold war politics, and the inherent intractability of quantum mechanics makes for a fascinating history with Rosenfeld at the eye of the storm in the 1950s. Bohm's theory especially was in the crosshairs. I would say that Gottfried's orthodoxy is very much in line with Rosenfeld.

Gottfried also makes a somewhat technical requirement of an acceptable measurement apparatus:

> "If **A** is properly designed, it is an amplifier whose output is a superposition of states of the detecting systems, each of which has no overlap in configuration space (in this example, the space of the $\mathbf{x}_m$) with any of the others. To be more precise, if **O** is any observable related to the atomic centre-of-mass and/or the detectors, and if $\mathbf{m} \neq \mathbf{m}'$, the matrix element $< \mathbf{R}, \mathbf{x_m} | \mathbf{O} | \mathbf{R}', \mathbf{x_{m'}} >$ is assumed to tend rapidly to zero when the separation between the coordinates in the bra and ket grows to macroscopic proportions. We call such observables macroscopically local. For any such **O** this apparatus has established a one-to-one correlation between the states of the micro-system **S** and macroscopically distinguishable states of the apparatus. If A cannot produce such an outcome it is a contraption, however useful or intriguing, that does not accomplish the measurement task, but such a failure would put no onus on quantum mechanics."

The existence of such apparata are proven by experimental construction. They act like a bridge between the microworld and the macro world. They take the place of the Heisenberg cut in the statistical interpretation and Gottfried's analysis. But, one might wonder if the existence of such apparata can be proven from the unitary quantum evolution of the atoms that make them up alone, without resorting to experience. Perhaps it is this hint of circularity that prompted Gottfried to describe his analysis as a solution to the measurement problem FAPP. His arguments are similar to decoherence arguments, and as such may have the same problems as described in [27].

Gottried illustrates his argument by considering an ensemble of neutral atoms passing through a Stern-Gerlach apparatus. He introduces a density matrix formalism to describe the system. This formalism mixes the classical probability elments in the apparatus with the quantum probability amplitudes. He argues that an initial pure state of atom plus apparatus is indistinguishable from a "butchered" density matrix with certain small interference terms set to exactly zero. Substituting the true density matrix with this butchered one makes the system look like a classical random system, like a roulette wheel, with positive probabilities for different outcomes. Bell objected to this crucial step in Gottfried's argument. He felt that it amounted to sweeping the Heisenberg cut under the rug. It has been suggested by Weinberg that the density matrix operator should be taken as the representative theoretical object for defining a state, rather than the Hilbert space vector [28, 29]. Gottfried's explanation of the measurement process in terms of density matrices is undoubtedly one of the best in the physics literature, and despite Bell's objections, deserves a prominent place in the historical development of the foundations of quantum mechanics. A review of the Bell-Gottfried debate is found in [30].

## The revival of local hidden variable theories

Despite the results of the Bell experiments, research has continued on realistic hidden variable theories, ignoring the nonlocality implications. The ultimate loophole in the Bell experiments is the possibility that free will may not exist. The universe might be completely deterministic so that the random choice required in a Bell experiment cannot be truly random [31]. The most famous example of this idea is the cellular automata theory of Gerard 't Hooft [32]. A somewhat similar theory has also been proposed by Stephen Wolfram [33]. Just as a rising tide raises all ships, if the universe is governed by a superdeterministic algorithm, then a stochastic interpretation of the universe in terms of particles moving along stochastic trajectories could be an accurate picture of quantum phenomena. Just as a random walk can be simulated in terms of algorithmic random number generators, so too might the motions of particles be random walks whose trajectories are generated by the universe's cellular automata. Before getting carried away with this idea, we must keep in mind that there is no existence proof that such cellular automata algorithms exist, and certainly none have been found that can replicate the standard model of particle physics along with all the quantum phenomena. However, the possible existence of such algorithms gives a green light to research in stochastic models of quantum mechanics. A particularly large set of possible hidden variable models including Bohmian mechanics is given by generalized stochastic mechanics [15] (and references therein). Of necessity, a cellular automaton describing our universe would obviously have to include what could be interpreted as spontaneous wave function collapse. The game is afoot!

Mark Davidson
Palo Alto, California, USA
2025

## References

[1] Kurt Gottfried. Does quantum mechanics carry the seeds of its own destruction? *Physics World*, 4(10):34, October 1991.

[2] Guido Bacciagaluppi and Antony Valentini. *Quantum Theory at the Crossroads: Reconsidering the 1927 Solvay Conference*. Cambridge University Press, Cambridge, 2009.

[3] Flavio Del Santo. An unpublished debate brought to light: Karl Popper's enterprise against the logic of quantum mechanics. *Studies in History and Philosophy of Science Part B: Studies in History and Philosophy of Modern Physics*, 70:65–78, May 2020.

[4] Albert Einstein. Physics and reality. *Journal of the Franklin Institute*, 221(3):349–382, March 1936.

[5] Fritz Rohrlich. Schrödinger and the interpretation of quantum mechanics. *Foundations of Physics*, 17(12):1205–1220, December 1987.

[6] J. F. Clauser and A. Shimony. Bell's theorem. Experimental tests and implications. *Reports on Progress in Physics*, 41(12):1881, December 1978.

[7] John F. Clauser, Michael A. Horne, Abner Shimony, and Richard A. Holt. Proposed Experiment to Test Local Hidden-Variable Theories. *Physical Review Letters*, 23(15):880–884, October 1969. Publisher: American Physical Society.

[8] Stuart J. Freedman and John F. Clauser. Experimental Test of Local Hidden-Variable Theories. *Physical Review Letters*, 28(14):938–941, April 1972. Publisher: American Physical Society.

[9] Alain Aspect, Jean Dalibard, and Gérard Roger. Experimental Test of Bell's Inequalities Using Time-Varying Analyzers. *Physical Review Letters*, 49(25):1804–1807, December 1982. Publisher: American Physical Society.

[10] David Bohm. A Suggested Interpretation of the Quantum Theory in Terms of "Hidden" Variables. I. *Physical Review*, 85(2):166–179, January 1952.

[11] K. Berndl, M. Daumer, D. Dürr, S. Goldstein, and N. Zanghi. A survey of Bohmian mechanics. *Il Nuovo Cimento B (1971-1996)*, 110(5):737–750, May 1995.

[12] Edward Nelson. *Dynamical Theories of Brownian Motion*. Princeton University Press, Princeton, NJ, second edition, 2001 edition, February 1967.

[13] L. de la Peña and A. M. Cetto. Does quantum mechanics accept a stochastic support? *Foundations of Physics*, 12(10):1017–1037, October 1982.

[14] Mark Davidson. A generalization of the Fényes-Nelson stochastic model of quantum mechanics. *Letters in Mathematical Physics*, 3(4):271–277, July 1979.

[15] Mark Davidson. Deriving Heisenberg's operator algebra from a classical model of stochastic mechanics. Zenodo preprint archiv, doi = 10.5281/zenodo.11456201, June 2024. To be published in a celebratory volume in july 2025 by World Scientific Publications, Singapore.

[16] E. Madelung. Quantentheorie in hydrodynamischer Form. *Zeitschrift für Physik*, 40(3):322–326, March 1927.

[17] Samuel Colin, Thomas Durt, and Ralph Willox. de Broglie's double solution program: 90 years later, March 2017. arXiv:1703.06158.

[18] A. Einstein, B. Podolsky, and N. Rosen. Can Quantum-Mechanical Description of Physical Reality Be Considered Complete? *Physical Review*, 47(10):777–780, May 1935.

[19] Basil J. Hiley. Bohm's Approach to the EPR Paradox. In Daniel Greenberger, Klaus Hentschel, and Friedel Weinert, editors, *Compendium of Quantum Physics*, pages 55–58. Springer, Berlin, Heidelberg, 2009.

[20] John Bell. Against 'measurement'. *Physics World*, 3(8):33, August 1990.

[21] John Von Neumann. *Mathematical Foundations of Quantum Mechanics*. Princeton University Press, February 2018.

[22] Michel Paty. The Nature of Einstein's Objections to the Copenhagen Interpretation of Quantum Mechanics. *Foundations of Physics*, 25(1):183–204, 1995.

[23] Tuck C. Choy. A centennial reappraisal of Heisenberg's Quantum Mechanics with a perspective on Einstein's Quantum Riddle, August 2023. arXiv:2308.04199.

[24] Kurt Gottfried. *Quantum mechanics - Vol.1: Fundamentals*. January 1966. Publication Title: New York: Benjamin ADS Bibcode: 1966qume.book.....G.

[25] L. E. Ballentine. The Statistical Interpretation of Quantum Mechanics. *Reviews of Modern Physics*, 42(4):358–381, October 1970. Publisher: American Physical Society.

[26] Anja Skaar Jacobsen. *Leon Rosenfeld: Physics, Philosophy, And Politics In The Twentieth Century*. World Scientific, January 2012. Google-Books-ID: d57FCgAAQBAJ.

[27] Stephen L. Adler. Why decoherence has not solved the measurement problem: a response to P.W. Anderson. *Studies in History and Philosophy of Science Part B: Studies in History and Philosophy of Modern Physics*, 34(1):135–142, March 2003.

[28] Steven Weinberg. Quantum mechanics without state vectors. *Physical Review A*, 90(4):042102, October 2014.

[29] Steven Weinberg. What happens in a measurement? *Physical Review A*, 93(3):032124, March 2016. Publisher: American Physical Society.

[30] M. A. B. Whitaker. Can the Statistical Interpretation of Quantum Mechanics be Inferred from the Schrödinger Equation?—Bell and Gottfried. *Foundations of Physics*, 38(5):436–447, May 2008.

[31] Sabine Hossenfelder. Superdeterminism: A Guide for the Perplexed, October 2020. arXiv:2010.01324.

[32] Gerard 't Hooft. *The Cellular Automaton Interpretation of Quantum Mechanics*, volume 185 of *Fundamental Theories of Physics*. Springer International Publishing, Cham, 2016.

[33] S. Wolfram. *A new kind of science*. Wolfram Media, 2002.

# Chapter 9

## Unprincipling Heisenberg's uncertainties

Jean-Marc Lévy-Leblond
Professor Emeritus, University of Nice

**Abstract**. The first part of this paper is a critical essay on the conceptual meaning of the so-called « Uncertainty principle ». It will be stressed that there is no independent principle underlying the Heisenberg's inequalities [1] as they derive from the fundamental theoretical structure of quantum theory. More to the point, it will be emphasized that the term « uncertainty » is a misnomer, leading to subjective interpretations about the limits of our knowledge. Recognizing at face value the conceptual nature of quantum physical quantities, as endowed with a spectrum of numerical values, one should rather call « dispersions », « extensions » or « spectrum widths » the magnitudes entering Heisenberg's type inequalities. A historical inquiry about the origins of the conventional terminology, from German to other languages, will reveal the roots and extent of the confusion.
The second part of the paper will review and assess many of the various formal definitions leading to Heisenberg's type inequalities, well beyond the pairs of physical quantities, such as the position and velocity of a particle and the customary use of quadratic dispersions. Some still open issues, such as the time-energy inequality or the angle-phase inequality will be briefly discussed.
A sobering conclusion will assess the concrete uses of Heisenberg's inequalities and stress that their genuine relevance and indisputable importance lie mainly at the heuristic and qualitative level, thus somewhat reducing the interest of the continuous discussions concerning their mathematical validity and philosophical interpretation.

### I. Dismissing the "Principle of Uncertainty"

One of the most troubling assertions of quantum theory is commonly known as Heisenberg's "uncertainty principle"; it refers to the impossibility of simultaneously assigning a precise numerical value to certain pairs of physical properties of a quanton (a quantum particle), such as its position and momentum. This state of affairs was highlighted by Heisenberg in 1927, through the analysis of a thought experiment in which an electron's state is determined by "illuminating" it with a photon[2]. Heisenberg showed that the coexistence of wave and corpuscular characteristics for both the

[1] Pace Trump who advocates the banning of the word "inequalities": https://groundreport.in/latest/full-list-of-words-banned-by-trump-including-climate-change-8839026.
[2] W. Heisenberg, "Uber den anschaulich Inhalt der quantentheoretischen Kinematik und Mechanik", *Zeitschr. Physik*, 43 (1927), 3.


To appear in a forthcoming collective volume on the 100th birthday of Quantum Mechanics, T. Choy ed., World Scientific, 2026 photon and the electron prevented precise simultaneous measurement of the electron's position and velocity. It soon became clear that this thought experiment, known as the "Heisenberg microscope", was only a special case of a general phenomenon in quantum physics. Any measurement of the position and velocity of a particle in a given state regardless of the method used, is thus limited: the numerical values resulting from such a measurement are defined only with irreducible discrepancies, the product of which is always greater (modulo a certain numerical factor) than the fundamental characteristic constant of quantum theory, that is, Planck's constant, which is the standard of quantum orders of magnitude. Moreover, many other pairs of physical quantities, such as the components of the angular momentum of an object and its spatial orientation, or the energy involved in a phenomenon and its duration, are subject to similar constraints. The generality of this situation has been interpreted as a fundamental "uncertainty principle". But this formulation is a major example of the epistemological slippage caused by semantic wanderings resulting from uncontrolled linguistic transfers.

In past decades, it has become clear that the rather muddled meaning of Heisenberg's seminal heuristic paper is due to the confusion of two questions, concerning either the values of physical properties for a given state of a quantum system (often referred to as the "preparation" of the state), or the values resulting from a joint measurement of these properties, leading to "disturbances" of these values. The second question, depending on the general and vexed theory of measurement in quantum theory, is certainly more involved, as it needs the consideration of measuring apparatus, statistical noises, etc.[3] This problem has been thoroughly investigated, leading to generalized forms of Heisenberg's inequalities[4]. However, since these developments, sophisticated as they may be, still rely on the usual quantum formalism, we will stick here to a discussion of the Heisenberg inequalities as they concern the understanding of a quantum state per se.

**Misled comments**

As it seems to impose an intrinsic limitation on our knowledge capacity, the "uncertainty principle" has been the subject of countless misapplications. Outside physics proper, unjustified recourse to this so-called principle has developed in all fields, from epistemology to metaphysics, and from aesthetics to politics. Physicist James Jeans wrote in 1930:

> "Heisenberg shows us that nature abhors precision and certainty above all else. (...) The image of the universe offered by the new physics leaves more room for

[3] For a clear and exhaustive discussion, see P. Busch, T. Heinonen & P. Lahti, "Heisenberg's Uncertainty Principle", *Phys. Rep.* 43 (2007), 155

[4] See in particular M. Ozawa, "Uncertainty Relations for Joint Measurements of Noncommuting Observables", *Phys. Lett. A* 320 (2004) 367, and "Universal Uncertainty Principle in Measurement Operator Formalism », *J. Opt. B* **7** (2005), S672.

> life and consciousness than did the old mechanistic image"[5].

Or, from a once illustrious author:

> "An atom is 'free' within the limits of Heisenberg's uncertainty principle (...). So when a message of extrasensory perception, in the form of mindons, psitrons or whatever you like, touches a neuron in unstable equilibrium, it operates at the level of quantum uncertainty and can, if I may say so, work miracles."[6]

More recently, in 2000, the newspaper *Le Monde* jokingly invoked the uncertainty principle, which would prevent "the speed and position of the President of the Republic (...) from being known simultaneously": Jacques Chirac ou la politique quantique". But politicians themselves do not shy at using the uncertainty principle at their own ends: Indian Home minister Rajnath Singh thus advanced in 2014 that the Heisenberg Uncertainty Principle was based on the Vedas[7].

Salvador Dali's burlesque provocation is particularly noteworthy:

> "And for me, the enraged paroxist of imperialist precisions, nothing in the world seems so sweet, pleasant and restful, and even graceful, as the transcendental irony implied by Heisenberg's uncertainty principle"[8].

Such wild extrapolations flourish today more than ever on the internet. Here are but two recent examples, the first one in commercial management:

> "One fascinating aspect of the uncertainty principle is its application beyond the physical sciences. In business or crisis management, it can be used as an analogy for understanding and managing uncertainty. For example, when a company is trying to forecast its profits, greater precision in estimating one factor (such as sales) can lead to greater uncertainty in other areas (such as costs)."[9]

And the second in design esthetics:

> "In the Uncertainty Principle, we find a parallel in the design principle of ambiguity. Ambiguity in design is the intentional introduction of elements or details that leave room for interpretation or multiple meanings. It's a deliberate departure from clarity and the expected, inviting viewers to engage more deeply with the piece."[10]

It is somewhat depressing to record all these recent blunders if one remembers the most salubrious warning issued by at least one cautious physicist, P. W. Bridgman, as early as 1929:

> "The immediate effect [of the uncertainty principle] will be to let loose a veritable intellectual spree of licentious and debauched thinking. (…) [This idea] will be made the basis of an orgy of rationalizing. It will be made the substance of the soul; the spirits of the dead will populate it; God will lurk in its shadows; the principle of vital processes will have its seat there; and it will be the medium of

---

[5] James Jeans, *The Mysterious Universe*, Cambridge Univ. Press (1930), 26.

[6] Arthur Koestler, *Impact of Science on Society* n° 24 (« The Parasciences »), Unesco, 1974, p. 281.

[7] https://timesofindia.indiatimes.com/india/Heisenbergs-theory-of-uncertainty-based-on-Vedas-Rajnath-Singh-says/articleshow/45163343.cms

[8] Salvador Dali, *Journal d'un génie*, Gallimard (1994), 59

[9] https://jfgouyet.fr/le-principe-dincertitude-de-heisenberg-comprehension-et-impacts/

[10] https://orrbitt.com/news/quantum-aesthetics-designing-uncertainty-principle/

telepathic communication. One group will find [in it] the solution to the age-long problem of the freedom of the will; and on the other hand, the atheist will find the justification of his contention that chances rule the universe."[11]

**Unfounded uncertainties**

It is by now due time to acknowledge that the "uncertainty principle" does not deserve this name at all: it is not a principle, and it does not deal with uncertainties! If we place ourselves firmly within the established framework of quantum theory - which Heisenberg obviously could not have done in 1927, since he was in the very process of developing this theory -, it becomes clear that the inescapable fuzziness about the numerical values of measurements of certain pairs of quantities is in no way the result of experimental uncertainties that would express a forced limitation of our knowledge. Rather, it must be understood as the consequence of an essential mutation in the very notion of physical quantity, consubstantial with quantum theory. In quantum theory, a property is generally characterized not by a well-defined single numerical value, but by a spectrum, i.e. a range of numerical values. Note that this is already the case for the notion of frequency in classical wave physics: a sound in general does not have a single "pitch", but a tone quality consisting of a whole spectrum of sound frequencies. The simplest description of this *timbre* is based on its width, i.e. the difference between the lowest and highest frequencies in the composition of this sound. In quantum theory, this situation applies to any physical property, which is thus characterized by a spectrum width or dispersion; this is not an "uncertainty" expressing the impossibility of knowing a defined quantity, but rather an intrinsic numerical extension (that depends on the physical state of the object under consideration). Furthermore, for certain pairs of properties, in a given state of the system considered, the widths of their numerical spectra are correlated and cannot be simultaneously reduced to zero, as their product is inferiorly limited. This is usually expressed by a formula such as

$$\delta A.\delta B \succcurlyeq \hbar, \qquad (1)$$

where $\delta A$ and $\delta B$ denote the numerical dispersions, suitably defined (see part II), of two such quantities $A$ and $B$, the symbol $\succcurlyeq$ meaning "greater than, up to a dimensionless factor of order unity", and where $\hbar$ is the fundamental quantum constant (reduced Planck constant). As we shall see later, various other expressions are possible, all leading to inequalities with the same general significance but various contexts of validity. By far the most frequent use of the general inequality (1) refers to the position $X$ and momentum $P$ respectively of a quanton, so that it becomes

$$\delta X.\delta P \succcurlyeq \hbar. \qquad (2)$$

The generalization to three dimensional situations is immediate, so that for most of this paper we will stick to one-dimensional expressions.

[11] P. W. Bridgman, "The New Vision of Science", *Harper's Magazine* 158 (1929) 443.

**A baseless principle**

To sum up the situation as briefly as possible, if we *cannot* know exactly the position or velocity of a particle governed by quantum theory, it is because it generally *does not have* a well-defined position and velocity: there simply do not exist any states for this particle in which each of these two quantities would take on a unique numerical value. This is the essence of what should be more soberly called "Heisenberg inequalities" - all the more so since this is not an independent "principle", but a consequence of the basic postulates of quantum theory as they are understood today. Indeed, these inequalities are rigorously demonstrated within the mathematical formalism of the theory. In the hilbertian formalism, the mere fact that the operators representing two physical properties do not commute prevents them from possessing the same set of proper states. This result is specifically quantum, thus invalidating any attempt to generalize it beyond its area of validity, namely theoretical physics . On a macroscopic scale, in classical physics, Heisenberg's inequalities are usually irrelevant, since they are generally masked by experimental inaccuracies far greater than the intrinsic quantum dispersions. As for proposals to import Heisenberg-style "uncertainties" in various fields of the social sciences, they are totally unfounded.

**Uncertain terminology**

It is worth emphasizing that in common parlance the term "uncertainty" — as "certainty" — refers to some psychological disposition and is of an essentially subjective nature. *We* are certain or uncertain about something. Thus, characterizing the quantum dispersion of some physical property as an "uncertainty" merely expresses the fact that *we* do not know the exact value of this property, while inducing us to suppose that such an exact value indeed does exist. Such is the meaning of the term when applied to the empirical measurements of physical quantities: "error bars" are used on graphs to indicate the reliability of a reported result. They give a general idea of how precise a measurement is, or conversely, of how far from the recorded value the true (error free) value might be. But, as we have stressed, in quantum theory, physical quantities of a system in general cannot be attributed such a single numerical value.

As it happens, the unfortunate terminology of "uncertainties" has a complex historical origin, linked to physicists' poor critical awareness of linguistic matters[12]. In his

[12] For an interesting exchange on the question, see:
W. D. Harris, « When did the Indeterminacy Principle become the Uncertainty Principle ? », *Am. J. Phys.* 65 (1997), 461.
D.C. Cassidy, « When did the Indeterminacy Principle become the Uncertainty Principle ? », *Am. J. Phys.* 66 (1998), 278.
F. Balibar & J.-M. Lévy-Leblond, « When did the Indeterminacy Principle become the Uncertainty Principle ? », *Am. J. Phys. 66* (1998), 279.

founding article in German, Heisenberg used the term "Ungenauigkeit" = "imprecision" (see the common German adverb "genau" = "exactly"), or "uncertainty", including its customary use for experimental measurements in physics. He and other contemporaries also used the less abstract term "Unsicherheit", which has the same meaning. But Heisenberg also put forward a much more appropriate term, "Unbestimmtheit", from "bestimmen" = "to determine", which has a stronger philosophical connotation (it may be found in Hegel), and can be translated as "indefiniteness", "indeterminacy" or even better as something like "indeterminitude". It is this word, "Unbestimmtheit", that came to the fore in German, where one speaks of "Heisenberg'schen Unbestimmtheitrelationen". In modern German, however, another term has become widespread: "Unschärferelationen", from "Unschärfe" = "vague, imprecise".

In French, "Unbestimmtheit" gave rise to a similar formulation in terms of "indétermination", common in the 1930s, for example with de Broglie and Langevin. Anglo-American language, on the other hand, settled on "uncertainty". In the context of the 1950s, with science now dominated by the Anglo-Saxon world, it was this term that came back into French in the form of "incertitude", and ended up imposing itself for a long time, which we can only regret. However, it seems that the terminology is tending to evolve, as shown by several recent textbooks and by the entry "Uncertainty principle" on Wikipedia: "In quantum mechanics, the uncertainty principle or, more correctly, the principle of indeterminacy, etc.".

The situation varies according to the languages considered:

— as far as Romance languages are concerned, Italian has stuck to the term "indeterminazione" since the 1930s, Spanish uses both "indeterminacion" and "incertidumbre", while Portuguese is content with "incerteza".

It is worth extending the sampling. Thus, we find:

— in Greek, "αβεβαιότητας" (*abebaiotètas*), meaning uncertainty, literally "that which cannot be attested".

— In Russian, "неопределенность" (*neopredelennosti*), derived from "определять" = "to define, to put an end to", thus inducing the interesting idea of an "indefiniteness" for the values of physical quantities.

— in Arabic, one finds several expressions among which عدم اليقين (*adam al yakin*), "lack of certainty", or عدم اليقين (*el-yaqin*), "uncertainty", formed on the root of the verb يقن (*yakana*), "understand".

— in Hebrew, האי-ודאות (*i-vadaout*), "uncertainty", although coming from ודא = "to verify", can be understood as something like "unverifiability" (see Greek above).

— in Japanese (*fukakuteiseig*), rather "indeterminacy".

G. Battimelli, « When did the Indeterminacy Principle become the Uncertainty Principle ? », *Am. J. Phys.* 66 (1998), 280.

**But positive content**

Behind the suggestive semantic distinctions shown by the diversity of these denominations, lies in fact a fundamental linguistic problem. If most of these terms belong to the negative (in-, не-, α-, عدم, אי-), it is because the proper quantum property concerned is specifically formulated by the negative, in opposition to its implicitly summoned classical counterpart. In classical mechanics, physical quantities - position, momentum, etc. - are in general determined and defined by unique numerical values. But it would be desirable to be able to explain and formulate quantum theory in its own right, without reference to the less valid classical theory it replaces. After all, the revolutionaries of the XVIIIth century chose to claim freedom, not non-servitude! Here, it would suffice to emphasize the characteristic feature of physical quantities in quantum theory, namely that their numerical values constitute a spectrum, of a certain extension. We could then speak of "Heisenberg's spectral inequalities", which would certainly lend itself to less questionable exegeses.

In order to dispel the somewhat ominous aura of Heisenberg's inequalities, let us insist that they are far from having a solely negative content. Insofar as they express, in a particularly simple and compact way, the specific character of quantum notions and their distinction from classical ones, these inequalities can be used effectively to develop a qualitative and heuristic approach to many problems in quantum physics, thus helping not to limit, but to extend our knowledge[13]. It could be said that Heisenberg's inequalities operate as a sort of free pass allowing us to cross the border between classical and quantum physics, without too much damage, at least for a short distance.

**Epistemological overhaul and linguistic reform**

Of course, this update of ideas concerning Heisenberg's so-called "Uncertainty Principle" is not intended to belittle the tremendous intellectual work accomplished by the founders of quantum theory in the 1920s. Misinterpretations and abuses of language are inevitable when a new theory is born, in science as elsewhere. How could we extemporaneously produce correct and valid expressions from ideas that turn out to be erroneous, and escape without a moment's hesitation the restraints of the old formulations that we are trying to overcome? The new can only be built from the old. And, as Bachelard has shown, every scientific revolution must be followed by a phase - sometimes a long one - of epistemological reform, enabling us to grasp the new concepts and express their intrinsic meaning. It remains to be discussed why modern physics, so brilliant in its revolutionary contributions, has been so ineffective in its recasting tasks.

---

[13] J.-M. Lévy-Leblond & F. Balibar, *Quantics : Rudiments of Quantum Physics*, North-Holland, 1990, Chapter 3.

I would venture to suggest that the dominating use of English in the modern physical sciences is one of the reasons for their epistemological backwardness. To the few examples given above of the variety of formulations used in different languages to state the conceptual meaning of Heisenberg's spectral inequalities, we could add many more concerning numerous problematic terms[14]. All of which goes on to show that maintaining and, better still, voluntarily developing as much multilingualism as possible in scientific activities would be a fruitful source of semantic richness, helping to deepen our understanding.

## II. Inequalities galore

The Heisenberg inequality, after its rather informal appearance in 1927, was at first the subject of a few papers establishing for it rigorous proofs and formulas, but it also gave rise to a vast flurry of conceptual discussions on its validity and meanings. It constituted, so to speak, a fixation abscess of the debate over the foundations of quantum theory which raged for more than half a century and still goes on today. It is less appreciated that in the last decades, the Heisenberg inequalities were revisited on the formal level, giving rise to many variations on their expressions and mathematical bases.[15]

Heisenberg, in his seminal 1927 paper, introduces his inequalities in order to develop an *anschauliche Inhalt* ("vivid contents") of the nascent quantum theory. The crucial point is that he does not propose a proof of the inequalities, would it be only because he does not give a precise definition of the "uncertainties" he refers to and only relies on a qualitative heuristic argument taken from an ad hoc description of a microscope experiment. To present day eyes, this argument is hardly convincing and may even be considered as plainly wrong[16]. Furthermore, there is but a vague reference to the fact that the "uncertainties" would arise from an "uncontrollable" disturbance in the measurement process, an idea which would become gospel talk in the following years and decades. It was most clearly disposed off by A. Peres:

> "The only correct interpretation [of the "uncertainty relations" for $x$ and $p$] is the following. If the same preparation procedure is repeated many times, and is followed either by a measurement of $x$ or by a measurement of $p$, the various results obtained for $x$ and $p$ have standard deviations $\Delta x$ and $\Delta p$, whose product cannot be less than $\frac{\hbar}{2}$. There never is any question here that a measurement of $x$ 'disturbs' the value of $p$ and vice-versa, as sometimes claimed. These measurements are indeed incompatible, but they are performed on different

[14] J.-M. Lévy-Leblond, "Quantum Words for a Quantum World", in D. Greenberger, W. L. Reiter, A. Zeilinger (eds), *Epistemological and Experimental Perspectives on Quantum Physics*, Vienna Circle Institute Yearbook, vol 7, Springer (1999).
[15] See a large collection in V. D. Dodonov, « Variance-Based Uncertainty Relations: A Concise Review of Inequalities Discovered Since 1927 », *Quantum Reports* 7 (2025), 34.
[16] R. F. Werner & T. Farrelly, "Uncertainty from Heisenberg to Today", *Found Phys* 49 (2019), 460.

particles (all of which were identically prepared) and therefore these measurements cannot disturb each other in any way."[17]

**Quadratic dispersions: Kennard, Robertson, etc.**

It was left to E. H. Kennard to give a mathematically reliable definition of the dispersions $\Delta x$ and $\Delta p$ as standard quadratic deviations and to give a correct, although cumbersome, mathematical proof of the inequality, in the framework of Schrödinger wave mechanics, relying on wavefunctions [18]. Note that Heisenberg was rather deprecatory about Kennard's work. Kennard's proof, dealing with wavefunctions, in fact is but a lemma in Fourier analysis. As such, its meaning, not to say its statement, was certainly known since the beginning of the XXth century, at least qualitatively, by many mathematicians interested in harmonic analysis (Wiener being doubtlessly one of them, as was Weyl) and even before, by physicists working in optics (such as Rayleigh). It is interesting to note that many strictly mathematical papers on the subject appeared only by the end of the XXth century. They have been thoroughly reviewed by G. B. Folland and A. Sitaram[19] as well as more recently by A. Wigderson and Y. Wigderson[20].
Kennard's proof was soon superseded by Robertson[21], who, working in a more general algebraic set-up, proposed the standard modern inequality. Consider any two physical properties $A$ and $B$ of a quantum system, represented by self-adjoint operators acting on the Hilbert space of states. Noting $[A, B]$ the commutator of $A$ and $B$, a simple adaptation of the Cauchy-Schwartz inequality yields:

$$\Delta A.\Delta B \geq \tfrac{1}{2} | <[A,B]> | . \tag{3}$$

It is worth noting that Weyl simultaneously obtained this inequality, although he does not cite Heisenberg (while thanking Pauli for a clue)[22]. It is the incompatibility of the physical properties, that is, the nonexistence of a common complete set of eigenstates for $A$ and $B$, which entails the existence of a lower bound for the product of their dispersions.
Coming back to the elementary Robertson's inequality (3) and considering the standard operators for position $X$ and momentum $P$ along with their canonical commutation relation

$$[X, P] = i\hbar I, \tag{4}$$

we obtain a rigorous statement of the usual Heisenberg inequality:

$$\Delta X . \Delta P \geq \frac{\hbar}{2}. \tag{5}$$

[17] A Peres, *Quantum Theora y : Concepts and Methods*, Kluwer, 1993.
[18] E. H. Kennard, "Zur Quantenmechanik einfacher Bewegungstypen", *Zeitschr. Physik* 44 (1927), 326.
[19] G. B. Folland & A. Sitaram, "The Uncertainty Principle: A Mathematical Survey", *J. Fourier Anal. Appli.* (1997), 207.
[20] A. Wigderson & Y. Wigderson, "The Uncertainty Principle: Variations on a Theme", *Bull. Am. Math. Soc.* 58 (2021), 225.
[21] H. P. Robertson, « The Uncertainty Principle », *Phys. Rev.* 34 (1929), 163.
[22] H. Weyl, *Gruppentheorie und Quantenmechanik*, Hirzel, 1928.

Schrödinger[23] also added his touch by generalizing Robertson's formula (3) to the following:

$$\Delta A.\Delta B \geq \tfrac{1}{2}\sqrt{(<\{A,B\}> -2 <A><B>)^2 + |<[A,B]>|^2}. \qquad (6)$$

where the anticommutator $\{A,B\} = AB + BA$ can be interpreted as expressing the correlation between $A$ and $B$[24].

Robertson's derivation of inequality (3) is susceptible of simple but fruitful generalizations[25]. The momentum $P$ as an element of the Lie algebra operates as the derivation operator on a function of the position $X$:

$$[F(X), P] = iF'(X) \qquad (7)$$

(where frow now on we choose units such that $\hbar = 1$).

This leads to the family of generalized inequalities:

$$\Delta F.\Delta P \geq \tfrac{1}{2}|<F'>| \qquad (8)$$

By an adequate choice of the function $F$, we may obtain interesting bounds on the local values of the wavefunction. Let us give but one example. Taking for $F$ the Heaviside step function

$$H_{x_0}(x) = \begin{cases} 0 & \text{if } x < x_0 \\ 1 & \text{if } x > x_0 \end{cases}, \qquad (9)$$

we leave it to the reader to prove that:

$$|\varphi(x)|^2 \leq 2\Delta P\{\mathcal{P}(x)[1-\mathcal{P}(x)]\}^{\frac{1}{2}} \leq \Delta P, \qquad (10)$$

where $\mathcal{P}(x)$ is the probability of localization on the half axis $[0,\infty[$.

The inequality (8) can be interpreted in an instructive general way. Indeed the quantity defined by

$$\Delta_F(X) := \frac{\Delta F}{|<F'>|} \qquad (11)$$

is a length associated with the space function $F$, giving the interval over which it changes by its dispersion $\Delta F$ through a displacement with its average slope $|<F'>|$. In other terms, it may be considered as a *characteristic variation length* of the property $F$. It obeys a generalized Heisenberg inequality:

$$\Delta_F(X).\Delta P \geq \tfrac{1}{2}. \qquad (12)$$

Similar arguments of course may be symmetrically developed concerning the momentum $P$.

**Quadratic dispersions and the triviality problem: Maccone and Pati**

[23] E. Schrödinger, "Zum Heisenbergschen Unschärfeprinzip", *Sitz. Preuss. Akad. der Wissenschaften, Physik.-math. Klasse,* 14 (1930), 296.

[24] J.-M. Lévy-Leblond, "Correlations of Quantum Properties and the Generalized Heisenberg Inequality", *Am. J. Phys.* 54 (1986), 135.

[25] W. G. Faris, "Inequalities and Uncertainty Principles", *J. Math. Phys.* 19 (1978), 461.
J. F. Price, "Position versus Momentum", *Phys.Lett.* 105A (1984), 343.
J.-M. Lévy-Leblond, "Local Heisenberg Inequalities", *Phys. Lett.* 111 (1985), 353.

We are so accustomed to using the Heisenberg-Robertson inequality in its canonical form (5) that we do not realize that in the general case (3) it does not fully capture the concept of incompatible properties, because it can be trivially true. That is, the right side of the inequality may be null even for two non-compatible properties $A$ and $B$, if the state is an eigenstate of the commutator. In such a case, the Heisenberg-Roberston inequality yields no information about their incompatibility. As an elementary example, consider the case of a spin ½ and take:

$$A = S_x \qquad B = S_y \qquad |\varphi> = \frac{1}{\sqrt{2}} \begin{pmatrix} 1 \\ e^{i\alpha} \end{pmatrix} \tag{13}$$

One computes easily:

$$\Delta A = \tfrac{1}{2}|\sin\alpha| \qquad \Delta B = \tfrac{1}{2}|\cos\alpha| \qquad |<[A,B]>| = 0 \tag{14}$$

The right side of inequality (3) thus is null, despite the non-compatibility of $A$ and $B$, thus depriving the inequality of any relevance. The Schrödinger improvement does not help.

Almost a century after Heisenberg's work, Maccone and Pati proved two inequalities on the *sum* of the variances [26]:

$$\Delta A^2 + \Delta B^2 \geq |< [A, B] >| + |\langle\varphi|A \pm iB|\varphi^{\perp}\rangle|^2 \tag{15}$$

$$\Delta A^2 + \Delta B^2 \geq \tfrac{1}{2}|\langle\varphi|A + B|\varphi^{\perp}\rangle|^2 \tag{16}$$

where $\varphi^{\perp}$ is an arbitrary state orthogonal to $\varphi$. For the example given above, the right sides of (15) and (16) now saturate the inequalities.

These inequalities have been investigated experimentally and proved indeed to be stronger than the usual ones (5) [27].

**Slim dispersions: Hilgevoord and Uffink widths**

It is surprising that it took such a long time to realize — at the level of common physicists' knowledge — that the evaluation of dispersions through mean square root deviations was not the only possible, nor often the best one. Indeed, consider a wavefunction with two peaks of small width $a$, separated by an interval of length $l \gg a$, the value of the function being quite small outside the peaks. It is easy to check that the quadratic dispersion is of the order of the separation $l$, while the intrinsic width of both peaks is of order $a$. Of course, the last is a much better estimate of the proper fuzziness of the position value because it takes into consideration the fine structure of the wavefunction while the quadratic dispersion only estimates its overall span. There is even worse, as some simple wavefunctions, currently used in exercises and examples, do not have finite values for their quadratic dispersions. Consider a wavefunction the graph of which in cartesian coordinates is a rectangle. Its Fourier

[26] L. Maccone & A. K. Pati, "Stronger Uncertainty Relations for all Incompatible Observables", *Phys. Rev. Lett.* 113 (2014,) 261401.

[27] K. Wang, X. Zhan, Z. Bian, J. Li, Y. Zhang, P. Xue, "Experimental Investigation of the Stronger Uncertainty Relations for all Incompatible Observables", *Phys. Rev.* A 93 (2016), 052108.
W. Ma & al. "Experimental Test of Heisenberg's Measurement Uncertainty Relation Based on Statistical Distances", *Phys. Rev. Lett.* 116 (2016), 160405.

transform is a sinc function, the second moment of which diverges, leading to an infinite quadratic dispersion in momentum. This simple model for the diffraction of a quanton by a slot thus cannot be understood by relying on the standard Heisenberg inequality.

More than half a century after Kennard and Robertson, Uffink and Hilgevoord[28] pointed out that in fact we need a more refined characterization of the extension of a wavefunction, distinguishing its global width from its finer peak structure. For a given normalized wavefunction $\varphi$, they define its "bulk width" at level $\alpha$ ($0 < \alpha < 1$) by the expression:

$$W_\alpha(X) = \min\left\{|b-a| \,\Big|\, \int_a^b |\varphi(x)|^2 dx = \alpha\right\}, \tag{17}$$

that is, the smallest spatial extent of the support which captures the proportion $\alpha$ of the localization probability. They go on to define the "mean peak width" at level $\beta$ ($0 < \beta < 1$) by:

$$w_\beta(X) = \min\{c \,|\, \int \varphi(x)\overline{\varphi(x+c)}dx = \beta\}, \tag{18}$$

that is, the minimum shift which respects a covering proportion $\beta$ between the displaced function and the original; this quantity is manifestly sensitive to the detailed peak structure of the function.

Homologous expressions can be given to the dispersions in momentum by using the Fourier transform $\hat{\varphi}$ of the wavefunction.

These dispersions are finite for any wavefunction. They can be related to the quadratic dispersions by simple inequalities:

$$W_\alpha(X) \leq 2(1-\alpha)^{-\frac{1}{2}}\,\Delta X \tag{19}$$

$$w_\beta(X) \geq (1-\beta^2)^{\frac{1}{2}}\,(\Delta P)^{-1}, \tag{20}$$

showing that they definitely give a more detailed account of the wavefunction structure. Finally, these Uffink and Hilgevoord dispersions themselves do obey Heisenberg's type inequalities:

$$W_\alpha(X) w_\beta(P) \leq 2\left[\frac{2(2\alpha-\beta+1)}{\alpha}\right]^{\frac{1}{2}} \hbar. \tag{21}$$

This plainly justifies the heuristic use of Heisenberg's inequalities in cases where the quadratic dispersions do not.

**The time-energy inequality**

In many textbooks and popularization papers, a time-energy Heisenberg inequality is asserted under the heuristic form:

$$\delta E . \delta t \gtrsim \hbar \tag{22}$$

by sheer analogy with the space-momentum case.

[28] J. B. M. Uffink & J. Hilgevoord, "Overall Width, Mean Peak Width and the Uncertainty Principle", *Phys. Lett.* 95A (1983) 474 ; "New Bounds for the Uncertainty Principle", *Phys. Lett.* 105A (1984), 276 ; "Uncertainty Principle and Uncertainty Relations", *Ann. Phys.* 15 (1985), 925.

However, the meaning of the time dispersion $\delta t$ is far from clear and no immediate rigorous definition of it is usually given. Indeed, in quantum theory, time is not a property of the system, unlike its position or energy. It is not represented by an operator which would constitute such a property. It is an external parameter the change of which marks the evolution of the system. A vast body of literature has been devoted for a full century to the problem of stating a relevant time-energy inequality. Still, a solution was set up by L. I. Mandelstam and I. E. Tamm back in 1945[29], that is, eighty years ago — although it is poorly known. The idea is to consider the time evolution of a given property of the system, say $G$. As is well-known, the Hamiltonian $H$ generates its time evolution, so that one may write

$$[G, H] = i\dot{G}, \qquad (23)$$

where the quantity $\dot{G}$ represents the time rate of evolution of the property $G$, in close analogy with the expression (8) for the spatial dependence of $F$. A Heisenberg-Robertson inequality readily follows in the form:

$$\Delta_G(T).\Delta H \geq \tfrac{1}{2}. \qquad (24)$$

where we have defined

$$\Delta_G(T) := \frac{\Delta G}{|<\dot{G}>|}. \qquad (25)$$

This quantity expresses the order of magnitude of the time necessary for the quantity $G$ to change by an amount equal to its dispersion $\Delta G$ at its average rate of evolution $|< \dot{G} >|$. It is thus appropriate to consider it as a *characteristic time of evolution* of $G$. As shown by Mandelstam and Tamm, this inequality applies to most specific problems where a heuristic inequality such as (22) is applied, such as the relationship between the lifetime of an unstable particle and its energy width. It must yet be emphasized that a precise interpretation of (24) nevertheless requires some caution[30]. The question also has been the object of more sophisticated approaches[31].

**Angle-angular momentum and phase-number inequalities**

In classical physics, angle and angular momentum form a standard pair of physical properties obeying a canonical Poisson bracket. But it was pointed out very early that an operator representing angular position and entertaining a canonical commutation relation with the angular momentum does not exist within quantum theory[32]. As a

---

[29] L. I. Mandelstam and I. E. Tamm, "The Uncertainty Relation between Energy and Time in Nonrelativistic Quantum Mechanics" [English version], *J. Phys. (USSR)* 9 (1945), 249.

[30] D. D. Georgiev, "Time-Energy Uncertainty Relation in Nonrelativistic Quantum Mechanics", *Symmetry* 16 (2024), 100.

[31] P. Pfeiffer & J. Fröhlich, "Generalized Time-Energy Uncertainty Relations and Bounds on Lifetimes of Resonances", *Rev. Mod. Phys.* 67 (1995), 759.

[32] P. A. M. Dirac, "The Quantum Theory of the Emission and Absorption of Radiation", *Proc. R. Soc. London*, Ser. A 114 (1927), 243. See the review by P. Carruthers & M. M. Nieto, "Phase and Angle Variables in Quantum Mechanics", *Rev. Mod. Phys.* 40 (1968) 11, and M. Nieto, "Quantum Phase and Quantum Angle Operators: some Physics and some History", *Physica Scripta* T48 (1993) 5.


To appear in a forthcoming collective volume on the
100th birthday of Quantum Mechanics, T. Choy ed., World Scientific, 2026

result, the quest for a Heisenberg type inequality between angular position and angular momentum has not received a generally accepted solution — even though such an inequality is readily admitted in heuristic discussions. The same situation prevails in quantum field theory concerning the phase and number quantities. Since many works continually appear on the subject[33], I will refrain here from reviewing the state of the art, while mentioning that a possible issue consists in enlarging the mathematical framework of quantum theory by admitting non-hermitian operators as bona fide mathematical representations of physical quantities[34].

**Information-theoretic inequality**

For a probability density $\rho(x)$, its entropy, which is an information-theoretic measure of its distribution, has been proposed by Shannon in his seminal work[35]:

$$I(\rho) = -\int \rho(x) \log \rho(x)\, dx. \tag{26}$$

Shannon showed that it is logarithmically bounded by the quadratic dispersion:

$$I(X) \leq \log\left[(2\pi e)^{\frac{1}{2}}\Delta(X)\right]. \tag{27}$$

A quite esthetic entropic Heisenberg's type inequality can then be proved (through subtle analytic arguments)[36]:

$$I(X) + I(P) \geq \log\left({}^{e}/_{2}\right), \tag{28}$$

which may be shown to be sharper than the Heisenberg-Robertson inequality.

**The Heisenberg-Pauli inequalities**

Consider a system of one-dimensional particles, classical ones to start with. The state of the system can be represented at a given time, by a collection of points in the two-dimensional one-particle phase-space $(x, p)$. Contemplate now a system of quantons (quantum 'particles'). Denoting by $\delta X$ and $\delta P$ respectively the extensions of an individual cell in position and momentum respectively (in order of magnitude), the minimal area of the cell constitutes but a rewriting (and an interpretation) of the Heisenberg inequality (2). The collective state then consists of an ensemble of such cells. Let us suppose now that the quantons are identical fermions. The effect of Fermi statistics may be expressed by the Pauli exclusion principle which requires that no two

[33] For instance :
J.-P. Gazeau & F. H. Szafraniec, "Three Paths toward the Quantum Angle Operator", *Ann. Phys.* 375 (2016) 16
P. Busch, J. Kiukas, R.F. Werner, "Sharp Uncertainty Relations for Number and Angle", *arXiv:*1604.00566v2 [quant-ph] (2018).
[34] J.-M. Lévy-Leblond, "Who is afraid of Nonhermitian Operators? A Quantum Description of Angle and Phase", *Ann. Phys.* 101 (1976), 319.
[35] C. Shannon, "A Mathematical Theory of Communication", *Bell System Tech. J.* 27 (1948) 379 and (1949), 623.
[36] W. Becker, "Inequalities in Fourier Analysis", *Ann. Math.*102 (1975), 159.
I. Bialynicki-Birula & J. Mycielski, "Uncertainty Relations for Information Entropy in Wave Mechanics", *Commun. Math. Phys.* 44 (1975), 129.

individual states be identical. It is straightforward to translate this constraint as requiring the cells now to be disjoint. One sees vividly how the Pauli principle requires the collective state of the system to have a much greater minimal extension in phase-space. Indeed, for a system with quantons *not* obeying the Pauli principle, the inequality (2) for the individual regions in phase-space applies as well to the total region, since nothing prevents the individual states to be one and the same. For identical fermions however, the total region, consisting of non-overlapping cells, must have a minimal area at least equal to $N$ times that of an individual cell, leading to an inequality characterizing the collective extensions in position and momentum of a system with $N$ identical fermions in one dimension:

$$\delta X.\delta P \succ N\hbar \qquad \text{(1-D)} \qquad (29)$$

I call it the 'Heisenberg-Pauli inequality'. It can easily be generalized to three (or more) dimensions. The individual cells now have a minimal volume $\hbar^3$ and the collective state with $N$ identical fermions must occupy a region of volume at least $N\hbar^3$. Hence the Heisenberg-Pauli inequality in three dimensions:

$$\delta R.\delta P \succ N^{\frac{1}{3}}\hbar \; \text{(3-D)} \qquad (30)$$

It should be emphasized that this heuristic derivation may be given a fully rigorous form, and that an inequality similar to (30) can be proved formally[37]. The best way to interpret the Heisenberg-Pauli inequalities (29) and (30), is to compare them to the standard Heisenberg inequality (2), and to consider that the effect of the Pauli principle is to replace the fundamental quantum constant $\hbar$ by an effective fermionic constant, $\hbar_{ferm}^{(D)}=N^{\frac{1}{D}}\,\hbar$, where $D$ is the dimensionality of space (which shows, by the way, that the effect of the exclusion principle is all the more powerful, the lower is the dimensionality of space).

The inequality (30) may be fruitfully applied in heuristic studies of many-fermion systems, such as atoms with a large number $Z$ of electrons, for which it explains the $Z^{7/3}$ dependency of the ground-state energy. More than that, it underlies the saturation of Coulomb forces in bulk matter and thus guarantees its stability[38].

**An inverse Heisenberg inequality**

It is common practice in elementary quantum theory, when trying to evaluate the ground state energy of a quanton in a simple potential well, to use the heuristic Heisenberg inequality in the form of an approximate equality:

$$\delta A.\delta B \approx \hbar. \qquad (31)$$

[37] J.-M. Lévy-Leblond, "Generalized Uncertainty Relations for Many-Fermion Systems", *Phys. Lett.* 26A (1968), 540.

[38] J.-M. Lévy-Leblond, "Quantum Phenomena at Large", in coll., *Advances in Quantum Phenomena*, E. Beltrametti & J.-M. Lévy-Leblond eds, PlenumPress (1995), 281.

Up to a numerical constant of order unity, this gives in general correct order-of-magnitude estimates. It suggests that, for the situation considered, there might exist a rigorous inverse Heisenberg inequality:

$$\Delta X . \Delta P \leq K\hbar \qquad (32)$$

with a numerical coefficient $K$. Some familiarity with the Schrödinger equation leads one to believe that this inequality is saturated with $K = \frac{1}{\sqrt{2}}$; this holds in the case of a Dirac delta potential, for which the wave function is a decreasing exponential (which can be seen as a complementary situation to that of the harmonic oscillator). Indeed, the inverse Heisenberg inequality (32) has been proved [39], not without some mathematical ingenuity, under certain technical conditions on the potential $V$, such as $V = -\lambda\delta + U, \lambda \geq 0, U$ being even, locally integrable, with $U, U''$ increasing and $U'$ decreasing on $(0, \infty)$, or $V$ convex and even, with $\sup V'' \leq 2 \inf V''$. A simpler and more general proof would certainly be desirable as well as an extension to the excited states.

## III. Some sobering conclusions

What are the Heisenberg inequalities good for? One cannot avoid recognizing that they have not, as such, led to fundamental new advances, either theoretical or experimental. True, measurements have verified the validity of some of their formal expressions. But these works constitute a rather small proportion of the vast and increasing body of literature on the subject. As we have seen in the second part of this essay, the quest still goes on for ever more diverse and sophisticated new formal expressions of the general idea. Meanwhile, it is somewhat disappointing to realize that the interpretation of the Heisenberg inequalities is still largely marred by the endless repetition of unfounded terminology and interpretations. Despite the admonitions exposed in the first part of this essay and other recent critical assessments[40], the (unjustified) "Principle" and (unfounded) "uncertainties", as well as references to "measurement disturbances", are still found in many specialized papers as well as in popularization books and articles.

It is probably too much to hope for a collective effort of physicists in reconsidering their terminology. Indeed, as aptly put by Englert, "it should be clear that the subject matter of uncertainty relations is more important for the teaching of quantum mechanics than for research at the frontier of knowledge". Let me yet remark that scientific activity does not proceed only at this frontier, and that we all need to approach our theoretical knowledge with care, relying on an understanding as sound and mastered as possible of its foundations. From that perspective, one will never assert too strongly the need to supplement and comfort our formal knowledge (the equations, so to speak) with a conceptual (that is verbal) grasp of its meanings. The

[39] J.-P. Kahane, J.-M. Lévy-Leblond, J. Sjöstrand, « Une inégalité de Heisenberg inverse", *J. Anal. Math.* 60 (1993) 135.

[40] B.-G. Englert, "Uncertainty Relations Revisited", *Phys. Lett.* A 494 (2024), 129278

Heisenberg inequalities thus find their main genuine relevance and indisputable importance at the heuristic and qualitative level (see an example in the Appendix). May we then hope that the present considerations will be useful for the students and lecturers in physics, the philosophers and historians of science — and the cultured public, if it still exists.

**Appendix. A heuristic example.**
In order to dispel the unfounded impression this essay might leave on the reader that I shun Heisenberg inequalities and their heuristic use, let me give here an example which I hope will be of some didactical interest. The quantum theory of angular momentum shows that the proper values of the scalar square $J^2$ are of the form $j(j+1)\hbar^2$ where $j\hbar$ is the highest proper value of a component, say $J_z$. A classical reasoning dealing with ordinary vectors would have led to the simpler expression $j^2\hbar^2$. So where does the extraterm $j\hbar^2$ come from? Start from the definition:

$$J^2 = J_x^2 + J_y^2 + J_z^2. \qquad \text{(A1)}$$

Hence, averaging in the common proper state of $J^2$ and $J_z$ with its maximal value $j$:

$$< J^2 >_j = < J_x^2 >_j + < J_y^2 >_j + < J_z^2 >_j. \qquad \text{(A2)}$$

That is, for the proper value of $J^2$:

$$(J^2)_j = < J_x^2 >_j + < J_y^2 >_j + j^2\hbar^2. \qquad \text{(A3)}$$

It is easy to show that $< J_x >_j = < J_y >_j = 0$, so that $< J_x^2 >_j = \Delta_j J_x^2$ and $< J_y^2 >_j = \Delta_j J_y^2$. Thus

$$(J^2)_j = \Delta_j J_x^2 + \Delta_j J_y^2 + j^2\hbar^2 \geq 2\Delta_j J_x \, \Delta_j J_y + j^2\hbar^2. \qquad \text{(A4)}$$

And now comes the crux of the argument. The commutation relation:

$$[J_x, J_y] = i\hbar J_z \qquad \text{(A5)}$$

leads through the Robertson's inequality (3) to

$$\Delta_j J_x \, \Delta_j J_y \geq \tfrac{1}{2}\hbar < |J_z| > = \tfrac{1}{2} j\hbar, \qquad \text{(A6)}$$

So that, finally, from (A4):

$$(J^2)_j \geq j\hbar^2 + j^2\hbar^2. \qquad \text{(A7)}$$

Of course, to prove that this inequality is saturated and has to be replaced by an equality, does require the full algebraic treatment of angular momentum in quantum theory. Nevertheless, the present argument shows quite clearly how the extra term comes from the non-commutativity of the components of the angular momentum

[Many thanks to Y. Rabkin, T. Tokieda, and to A. Abbes, F. Charfi, S. Horchani, N. Fatès, M. Hamdi, for their linguistic help, as well as to O. Darrigol, J.-P. Gazeau, A. Grinbaum, R. Nardone, A. Saint-Ours, F. Varenne for various exchanges on a first version of this essay, and to T. Choy, M. Davidson, H. Leschke, M. Ozawa for their criticisms and suggestions, to whom I must add the anonymous but keen proofreader.]

# Chapter 10

## Solvay 1927: Quantum at the crossroads

D. H. Delphenich
Spring Valley, OH USA 45370
feedback@neo-classical-physics.info

**Abstract:** Two main currents of progress in quantum theory had been established as of the 1927 Solvay conference on electrons and photons: the wave mechanics of de Broglie, Schrödinger, and others, and the operator mechanics of Born, Heisenberg, and Jordan. It was the latter approach that seemed to dominate the thinking of Heisenberg and Pauli's attempt at quantum electrodynamics, but in the long run it became clear that the problems with that approach were at the fundamental level and were never fully resolved.

## 1 Introduction

When the Solvay conference on electrons and photons convened in 1927 [1], as its title would suggest, the main issues involved devising an adequate theory that would account for the experimental results that had been obtained regarding the interactions of electrons and photons, primarily in the context of atomic structure. If the list of elementary particles under discussion seems quite incomplete by modern standards, keep in mind that although the proton had been discovered by Rutherford in 1917, the neutron would not be discovered by Chadwick until 1932, and all of the other elementary particles would come later than that.

The attendees represented the leading theoretical physicists of that era, including Hendrik Antoon Lorentz, Marie Curie, Niels Bohr, William Bragg, and a host of others. The presentations were made by Bragg, Arthur Compton, Louie de Broglie, Max Born, Werner Heisenberg, and Erwin Schrödinger, and were followed by spirited discussions between the presenters and attendees.

If one would like to examine the most-fundamental empirical roots of the basic problem confronting anyone who wishes to formulate a mathematical theory of quantum phenomena (i.e., the ones that pertain to atomic and sub-atomic processes) then one would be well-advised to consult the presentation by Arthur Compton. His basic thesis was that what quantum theory was confronting most fundamentally was the breakdown of the Maxwell-Hertz theory of electromagnetic radiation, which could not properly account for the experimental results that involved the interaction of photons (mostly X-rays in his talk) and electrons with atoms and isolated electrons. Indeed, one notices that empirically those phenomena eventually became the bulk of the successes of the quantum electrodynamics that came later. In particular, Compton discussed the experimental work that was concerned with the photoelectric effect and Compton scattering.

It is worth repeating the issues that Compton raised in his presentation:

---

[1] The title of this presentation is a paraphrase of the title of the book [1] by Guido Bacciagluppi and Antony Valentini with the title *Quantum Theory at the Crossroads; Reconsidering the 1927 Solvay Conference*, which includes English translations of the various talks that were presented, along with their own commentary on the nature of the problem of quantum theory. However, the similarity between the present discussion and their own largely ends there.

1. Is there an ether?

2. How are waves produced?

3. The photoelectric effect.

4. The scattering of X-rays and their recoil electrons.

5. Experiments concerned with interactions between isolated quanta of radiation and electrons.

The earliest work in the name of quantum physics was due to Planck's theory of why the experimental facts regarding blackbody radiation were at odds with the reigning theories of the late nineteenth century. One sees that, once again, one is dealing with the breakdown of the classical theory of electromagnetic radiation. Similarly, the mounting experimental data concerned with atomic spectroscopy also suggested that the classical theory of radiation could not account for the nature of the photons that were emitted by atomic electrons and that a better model for the structure of the energy levels of atomic electrons was needed. In particular, some fatal flaws in the Bohr planetary model were due to the fact that orbiting electrons, as accelerated charges, should emit radiation, and thus gradually lose energy until they merged with the atomic nucleus and the fact that the energy levels of atomic electrons had an inexplicable discreteness. In both cases, one was still looking at a breakdown of the classical theory of radiation.
However, a promising first step in the direction of a quantum theory of electrons, photons, and atomic structure was the 1924 thesis [2] by Louis de Broglie [2] on the wave-like nature of matter, which complemented the advances of Planck and Einstein regarding the association of kinetic energy and linear momentum with the photon (a term that was coined by the chemist G. N. Lewis [3]) by proposing that matter (i.e., electrons, at that time) would also have a wave-like character, and the association of energy $E$ and momentum $\mathbf{p}$ with the (radial) frequency $\omega$ and wave number $\mathbf{k}$ of the matter wave would be the same as for photons:

$$E = \hbar\omega, \qquad \mathbf{p} = \hbar\mathbf{k}\,. \tag{1}$$

The experiments of Davisson and Germer between 1923 and 1927 regarding electron diffraction seemed to confirm the validity of the matter wave conception of electrons quite dramatically. Indeed, one finds that electron diffraction patterns are qualitatively identical to corresponding X-ray diffraction patterns (see the figures in the book by Max von Laue [4]).

## 2 Two main schools of quantum mechanics

[2] There is an enduring legend regarding de Broglie's doctoral dissertation that needs to be debunked. The belief that it was somehow inadequate in its content (one version of the story says it was only four pages long) is probably Nazi propaganda left over from between the wars and is simple enough to refute nowadays since the dissertation is freely available on the internet in both its original French version and in English translation.

The Solvay conference of 1927 primarily defined the fundamental conflict between two approaches to the formulation of a theory of quantum phenomena, namely, matrix mechanics and wave mechanics. Thus, we shall briefly summarize the main points of those two approaches.

## 2.1 Matrix mechanics

In 1925, Werner Heisenberg published a paper on the quantum-theoretic reinterpretation of kinematical and mechanical relationships [5], which was followed by a paper by Max Born and Pascual Jordan [6], and then a paper by all three of them [7]. Together, they created a theoretical framework for quantum phenomena (but mostly atomic spectra) that came to be called *matrix mechanics*, although nowadays the matrices in question would appear as a consequence of evaluating a Hermitian operator $T$ on a complex Hilbert space of states between two states that belong to a basis for it, such as $| i >$ and $| j >$, to use the later Dirac notation, i.e., the matrix $T_{ij}$ would be obtained from $T$ in the form of:

(2) $$T_{ij} = < i \,|\, T \,|\, j > .$$

Hence, what Born, Heisenberg, and Jordan were calling matrix mechanics is now called *operator mechanics*. Of course, since the Hilbert space of quantum states is typically infinite-dimensional, so any basis for it would have an infinitude of elements (and not necessarily a denumerable infinitude), the matrix $T_{ij}$ would necessarily have an infinitude of rows and columns.
The physical interpretation that was given to the Hermitian operator was that it was the quantum representation of a physical "observable," such as linear or angular momentum, and as such, it represented a measurement of the observable. The eigenvalues $\lambda_i$ of the operator $T$ would then represent the result of measuring the observable when the quantum system is in an eigenstate $| i >$:

(3) $$T \,|\, i > = \lambda_i \,|\, i > .$$

In order to obtain a dynamical equation for the time evolution of a quantum operator $T$, one starts by representing the canonical observables of classical (point) mechanics, namely, the position coordinates $x^i$ and momentum components $p_i$ as Hermitian operators on a complex Hilbert space of functions of space and time using *canonical quantization:*

(4) $$x^i \to X^i = \text{scalar multiplication by the function } x^i , \qquad p_i \to P_i = \frac{\hbar}{i}\frac{\partial}{\partial x^i} .$$

Any other classical observable $f = f(x^i, p_j)$ will be represented by an "analogous" function of the quantum observables $F = F(X^i, P_j)$, although that is defined unambiguously only in the case where the function $f$ is a polynomial in the canonical coordinates. However, the linear space of functions on the classical phase space is a *commutative* algebra under point-wise multiplication,

whereas the linear space of Hermitian operators on the quantum phase space is *non-commutative* under composition of operators. However, the classical Lie algebra of functions on phase space is not the commutator bracket of the functions, but the Poisson bracket.
The Poisson bracket of two classical observables $f$, $g$ :

(5) $$\{f, g\} = \frac{\partial f}{\partial x^i}\frac{\partial g}{\partial p_i} - \frac{\partial f}{\partial p_i}\frac{\partial g}{\partial x^i}$$

then gets represented as the Lie bracket of the operators $F$, $G$ that represent the functions $f$, $g$:

(6) $$[F, G] = FG - GF\,.$$

In particular, the classical canonical commutation relations between the canonical observables $x^i$ and $p_i$ :

(7) $$\{x^i, x^j\} = 0\,, \qquad \{p_i, p_j\} = 0\,, \quad \{x^i, p_j\} = \delta^i_j$$

will become the quantum canonical commutation relations:

(8) $$\{X^i, X^j\} = 0\,, \qquad \{P_i, P_j\} = 0\,, \; \{X^i, P_j\} = i\hbar\,\delta^i_j\,.$$

Perhaps the most-fundamental Hermitian operator that one defines is the one that corresponds to the classical Hamiltonian:

(9) $$H(x^i, p_j) = T + U = \frac{1}{2m}\delta^{ij}\,p_i\,p_j + U(t, x)\,.$$

Upon canonical quantization, that will become the Hermitian operator:

(10) $$\mathsf{H}(X^i, P_j) = \frac{1}{2m}\delta^{ij}\left(\frac{\hbar}{i}\frac{\partial}{\partial x^i}\right)\left(\frac{\hbar}{i}\frac{\partial}{\partial x^j}\right) + U(t, x) = -\frac{\hbar^2}{2m}\Delta + U(t, x)\,.$$

The classical equation of motion for a classical observable $f = f(t, x^i, p_j)$ :

(11) $$\frac{df}{dt} = \frac{\partial f}{\partial t} + \frac{dx^i}{dt}\frac{\partial f}{\partial x^i} + \frac{dp_i}{dt}\frac{\partial f}{\partial p_i} = \frac{\partial f}{\partial t} + \frac{\partial H}{\partial p_i}\frac{\partial f}{\partial x^i} - \frac{\partial H}{\partial x^i}\frac{\partial f}{\partial p_i} = \frac{\partial f}{\partial t} + \{f, H\}$$

will then become the quantum equation of motion for the corresponding quantum observable:

(12) $$\frac{dF}{dt}=\frac{\partial F}{\partial t}+\frac{i}{\hbar}[\mathsf{H},F]\,.$$

Hence, a Hermitian operator that is not explicitly dependent on time will be constant in the motion of the quantum state if and only if it commutes with the Hamiltonian operator.

Because Heisenberg had been an assistant to Niels Bohr at his institute in Copenhagen at the time when Heisenberg began formulating matrix mechanics, it came to be referred to as the "Copenhagen interpretation" of quantum theory, and especially when combined with the statistical interpretation of quantum mechanics that was introduced by Max Born [8] in 1926 and served as the basis for his 1954 Nobel Prize. Although the Copenhagen interpretation certainly proved its utility in accounting for the energy levels of atomic electrons, which were assumed to be the basis for the structure of all atomic spectra, one has to accept that it is essentially a refinement of the theory of physical measurements that would be applicable to the measurements that one makes of quantum phenomena. One might argue that a theory of physical measurements does not have the universality in its scope to explain all physical phenomena. For instance, astronomical observations are basically passive measurements, in the sense that observing the state of a distant star cannot affect its state since the light that one observes might already be billions of years old.

## 2.2 Wave mechanics

Although De Broglie's theory of matter waves accounted for some aspects of quantum phenomena, such that the idea that the discreteness in the energy levels of atomic electrons was consistent with the nature of standing waves for the matter waves that represented the electrons, it still lacked a fundamental dynamical principle for the propagation of matter waves. That advance came in 1926 with a series of papers by Erwin Schrödinger [9].

The titles of the papers suggested that the basis for his approach to obtaining a wave equation for matter waves was rooted in the eigenvalue problem. That would certainly be consistent with the nature of standing waves, which are typically regarded as eigenfunction solutions of the wave equation in question. Of course, at that point in time, the concept of an eigenvalue problem was most intimately linked with the Sturm-Liouville equation:

(13) $$\frac{d}{dx}\left[p(x)\frac{dy}{dx}\right]+q(x)\,y=-\,\lambda\,w\,(x)\,y\,.$$

The eigenvalue problem that was associated with that equation involved finding the special values of $\lambda$ (the eigenvalue) for which solutions to that equation would exist.

However, another key idea that was driving his analysis was the *optical-mechanical* analogy. It came down to his suspicion that matter waves related to point-like matter in the same way that waves in general relate to individual rays, such as light rays. In other words, wave mechanics was to classical (point) mechanics as wave optics was to geometrical optics. As such, it drew rather crucially upon the Hamilton-Jacobi equation for the *principal function* $S(t,x^i)$, which is:

(14) $$\frac{\partial S}{\partial t} + H\left(t, x^i, \frac{\partial S}{\partial x^i}\right) = 0$$

in its time-varying form, and:

(15) $$H\left(x^i, \frac{\partial S}{\partial x^i}\right) = E$$

when $S(t, x^i)$ takes the form $E\,t - S(x^i)$.
Schrödinger obtained his time-independent (or stationary) equation for the complex-valued matter wave function $\psi(x^i)$:

(16) $$\frac{\hbar^2}{2m}\Delta\psi + [E - V(x^i)]\psi = 0 \qquad \text{or} \qquad \Delta\psi + \frac{2m}{\hbar^2}[E - V(x^i)]\psi = 0$$

before he obtained the time-dependent form:

(17) $$i\hbar\frac{\partial \Psi}{\partial t} = -\frac{\hbar^2}{2m}\Delta\Psi + V(t, x^i)\Psi$$

for the complex-valued space-time wave function $\Psi(t, x^i)$.
Note that one can represent equation (17) in operator form by introducing the canonical quantization of the total energy function $E = E(t, x^i, p_j)$ as the operator:

(18) $$\mathsf{E} = -\frac{\hbar}{i}\frac{\partial}{\partial t} = i\hbar\frac{\partial}{\partial t}.$$

In conjunction with the Hamiltonian operator $\mathsf{H}$ that was defined above in equation (10), the time-dependent Schrödinger equation will take the form:

(19) $$\mathsf{E}\,\Psi = \mathsf{H}\,\Psi.$$

As we will see shortly, when we go to the time-independent Schrödinger equation, that equation will become precisely an eigenvalue equation for the energy operator.
One gets the time-independent form by separating the time variable from the spatial coordinates when the potential energy function $V$ is itself time-independent. If one sets:

(20) $$\Psi(t, x^i) = T(t)\psi(x^i)$$

then the time-dependent Schrödinger equation above will split into two equations for the functions $T$ and $\psi$.

(21)
$$\frac{dT}{dt} = \frac{iE}{\hbar}T, \quad -\frac{\hbar^2}{2m}\Delta\psi + V\psi = E\,\psi,$$

in which the real constant $E$ has been introduced as the separation constant. The first equation defines the wave to be time-harmonic with a frequency of $E/\hbar$, while the second equation can be put into the form above:

(22)
$$\frac{\hbar^2}{2m}\Delta\psi + [E - V(x^i)]\psi = 0,$$

as well as the eigenvalue form:

(23)
$$\mathsf{H}\,\psi = E\,\psi.$$

Such a separation of variables is closely related to the concept of standing waves and eigenfunctions since the separation constant $E$ that must be introduced clearly takes the form of the energy eigenvalue for that standing wave. The time-independent Schrödinger equation, especially in its second form above, has a close kinship with the Helmholtz equation for the spatial wave function $u(x^i)$:

(24)
$$\Delta u + k^2(x)u = 0$$

of (linear) wave optics that comes about when one separates the time variable from the spatial ones in the (linear) wave equation for a space-time wave function $U(t, x^i)$:

(25)
$$\frac{1}{c(x)^2}\frac{\partial^2 U}{\partial t^2} = \Delta U.$$

For the time-independent Schrödinger equation, the corresponding wave number would take the form:

(26)
$$k(x) = \frac{1}{\hbar}\sqrt{2m[E - V(x^i)]}.$$

Note that the expression in the square brackets amounts to the kinetic energy $\frac{1}{2}mv^2$, so one can also say that:

(27)
$$k(x) = \frac{mv(x)}{\hbar}.$$

In one of Schrödinger's papers in the series [10], he showed that his wave-mechanical formulation of quantum mechanics was equivalent to the Born-Heisenberg-Jordan matrix mechanics. In particular, the energy eigenvalues that one deals with in either case are consistent with each other [3].

The Schrödinger equation was extended by Pauli in 1927 to include the magnetic dipole moment of the electron [11], which was discovered experimentally in 1922 by Otto Stern and Walter Gerlach [12]. That dipole moment was attributed to an "intrinsic angular momentum" or *spin* of the electron by George Uhlenbeck and Samuel Goudsmit in 1925 [13], although Arthur Compton had previously proposed it in the Journal of the Franklin Institute [14]. When Ralph Kronig proposed it at the Eidgenössische Technische Hochschule in Zurich (viz., the Swiss Federal Institute of Technology), he was immediately shot down by the redoubtable Wolfgang Pauli. However, Pauli's argument was still worth considering because it was based in the classical model of the extended electron and amounted to saying that if one used the prevailing experimental numbers for the relevant constants and assumed a spherical electron with the classical radius then the tangential velocity at the equator would have to exceed the speed of light.

The Pauli equation for spinning electrons was still non-relativistic (i.e., not Lorentz invariant), so further attempts were made to formulate a relativistic quantum wave equation. In 1926, Oskar Klein [15] and Walter Gordon [16] independently obtained the equation for free matter waves [4]:

$$\Box\Psi + k_0^2\,\Psi = 0 \qquad \left(\Box \equiv \frac{1}{c^2}\frac{\partial^2}{\partial t^2} - \delta^{ij}\frac{\partial^2}{\partial x^i\,\partial x^j}\right), \tag{28}$$

although the fact that the wave function $\Psi$ took its values in the complex number plane did not provide enough degrees of freedom to include a spinning electron, but only a charged, spinless one. In this case, the wave number $k_0$ is the *Compton wave number*, which relates to the rest mass $m_0$ of the particle by way of:

$$k_0 = \frac{m_0\,c}{\hbar}. \tag{29}$$

The relativistic theory of the spinning electron was finally achieved by Paul Dirac in 1928 [18] by essentially taking the "square root" of the Klein-Gordon equation, which involved the introduction of elements from the "Clifford algebra" of Minkowski space as coefficients in the equation in order to make it consistent with the Klein-Gordon equation upon "squaring" it. The details would take us too far afield at the moment, but they are discussed in any good text on relativistic wave mechanics.

One of the fundamental issues that had to be resolved at the roots of wave mechanics was how to physically interpret the quantum wave function $\Psi$. The starting point for any of the interpretations of it is the fact that its modulus-squared:

$$\|\,\Psi\,\|^2 = \Psi\,\Psi^* \tag{30}$$

[3] Strictly speaking, the rigorous proof of mathematical equivalence came later in 1931 by way of von Neumann, who showed the uniqueness of unitarily-equivalent representations of the canonical observables.

[4] The Russian Vladimir Fock also obtained the equation slightly later than Klein in the same year [17], but that fact is not always acknowledged in the Western literature on relativistic wave mechanics.

is a real-valued function of time and position, which might then be associated with some familiar physical concept.

Schrödinger himself assumed that when one is dealing with the wave function of an electron, it would be reasonable to associate that modulus-squared with an electric charge density. (This interpretation reemerged later with the Pauli-Weisskopf theory of the scalar meson [19], which related to the Klein-Gordon equation.) Initially, Louis de Broglie agreed with that conception of things. However, the Copenhagen school felt that it would be better to give a statistical interpretation to the quantum wave function that would be consistent with their quantum theory of measurement. Namely, as long as the wave function $\Psi$ was normalized, so that the spatial integral of $\| \Psi \|^2$ would be 1, that modulus-squared could serve as a probability density function. The crucial step is in assuming that the wave function must describe a *point* mass as its classical limit. Hence, the probability density function was assumed to describe the probability of finding that *point* in a differential volume of space, which is sometimes referred to as an "occupation probability" or "probability of existence." Schrödinger himself objected to that interpretation, as did Einstein, but de Broglie eventually decided to reluctantly accept it, although he occasionally returned to some of the more deterministic interpretations, which shall be discussed below. Nonetheless, at the Solvay Conference of 1927, in the course of the presentation by Born and Heisenberg, they felt that there was nothing too premature or bombastic about characterizing their theory of quantum phenomena by asserting:

> "We consider quantum mechanics to be a closed theory whose fundamental physical and mathematical assumptions are no longer susceptible of any modification."

Indeed.

## 3 Quantum electrodynamics

Although both the matrix (operator) mechanics of Born, Heisenberg, and Jordan and the wave mechanics of de Broglie, Schrödinger, *et al.*, continued to be discussed in the name of quantum theory, it seemed that operator mechanics gained somewhat more traction as the reigning explanation for quantum phenomena. In particular, its reliance upon the statistical interpretation of the quantum wave function and the quantum theory of measurement seemed to dominate all *a posteriori* attempts to develop a natural-philosophical foundation for the success of the theory.

Eventually, it became clear that quantum mechanics would have to be expanded to quantum field theory. In particular, quantum mechanics appeared to be effective at accounting for the bound states of elementary particles, at least as far as atomic electrons were concerned, and even the bound state energies of atomic nuclei seemed to admit a shell model that was analogous to the shell model for atomic electrons, although with a different expression for the potential energy. The issue was how one was to account for the interactions between elementary particles that did not involve bound states, such as high-energy scattering processes. It is important to note that the energy and momentum spectra in that regime are continuous, not discrete, so one might expect that some of the conceptual foundations of quantum theory that were eminently adapted to the problems of atomic structure and atomic spectra might start to show their weaknesses.

## 3.1 Second quantization

The definitive attempt at extending from quantum mechanics to quantum field theory came from Heisenberg and Pauli [20], who predictably attempted to use operators as the basis for that extension, i.e., they extended from operators that acted upon quantum states to fields of operators that depended upon time and position. However, it was no longer clear what a field of operators would act upon or whether a field of operators would produce a field of eigenvalues in so doing, and whether the concept of eigenvalue would need to be reevaluated in light of the fact that having a function of time and space for an eigenvalue does not really capture the essence of eigenvalues, which are typically scalar constants.
A first extension would be to extend the canonical commutation relations for the quantum operators that represent the position and momentum of point-masses. Indeed, the initial extension was to make the operators functions of position, but not time. If we write $P_\alpha$, $Q_\alpha$ as abbreviations for $P_\alpha(x^i)$, $Q_\alpha(x^i)$, and $P'_\alpha$, $Q'_\alpha$ for $P'_\alpha(x^i)$, $Q'_\alpha(x^i)$ then the commutation rules for the fields of canonical operators can be written as follows:

$$(31) \qquad [Q_\alpha, Q'_\beta] = 0, \qquad [P_\alpha, P'_\beta] = 0, \qquad [P_\alpha, Q'_\beta] = [P'_\alpha, Q_\beta] = \delta_{\alpha\beta}\, \delta(x, x').$$

Those expressions can then be referred to as the *equal-time* commutators of the fields of operators, and Heisenberg-Pauli did not define them for unequal times.
The equations of time evolution for the canonical fields that they defined were borrowed from Hamilton's equation of motion for point mechanics. One first integrates the Hamiltonian density $H$ for the canonical fields to get the total energy of the system:

$$(32) \qquad \bar{H} = \int H\, dV,$$

so the equation of motion for the fields will take the form:

$$(33) \qquad \dot{Q}_\alpha = \frac{2\pi i}{h}[\bar{H}, Q_\alpha], \qquad \dot{P}_\alpha = \frac{2\pi i}{h}[\bar{H}, P_\alpha].$$

One essential difference between the classical Hamiltonian and the quantum version is that since the operators do not all commute, one must also specify an order in which they are multiplied.

For the particular case of the electromagnetic interaction, Heisenberg and Pauli first tried using the electromagnetic potential 1-form $\Phi = \Phi_\alpha\, dx^\alpha$ and the corresponding field strength 2-form [5]:

$$F = d_\wedge \Phi = \tfrac{1}{2}(\partial_\alpha \Phi_\beta - \partial_\beta \Phi_\alpha)\, dx^\alpha \wedge dx^\beta \tag{34}$$

as the canonical fields. However, that proved to be unsuitable. They then settled upon using the electric field strength $\mathfrak{E}$ as the canonical position and the magnetic field strength $\mathfrak{H}$ as its corresponding conjugate momentum. The equal-time canonical commutation relations they then obtained were:

$$[\mathfrak{H}_i(x),\ \mathfrak{H}_k(x')] = 0\,, \qquad [\mathfrak{E}_i(x),\ \mathfrak{E}_k(x')] = 0\,, \tag{35}$$

$$[\mathfrak{E}_1(x),\ \mathfrak{H}_2(x')] = -\,[\mathfrak{E}_2(x),\ \mathfrak{H}_1(x')] = \frac{\hbar c}{i}\frac{\partial}{\partial x_3}\delta(x,x')\,, \tag{36}$$

in which $x$ and $x'$ are two points in space at equal times. In order to integrate the resulting field equations, they then resorted to statistical methods and focused on the transition probabilities, which seemed to be an approach that came to dominate all subsequent research.

Eventually, Heisenberg abandoned the concept of a force of interaction between interacting elementary particles and replaced it with the "exchange-particle" concept. That is, the interaction is *defined* by the exchange of an elementary particle, which has the effect of deferring the deeper issue of finding the quantum analogue of Maxwell's equations that would pertain to the quantum analogues of electric and magnetic fields by essentially taking the interaction of charged particles (and photons, for that matter, although that issue was not discussed at the time in question), which presumably happens continually over a macroscopic time interval, and approximating it by an interaction (i.e., an exchange of a particle) that takes place in an essentially-infinitesimal time interval. That approximation is also equivalent to passing from the Cauchy problem for the time evolution of the incoming particle fields at a finite initial time to the outgoing particle fields at finite time after the interaction to the *scattering approximation*, in which the initial time is pushed to $-\infty$, while the final time is pushed to $+\infty$. That also has the effect of linearizing the time evolution operator since the asymptotic fields at time $\pm\infty$ are, by definition, free (i.e., non-interacting) fields that presumably satisfy the free-field equations, such as the Dirac equation for fermion fields. As a result of that linearization, the methods of Fourier analysis, which shifts the theory from space-time to energy-momentum space, become more useful. Instead of constructing a time evolution operator on space-time that solves an as-yet-undetermined system of partial differential equations, one constructs momentum space propagators that solve a system of algebraic equations that relate to the Fourier transforms of the incoming and outgoing asymptotic fields.

---

[5] The notation $d_\wedge$ refers to the exterior derivative operator, which is a type of generalized curl, and the wedge $\wedge$ refers to the exterior product of the 1-forms, which is essentially an antisymmetrized tensor product of the covector fields. A good reference on such matters for physicists is the book by Frenkel [21].

Ultimately, the result of such a field theory of interaction in the scattering approximation is to produce scattering amplitudes that can be represented as the elements of a scattering matrix, i.e., the matrix of the linearized scattering operator. The moduli-squared of the amplitudes then represent transition probabilities for incoming particle states to emerge as the various possible outgoing states. Typically, the scattering amplitudes for specific interactions are usually computed by perturbation expansions, such as vertex expansions, in which one defines power series expansions in powers of the coupling constant for the interaction in question. (For electromagnetic interactions, the coupling constant is defined by the fine-structure constant, which is $e^2 / \hbar c$, in C.G.S. units.) The coefficients of the successive powers are typically obtained by graphical methods that really originated in some unpublished lecture notes by Baron Ernst Karl Gerlach von Stückelberg at the Swiss Federal Institute, although they are more commonly associated with Richard Feynman, who developed them some years later.

An immediate problem arose in the early manifestation of quantum electrodynamics, namely it led to unphysical infinities in such things as the mass and charge of elementary particles. Heisenberg originally proposed a process that he called the "subtraction of infinities," but Pauli, who was more rigorous when it came to mathematical matters, objected that such a process was mathematically meaningless. That led to a gradual falling-out between the two physicists that ultimately took the form of Heisenberg declining to even attend Pauli's funeral [6]. For several years that followed, it seemed as if all of the theoretical work that went into quantum electrodynamics was concerned with trying to make something more mathematically rigorous out of Heisenberg's subtraction of infinities. It eventually took the form of the various approaches to regularization and renormalization, such as Pauli-Villars regularization and dimensional regularization, but ultimately one could not get around the salient fact that in any form, one was still basically trying to perfect a "kludge," i.e., an error-correcting algorithm, when there was clearly a more-fundamental issue that had to be addressed, namely, how to alter the theory from the outset so the infinities would no longer appear.

One possible direction to pursue is to consider the possibility that it is not the values of the numbers themselves (e.g., energy) that are fundamental, but only the differences between them, just as potential energy is not as fundamental as potential differences. Perhaps that points to the mistake that was made in assuming that a 1-form, such as force, was exact, and therefore admitted a "potential" function, as opposed to a more "non-conservative," i.e., "thermodynamical," picture that starts with a 1-form that is not exact to begin with.

## 3.2 Path integration

Feynman himself eventually proposed an alternative approach to quantum mechanics and quantum field theory that was based in an extension of the concept of a path integral [23]. Once again, the purpose of the integral was to obtain a transition probability between incoming and outgoing states. The innovation that he introduced was that in addition to the classical extremal that connects two

---

[6] There is a most-absorbing and entertaining discussion of the history of early quantum electrodynamics in the book by A. I. Miller [22].

points in configuration space or phase space, one also considers the neighboring non-extremal paths, which are then weighted accordingly to reflect the extent to which they actually contribute to the transition probability.

The path integrals were eventually extended to functional integrals for fields, although in either case, since one was effectively integrating over infinite-dimensional spaces of curves or fields, there was an unavoidable problem with mathematical rigor in this approach as well that was based on the possible non-convergence of the integrals, so it was only a small advance over second quantization in that regard.

Ultimately, one evaluated the integrals by means of perturbation expansions of a different sort, namely, *loop expansions*, in which the power series was defined by powers of $\hbar$. Thus, the constant term in the power series ($\hbar = 0$) would represent the classical approximation or "tree-level" diagram. Successive terms would then correspond to Feynman diagrams with internal loops, and the one-loop diagrams, which would correspond to the first power of $\hbar$, would represent the semi-classical, or W.K.B. approximation. It should be noted that the expansions that one deals with in quantum field theory do not generally converge, and are therefore *asymptotic expansions*, which always have to be truncated after a finite number of terms.

Since the functional integrals that represent the transition probabilities cannot typically be evaluated in full generality, but only by perturbation expansions, one finds that they lead to the same Feynman rules and diagrams that were developed for second quantization. Indeed, to experimental particle physicists, it is often the Feynman diagrams themselves that are regarded as fundamental, and not the theory that produced them. However, the number of diagrams that must be evaluated in order to get successive coefficients in the expansions rapidly gets quite prohibitive and the evaluation of the diagrams must be automated by computer algorithms. For instance, obtaining the fourth-order coefficient in the Schwinger (vertex) expansion of the anomalous magnetic moment of the electron requires the evaluation of over twelve thousand diagrams.

## 4 Return to the fundamental problems

Since flaws persist in quantum electrodynamics, in any of its proposed forms, one must inevitably return to the roots of those flaws. It might seem most radical to return to the empirical phenomena that suggested that classical electromagnetism had reached its limits. However, since many empirical phenomena that related to electrodynamics at the atomic-to-subatomic level were discovered only after 1927, such as anti-particles and pair creation and annihilation, rather than take a naively atavistic approach to quantum theory and simply return to the discussion that prevailed at the time of Heisenberg and Pauli, this author feels that one should take a more "neo-classical" approach to the problem, i.e., start with the problems in the classical theory of electromagnetism and consider how the empirical successes of quantum electrodynamics that followed the first attempts to formulate that theory might suggest some expansion in scope of the classical theory. In particular, the concept of a loop expansion seems to give strong hints as to how one might generalize the first principles.

## 4.1 Extension of Maxwell-Hertz theory of radiation

Since one of the fundamental paradigm shifts that quantum field theory introduces takes the form of regarding matter as wave-like excitations of a vacuum ground state, one sees that before one can deal with the structure of elementary matter, one must first deal with the structure of the vacuum. There is then a fundamental problem that is concerned with the difference between the classical electromagnetic vacuum and the quantum electromagnetic vacuum.

One aspect of the quantum electromagnetic vacuum that gets a lot of attention is the concept of a *zero-point* field, which represents the extension of the quantum-mechanical concept of a non-zero ground state energy for the quantum harmonic oscillator to a corresponding non-zero ground state for the quantum electromagnetic wave field, which is imagined to be a continuous infinitude of coupled quantum oscillators in the same way that a classical electromagnetic wave is envisioned to be a continuous infinitude of coupled classical harmonic oscillators. Of course, the problem arises when one tries to add up (i.e., integrate) an infinitude of non-zero numbers, no matter how small, only to get an infinite result. However, a point that is often overlooked is that the energy eigenvalue of a quantum harmonic oscillator is a *total* energy, whereas one would expect that its field analogue would have to be an *energy density* that would give a total energy only when integrated over all of space.

A recurring theme of quantum electrodynamics is the ubiquitous appearance of corrections due to *vacuum polarization*, i.e., the formation of virtual particle-antiparticle pairs. When they occur in internal branches of Feynman diagrams, they represent the loops that define the loop expansions. For example, two internal loops would correspond to the creation and annihilation of two successive virtual pairs during the exchange of a photon. One can also think of the formation of virtual pairs as radiative corrections to the classical picture that amount to interactions with quantum vacuum. They also indicate the effects of renormalization, which would be unnecessary in the classical theory.

One useful attempt at modeling the quantum electromagnetic vacuum took the form of the Heisenberg-Euler model [24], which was essentially a one-loop extension of the classical electromagnetic action functional to include vacuum polarization. Although such "effective" Lagrangians and action functionals are probably not fundamental by themselves, they nonetheless point the way to possible refinements at the fundamental level that would imply them as approximations.

One of de Broglie's more intriguing attempts to extend classical mechanics took the form of his theory of the "hidden" thermodynamics of isolated particles [25]. In other words, the isolated point-particle was exchanging energy with a "hidden thermostat." Although that might seem far-fetched and contrived, a footnote in that treatise shed a more modern light on the nature of the problem; namely, he observed that other people were using the term "the quantum vacuum" instead of "hidden thermostat."

Another unavoidable complication to classical electromagnetism that emerges at the quantum level is that when one gets close to elementary source-charges, such as electrons, one expects that the strengths of the electric and magnetic fields that they create will approach the so-called *Schwinger limit*, which is the field strength at which a photon would polarize into a bound state of an electron and positron. (One might possibly model that process as a critical-point phase transition.) At any rate, one would expect that such field strengths could only be described by nonlinear field equations. The likely source of that nonlinearity would be the electromagnetic constitutive law for the quantum electromagnetic vacuum, which would probably have to account for vacuum polarization, such as the law that one gets from the Heisenberg-Euler model.

Furthermore, since the fundamental breakdown of the Maxwell-Hertz theory historically centered around the theory of electromagnetic radiation, one would necessarily have to consider the limitations of the classical theory. In particular, one typically deals with a linear theory of radiation since its reliance upon Green function techniques unavoidably ties it down to linear integral operators that would solve the linear differential equations for classical fields. Hence, one would need to develop a nonlinear theory of electromagnetic radiation as a first step. Similarly, much of the classical theory of radiation deals with asymptotic fields that become invalid as one approaches the sources of the fields.

## 4.2 Limitations of optical-mechanical analogy

Perhaps one of the key misunderstandings of early wave mechanics was the fact that the earliest forms of it always regard classical mechanics as synonymous with point mechanics. Consequently, since the wave functions of quantum mechanics were generally defined over finite regions of space, not points, the question always came down to how one could relate a spatially-extended wave function propagating through space like a wave front to a point moving along a single curve in space.

In the same year that the Solvay conference on electrons and photons was being held, Erwin Madelung [26] proposed a continuum-mechanical [7] interpretation of the stationary Schrödinger equation that proposed to interpret the modulus-squared of the Schrödinger wave function as essentially a "number" density for an extended mass, such that multiplying it by the mass scalar $m$ that appeared in the Schrödinger equation would define a mass density whose support would coincide with that of the wave function. It would then be unnecessary to identify a particular point in that region unless one wished to pass to the point approximation for the extended object, which would probably involve the center of mass of the density function. Rather than a single point moving along a curve, the propagation of the matter wave would sweep out a congruence of curves with a finite cross-section. The statistical interpretation of wave function would then become unnecessary, although one could re-introduce it by assuming that the normalized number density would represent a probability density function for finding a point-mass in the differential element of volume around each point. The continuum-mechanical picture was extended by others, including Takehiko Takabayasi, de Broglie and his students, such Jean-Pierre Vigier and Francis Halbwachs, and David Bohm and his school. For a more detailed discussion of the subject and its bibliography, one might consult the author's book on the subject [27].

---

[7] Madelung and others who followed called it a "hydrodynamical" interpretation, but since the stress tensor that one obtains for the medium does not actually have the form of any established hydrodynamical stress tensors, this author has always felt that it would be more cautious to simply call it a "continuum-mechanical" interpretation and investigate the properties of such a quantum medium. See the author's book [27].

Since the optical-mechanical analogy was the driving inspiration for Schrödinger's theory, one needs to consider how to extend that analogy to extended matter accordingly. Basically, since the trajectories of point mechanics were supposed to correspond to single light rays and their infinitesimal neighborhoods, one would have to consider congruences of light rays with finite cross-sections accordingly. Moreover, the picture of a single light ray and its infinitesimal neighborhood is also the essence of the geometrical optics approximation, one sees that one would have to go beyond that approximation in order to deal with congruences of curves with finite cross-sections.

## 4.3 Quantum corrections to classical mechanics

Of course, when one removes the geometrical optics approximation, one is back to wave optics. However, since the essential difference between the two types of optics is due to the fact that geometrical optics does not consider the interference effects of interacting waves, such as diffraction, in particular, one sees that if one looks at how one corrects geometrical optics for diffraction then one might get some idea regarding how to correct point mechanics for quantum effects.

It helps to know that loop expansions of quantum field theory have a close analogy to the diffraction corrections to geometrical optics. The main difference is that the powers in the asymptotic expansion for diffraction are powers of the wavelength (or really the reciprocal of the wave number), and not powers of $\hbar$.

Another path to extending classical mechanics to quantum mechanics is suggested by the fact that the aforementioned loop expansions tend to suggest that since the classical extremals are only the tree-level approximation to the full quantum expression, perhaps it is the least action principle itself that needs to be generalized. A promising candidate for that extension would be the principle of virtual work when one deals with non-conservative forces. The author has made some progress in that direction in his earlier publications [28], in which he defined the virtual work functional, which generalizes the first variation of an action functional to the cases in which no action functional would exist, such as non-conservative forces or imperfect or non-holonomic constraints. Along those lines, one can consider a concept that Helmholtz introduced [29] that he called the *kinetic potential*. As opposed to the usual force potential, which takes the form of a differentiable function of position $U(x)$ whose differential is minus a force 1-form:

$$F = -dU, \tag{36}$$

a (first-order) kinetic potential is a function $T = T(t, x^i, v^i)$ whose *variational derivative* is the force 1-form:

$$F = \frac{\delta T}{\delta x^i}, \qquad F = F_i(t, x^j, v^j, a^j)\,dx^i. \tag{37}$$

Hence, when the force $F$ takes the form of a conservative force, the definition of the kinetic potential reduces to the Euler-Lagrange equations. More generally, the force 1-form can take the form of an exact 1-form, such as $-dU$, plus inexact "quantum corrections," i.e., its normal form as a Pfaffian:

(38) $$F = -\, dU + \lambda_\alpha \, d\mu^\alpha \,.$$

In addition to quantum corrections to the least-action principle, one also encounters quantum corrections to conservation laws, such as the Ward-Takahashi identities of the quantum theory of gauge fields [30]. Basically, classical observables that had vanishing divergences correspond to quantum observables with non-vanishing ones. Since Noether's theorem associates conserved quantities with symmetries of the action functional, that also suggests that the underlying symmetries need to be "quantum corrected."

Of course, from the original viewpoint of the theory of electromagnetic radiation, one must always regard the wave that is radiated by an accelerated charge as something that takes away from its energy and momentum; hence, the term "radiation damping." That would already imply that one is dealing with a breakdown of the conservation of energy-momentum. In order to get a conservative system, one would have to include the radiated wave, along with the accelerated charge, although that would raise the dimensionality of the mechanical system to infinity, and with it, the analytical complexity of its dynamics.

## Summary

In conclusion, let us summarize the basis points of this presentation:

1. Just as special relativity (i.e., the relativity of velocity) grew out of the breakdown of Maxwell's theory in regard to electrodynamics of bodies moving with constant relative velocities and the crucial step related to the way that the propagation of electromagnetic waves was affected by the motion of the source, early quantum physics seemed to grow out of the breakdown of Maxwell's theory of electromagnetic radiation, which then relates to the acceleration of the source, so to the relativity of acceleration.

2. The limited information that early quantum theory had to build upon necessitated that certain aspects of quantum phenomena, such as discreteness, that were more intrinsic to the nature of bound states, such as the energy levels of atomic electrons and the discreteness in atomic spectra, were assumed to be more fundamental than they actually were. As a result, when quantum mechanics needed to move into quantum field theory, the fact that free and scattering states no longer had discrete spectra of energy and momentum did not seem to play as fundamental a role as it should have.

3. Another limiting assumption of early wave mechanics, as it was defined by de Broglie and Schrödinger, was their assumption that classical mechanics had to mean "point mechanics," whereas that only led to problems in interpreting the classical limit of quantum wave functions, which were spatially-extended, and thus defined congruences of rays with finite cross-sections. A more-promising approach is to first reduce the motion of the wave to the motion of an extended object, not a point, and then treat the reduction of the extended object to the point in the usual way (i.e., method of moments).

4. One possible source of unphysical infinities in the Heisenberg-Pauli approach to quantum electrodynamics might be the fact that the eigenvalues of quantum operators that related to bound states of atomic electrons represented *total energies, orbital angular momenta, spin,* etc., whereas in a field that is distributed through space, those things should become *densities* that turn into totals only upon integration over all space. Then again, the concept of a field operator producing an eigenvalue that is itself a function of space defeats the purpose of introducing eigenvalues.

5. Perhaps a promising direction to pursue for an alternative formulation of quantum electrodynamics as an extension of classical electrodynamics is to return to the concept of the "quantum theory of wave fields" that Heisenberg and Pauli had in mind in the titles of their articles [20], as well as Gregor Wentzel [31] (although his book was retitled as simply *Quantum Theory of Fields* in its English translation), and consider some of the lessons that were learned from quantum electrodynamics since that time, such as the effective field theories that were obtained from loop expansions. Above all, a shift in emphasis to expanding upon the nature of the quantum electromagnetic vacuum as an extension of the classical electromagnetic vacuum seems unavoidable.

## References [8]

[1] G. Bacciagluppi and A. Valentini, *Quantum Theory at the Crossroads; Reconsidering the 1927 Solvay Conference*, Cambridge University Press 2009.

[2] L. de Broglie, "Recherches sur la théorie des quanta," Thesis, Paris (1924); *Annales de Physique* (10) **3**, 22 (1925); English translation by A. F. Kracklauer, "On the theory of quanta," 2004, available online through Fondation Louis de Broglie.

[3] G. N. Lewis, "The conservation of photons," *Nature* **118**, 874 (1926).

[4] M. von Laue, *Materiewellen und ihre Interferenzen*, Geest and Fortig K. G., Leipzig, 1948 (2. Auflage).

[8] English translations of many of these articles can be found at the author's website: neo-classical-physics.info.

[5] W. Heisenberg, "Über quantentheoretische Umdeutung kinematischer und mechanischer Beziehungen," *Zeitschrift für Physik* **33**, 879 (1925).

[6] M. Born and P. Jordan, "Zur Quantenmechanik," *Zeitschrift für Physik* **34** 858 (1925).

[7] M. Born, W. Heisenberg, and P. Jordan, "Zur Quantenmechanik II," *Zeitschrift für Physik* **35**, 557 (1926).

[8] M. Born, "Zur Quantenmechanik der Stoßvorgänge," *Zeitschrift für Physik* **37** 863 (1926); "Quantenmechanik der Stoßvorgänge," *Zeitschrift für Physik* **38** 803 (1927).

[9] E. Schrödinger, "Quantisierung als Eigenwertproblem (erste Mitteilung)" *Annalen der Physik* (4) **79**, 361 (1926); "... (zweite Mitteilung)," *ibid.*, 489; "... (dritte Mitteilung)," *ibid.*, **80**, 437 (1926); "... (vierte Mitteilung)," *ibid.* **81**, 109 (1926; English translations in E. Schrödinger, *Collected Papers on Wave Mechanics*, American Mathematical Society Chelsea Publishing, Providence, RI, 1982.

[10] E. Schrödinger, "Über das Verhältnis der Heisenberg-Born-Jordanschen Quantenmechanik zu der meinen," *Annalen der Physik* (4) **79**, 734 (1926); also included in the collected translations cited in [8].

[11] W. Pauli, "Zur Quantenmechanik des magnetischen Elektrons," *Zeitschrift für Physik* **43**, 601 (1927).

[12] W. Gerlach and O. Stern, "Über die Richtungsquantelung im Magnetfeld," *Annalen der Physik* **74**, 673 (1922).

[13] G. Uhlenbeck and S. Goudsmit, "Spinning electrons and the structure of spectra," *Nature* **117**, 264 (1926).

[14] A. Compton, "The magnetic electron," *Journal of the Franklin Institute* **192**, 145 (1921).

[15] O. Klein, "Quantentheorie und fünf-dimenzionale Relativitätstheorie," *Zeitschrift für Physik* **37**, 895 (1926); English translation available in G. Ekspong (ed.): *The Oskar Klein memorial lectures*, Volume 1, World Scientific, Singapore, 1991, pp. 67-80.

[16] W. Gordon, "Der Comptoneffekt nach der Schrödingerschen Theorie," *Zeitschrift für Physik* **40**, 117 (1926).

[17] V. Fock, "Über die invariante Form der Wellen- und der Bewegungsgleichunen für einen geladenden Massenpunkt," *Zeitschrift für Physik* **39**, 226 (1926).

[18] P. A. M. Dirac, "The quantum theory of the electron," Proceedings of the Royal Society of

London A **117**, 610 (1928); Part II: *ibid.* **118**, 351 (1928).

[19] W. Pauli and V. Weisskopf, "Über die Quantisierung der skalaren relativistischen Wellengleichung," *Helvetica Physica Acta* **7**, 709 (1934); also published in translation in [20].

[20] W. Heisenberg and W. Pauli, "Zur Quantendynamik der Wellenfelder ," *Zeitschrift für Physik* **56**, 1 (1929); "Zur Quantentheorie der Wellenfelder II," *ibid.* **59**, 168 (1929).

[21] T. Frenkel, *The Geometry of Physics*, 3rd ed., Cambridge University Press, Cambridge, 2011.

[22] A. I. Miller, *Early Quantum Electrodynamics: a source book,* Cambridge University Press, Cambridge 1994.

[23] R. P. Feynman and A. R. Hibbs, *Quantum Mechanics and Path Integrals*, McGraw-Hill, NY (1965); emended by D. F. Styer, Dover, Mineola, NY 2010.

[24] H. Euler and W. Heisenberg, "Folgerungen aus der Diracschen Theorie des Positrons," *Zeitschrift für Physik*, **98** 714 (1936).

[25] L. de Broglie, *La thermodynamique de la particule isolée*, Gauthier-Villars, Paris 1964.

[26] E. Madelung, "Eine Überträgung der Diracschen Theorie des Elektrons in gewohnte Form," *Zeitschrift für Physik* **54**, 303 (1929).

[27] D. Delphenich, *Continuum-mechanical Models for Quantum Mechanics*, Neo-classical Press, Spring Valley, OH 2017; available online as a free PDF download at neo-classical-physics.info.

[28] D. H. Delphenich, "Integrability and the variational formulation of non-conservative mechanical systems," *Annalen der Physik* **18**, 45 (2009); "On the variational formulation of systems with non-holonomic constraints," *ibid.*, 649; "A generalization of Noether's theorem based on the virtual work functional," arXiv: 1109.0461 [math-ph].

[29] H. von Helmholtz, "Ueber die physikalische Bedeutung des Princips der kleinsten Wirkung," *Journal für die reine und angewandte Mathematik* **100**, 137 (1887).

[30] S. Pokorski, *Gauge Field Theories*, Cambridge University Press 1987.

[31] G. Wentzel, *Einfuhrung in die Quantentheorie der Wellenfelder*, Franz Deuticke, Vienna 1943; English translation *Quantum Theory of Fields*, Dover, Mineola, NY, 2003.

# Chapter 11

# On the Peierls Development of Heisenberg's interpretation of the wave-function – a pragmatic approach to Quantum Measurement

M G Burt*

Department of Physics, Durham University, Durham, DH1 3LE, UK

October 2025

**Abstract**

In 1929 Heisenberg introduced the idea that the wavefunction encapsulates the knowledge of the observer of a quantum system. Over many years his erstwhile student, Rudolf Peierls, successfully developed this idea in a few brief works. Here these works are examined, interpreted, and expanded to widen appreciation of Peierls's development of Heisenberg's elegant idea. The approach followed here is very much in the spirit adopted by Peierls who eschewed a 'rigorous axiomatic' and aimed at using 'logic at the level of the working physicist'. The fundamental tenet of his work is that the wavefunction or density matrix represents the knowledge of an observer and that two observers of the same system may well have different knowledge and will use different density matrices to describe it. Essential to the understanding of Peierls's approach is the demonstration, given here, that the density matrix generally can be expressed entirely in terms of probabilities of observable outcomes and that such probabilities are subjective in the first instance. The process by which the knowledge of two observers may be combined is discussed in detail for some simple cases and, by using the Bayes theorem, the progress to a common understanding of the correct density matrix for a given situation demonstrated. Peierls gave a criterion to ensure that the amalgamation of data from two observers can never lead to violation of the uncertainty principle : that the density matrices of two observers must commute. It is pointed

*academic visitor : email prof@burt.uk.com

out that it is essential that the density matrices of two observers commute under all circumstances no matter how inaccurate the data or fanciful the beliefs used to construct them. Once this is appreciated it can be shown that Peierls's density matrix commutation criterion is the generalisation of the standard result in quantum mechanics that two observables must commute if they are to be known precisely and simultaneously, but now extended to imperfectly known mixed states rather than a single pure state. Peierls's second criterion for two observers to have compatible density matrices, that their product is non zero, is discussed and clarified.

# 1 Introduction

Heisenberg made a third great breakthrough in the early days of Quantum Mechanics beyond his two other monumental contributions, namely matrix mechanics(1925) and the uncertainty principle(1927): and that third great breakthrough(1929) was the interpretation of the wavefunction as representing an observer's knowledge of a quantum system (such knowledge being constrained, of course, by the uncertainty principle). This insight appeared in a series of lectures at the University of Chicago in the spring of 1929 and was published as a book by the University of Chicago Press in 1930. My source is the Dover edition of 1949 [1]. I refer especially to pages 16 and 36. I describe this interpretation as a great breakthrough because, by introducing the observer and their knowledge, it arguably circumvents the infamous measurement paradox, as described on pages 26 to 28 in *The Ghost in the Atom* [2], that plagues the standard interpretation, see e.g. [3], in which the measurement apparatus is deemed to be classical and the observer plays no role.

Rudolf Peierls obtained his doctorate under Heisenberg in 1929 and was subsequently Pauli's research assistant for 3 years [4]. But, it would seem that only much later in his career did he start to write about quantum measurement and develop, at least for a nonspecialist audience, this interpretation of Heisenberg's - that the wavefunction encapsulates the observer's knowledge. Indeed, in the proceedings of the symposium for his 80th birthday celebrations [5] there is no chapter on his quantum measurement work, though there is mention, see page 37 in Dalitz's preface, of his contribution [6] to the *Symposium on the Foundations of Modern Physics* held in Helsinki in 1985. However, his best known work on this topic is probably his response *in defence of 'measurement'* [7] in Physics World to Bell's iconoclastic article *against 'measurement'* [8] (both reprinted in this volume), and which appeared only a few years before his death in 1995. The striking and very welcome aspect of this article [7] is the declaration Peierls makes that he would not aim at a rigorous axiomatic, but only at the level of logic of the working physicist. As indicated by my title, I shall follow that pragmatic lead. This pragmatic 'working physicist' approach also characterises his other publications quoted in *in defence of 'measurement'* [7] namely in *Symposium on the Foundations of Modern Physics* [6] already mentioned above and the short sections in his books *Surprises in Theoretical Physics* and *More Surprises in Theoretical Physics* [9, 10].

Taking all the works by Peierls mentioned above together, one is struck not only by their clarity, but also by their brevity considering the profundity of the subject. But this is not perhaps surprising given both his prodigious output across physics generally, as evinced by his books mentioned above [9, 10], and his understandable desire to avoid overloading the nonspecialist reader with detail. Nonetheless, he gives a coherent overview of the quantum measurement problem for the nonspecialist and it is difficult not to resist studying it more closely to appreciate the subtleties and make sure one has really understood the material. Indeed, understanding the material is not as straightforward as it may appear: the eminent theoretician David Mermin has questioned the criterion that Peierls gave [6, 7, 10] that the density matrices of two observers should commute, and even claimed [11] to have found a counterexample. This emphasises the need to augment and expand the work of Peierls to avoid such misunderstandings and to demonstrate the perspicacity and also the fruitfulness of Heisenberg's original and elegant idea.

Before embarking on our study of Peierls's work, I would like to mention the work of van Kampen; it provides a somewhat different but valuable perspective. While Peierls uses the Stern-Gerlach experiment for illustration, van Kampen uses the somewhat more complicated double slit experiment. Fortunately van Kampen has published an engaging article [12] for the general reader (including the working physicist!), an article which summarises the work in his more erudite papers. The article is written in Dutch, but it is such a gem (well, for the ultra pragmatist, that is) that I have translated it (see 'Quantum Measurement Theory as simply explained by N G van Kampen' arXiv:2206.03219v2). I encourage the reader to take a look!

# 2 Preliminary Comments - Knowledge

We first need to familiarise ourselves with the broad meaning of knowledge as used by Peierls. For instance, in the final column of *in defence of 'measurement'* he uses information synonymously with knowledge and also by implication includes belief as a class of knowledge '....they may have seen part of the results of another physicist.'[7]. We also need to note that any knowledge may be inaccurate: in *More Surprises in Theoretical Physics* we find on page 10 '.... one [observer] may have only an incomplete or inaccurate view of the measuring instrument.'[10]. And further, knowledge need not come from measurements : on page 1 of *in defence of 'measurement'*[7], we find "..., the initial values [of the density matrix] represent knowledge usually obtained from observations. (They are not always measurements; for example, if an atom has been for a reasonable time in free space we know that it must be in its ground state.)"

We further need to appreciate, and this is essential, that the knowledge referred to by Peierls is subjective in the first instance, but becomes common and nominally objective (or at least less subjective) as the scientific process proceeds. This subjectivity can be seen among the quotations in the paragraph above, but much more clearly in Peierls's response to the question "Whose knowledge?" on page 32 of *Surprises in Theoretical Physics*[9]: "One question we may ask at this point is : Whose knowledge does the wavefunction represent? Do we have to write a separate wavefunction for every observer who participates in an experiment, and perhaps for every reader of the resulting paper? The answer is that in most practical experiments, as in the example of Stern-Gerlach, the later links in the chain leading from the quantum system to the reading of an instrument, are entirely classical. Many observers may therefore look at the same dial without affecting the state of affairs; they can pass information about what they have seen, and verify each other's statements. In these circumstances we are then dealing with common knowledge, and no further distinction need be made."

The emergence of common knowledge as evinced from the above quotation is related to the data emerging from a single experiment. But, it is more usual for common knowledge to have a wider meaning i.e. to also come from similar experiments done in different labs and we will certainly include such in our considerations.

The broad interpretation of knowledge as used by Peierls also extends to the time lapsed before it is acquired. In *in defence of 'measurement'*[7],

Peierls states that the observer does not have to be contemporaneous with the event, quoting the Cosmic Microwave Background. Quantum Mechanics provides a means for generating a probability distribution for possible outcomes. It is only when the observer becomes aware of a definite outcome that their probability distribution collapses to a certainty. This is a perfectly ordinary classical feature. Associated with this classical collapse of the probability, the observer has to update their wavefunction; this feature is often referred to as the collapse of the wavefunction, though it would perhaps be more helpful to call it a modification of the wavefunction acknowledging that it is the change brought about by the interaction with the 'apparatus'.

# 3 Overview

The essential feature of the Peierls interpretation, as already stressed, is that the wavefunction or more generally the density matrix expresses the knowledge of an observer. Of course, although Peierls does not discuss this directly, it is obvious that this can only be describing the first step in the scientific process. With any disparity in the knowledge of a group of observers, more experiments are performed until a consensus is reached among the scientific community as to the correct density matrix to be assigned to a given system. This process is carried out, if not explicitly, then implicitly, using the Bayes theorem to modify the probability of a hypothesis being correct in the light of data collected. The question immediately arises as to how this process can be applied to quantum theory which famously involves not only probabilities, but also probability amplitudes; indeed, the off diagonal elements of the density matrix are often referred to as containing phase information which has a probability amplitude 'ring' to it. However, it will be shown ( section 4 ) how to express the elements of any density matrix of finite dimension entirely in terms of the expectation values of observables of the system, which in turn can be expressed entirely in terms of the probabilities of the various possible outcomes of measurement of those observables. All information about probability amplitudes, so crucial to understanding quantum phenomena in general, is contained in a density matrix so constructed. Such probabilities as appear in the density matrix can be modified in the light of new data using the Bayes theorem and with sufficient high quality observations a consensus can be achieved as to the correct values to be assigned to them, an example of which is given in section 5.

It is within this general framework that we examine (section 6 ) Peierls's requirement that the density matrices of two observers of the same system must commute. It is shown that this requirement is equivalent to the familiar one that the observables used by the two observers must commute if their measurements are never to interfere with each other, but now extended to imperfectly known mixed states rather than a single pure state. It will be seen that, no matter how inaccurate the data supplied by the two observers, if the commutation criterion is always satisfied, their combined density matrix will always have non-negative eigenvalues and hence can never violate the uncertainty principle as will be shown explicitly. On the other hand, we will see that if Peierls's requirement is not met, then, with sufficiently inaccurate data (inaccuracy which is not due to the uncertainty principle), violations of the uncertainty principle, i.e. density matrices with one or more negative eigenvalues, are possible. Of course, with sufficiently accurate data it is impossible to violate the uncertainty principle since no physically or mathematically realizable state can do so.

In section 6 we will also clarify Peierls's other criterion that the product of the density matrices of two observers must not be zero in order to avoid contradiction. We show generally that if the Peierls product criterion is not obeyed, i.e. the product of the observers' density matrices does in fact equal zero, then no amount of further experimental information along with the subsequent use of the Bayes theorem will remedy the situation. In section 7 we will re-examine some of Peierls's statements in the light of our work and demonstrate how we can clarify and augment them.

Throughout this article the points will be illustrated using a concrete example, the Stern-Gerlach experiment, which focuses on the magnetic moment due to the outer electron of the silver atom. In the first instance, we assume that two observers, in separate labs, set up identical sources of silver atoms, investigate the polarization properties of their respective beams and then compare results. But we will also consider the case of two observers making sequential measurements on the same source. And to illustrate the theoretical concepts discussed we will imagine that we have an apparatus capable of measuring the nuclear magnetic moment of the atoms as well. Although it is far fetched in practice, it is possible in principle and will be a very useful illustrative aid.

## 4 The Density Matrix expressed in terms of Probabilities

We consider here the two dimensional (i.e. the spin 1/2) case, leaving the general case of higher dimensional state spaces to Appendices A and B. We will show that all the elements of the density matrix, $\rho$ , can be written in terms of probabilities of the outcome of measurements of observables, in this simple case the $x$, $y$ and $z$ components of the spin. Consider first, $\langle\sigma_z\rangle$, the quantum mechanical expectation value (or rather twice that value) of the $z$ component of the spin, represented by the appropriate Pauli matrix. We have

$$\langle\sigma_z\rangle = \mathrm{Tr}[\rho\sigma_z] = \rho_{1,1} - \rho_{2,2} \tag{1}$$

But we also have from the normalization of $\rho$

$$\mathrm{Tr}[\rho] = \rho_{1,1} + \rho_{2,2} = 1 \tag{2}$$

and so

$$\rho_{1,1} = (1 + \langle\sigma_z\rangle)/2 \tag{3}$$

and

$$\rho_{2,2} = (1 - \langle\sigma_z\rangle)/2 \tag{4}$$

Turning our attention to the $x$ component of the spin

$$\langle\sigma_x\rangle = \mathrm{Tr}[\rho\sigma_x] = \rho_{1,2} + \rho_{2,1} \tag{5}$$

and finally for the $y$ component,

$$i\langle\sigma_y\rangle = \mathrm{Tr}[\rho i\sigma_y] = -\rho_{1,2} + \rho_{2,1} \tag{6}$$

so that

$$\rho_{1,2} = (\langle\sigma_x\rangle - i\langle\sigma_y\rangle)/2 \tag{7}$$

and

$$\rho_{2,1} = (\langle\sigma_x\rangle + i\langle\sigma_y\rangle)/2 \tag{8}$$

the complex conjugate as it must be. The above results (equations (3) (4) (7) and (8)) can be combined into the single matrix equation

$$\rho = (\frac{1}{2})(I + \langle\sigma_x\rangle\sigma_x + \langle\sigma_y\rangle\sigma_y + \langle\sigma_z\rangle\sigma_z) \tag{9}$$

and its extension to all finite dimensions is given in Appendix A .

Now, to establish that these results for the elements of $\rho$ can be expressed in terms of probabilities of results of measurements, one first notes that each $\sigma_\alpha$, where $\alpha = (x, y, z)$, has eigenvalues $\pm 1$. And hence, if we evaluate the expectation value, $\langle\sigma_\alpha\rangle$, of $\sigma_\alpha$ using its eigenvectors, then we obtain

$$\langle\sigma_\alpha\rangle = p^{(\alpha,+1)} - p^{(\alpha,-1)} \tag{10}$$

where $p^{(\alpha,+1)}$ and $p^{(\alpha,-1)}$ are the probabilities for measuring eigenvalues $+1$ and $-1$ respectively for $\sigma_\alpha$. From $p^{(\alpha,+1)} + p^{(\alpha,-1)} = 1$ we readily find that $\rho_{1,1}$ and $\rho_{2,2}$ have particularly simple forms

$$\rho_{1,1} = p^{(z,+1)} \tag{11}$$

and

$$\rho_{2,2} = p^{(z,-1)} \tag{12}$$

while $\rho_{2,1}$ and $\rho_{1,2}$ can be written as

$$\rho_{2,1} = ((p^{(x,+1)} - p^{(x,-1)}) + i(p^{(y,+1)} - p^{(y,-1)}))/2 \tag{13}$$

and

$$\rho_{1,2} = ((p^{(x,+1)} - p^{(x,-1)}) - i(p^{(y,+1)} - p^{(y,-1)}))/2 \tag{14}$$

So we have shown that the density matrix for a spin 1/2 system can be expressed entirely in terms of the probabilities of the results of measurements. Even the off diagonal elements that provide the phase relationship between spin up and spin down states for the $z$ direction are determined by the probabilities of measurement results, those for the spin along the $x$ and $y$ directions. Consequently, all the elements of the density matrix can be updated as new information/measurement results become available using the Bayes theorem. We now readily see why complete ignorance is represented essentially by the unit matrix [7]. In the absence of any information, by The Principle of Indifference [13] we must have $p^{(\alpha,+1)} = p^{(\alpha,-1)} = 1/2$ for each and every $\alpha$ giving

$$\rho = (1/2)\begin{pmatrix} 1 & 0 \\ 0 & 1 \end{pmatrix} \tag{15}$$

Also, if we only have results for the measurement of $\sigma_z$, then the density matrix must be diagonal. The result that the density matrix for a spin 1/2 system can be expressed entirely in terms of the probabilities of the results of measurements can be extended to any finite number of dimensions as shown in Appendix B.

# 5 Constructing a Density Matrix from Data

Often the density matrix for a system is discussed as a purely theoretical construct. But we need to appreciate that it can be derived directly from suitable data. We will now do this for the simplest of cases, the Stern-Gerlach experiment with an inhomogeneous field in the $z$ direction. We will assume the beam is sufficiently weak that the spin 1/2 atoms pass through one at a time so that the detectors can act as counters and we record data $D(n_\uparrow, n_\downarrow)$ where $n_\uparrow$and $n_\downarrow$ are the number of counts corresponding to spin up and spin down respectively. From our considerations in the previous section we know that the density matrix is determined by $\langle\sigma_z\rangle$, the quantum mechanical expectation value, but, of course, the data do not determine $\langle\sigma_z\rangle$ precisely and the best we can do is compute a probability density distribution for the possible values, $\sigma$, of $\langle\sigma_z\rangle$. We can do this using the Bayes theorem. Normally one would consider no prior knowledge or prejudice and by The Principle of Indifference [13] use a constant, 1/2, for the prior probability distribution since $-1 \leq \sigma \leq 1$ and all values are equally likely. But it will be useful to consider the general case and assume a nonuniform prior probability density distribution, $p_0(\sigma)$.Using the Bayes theorem,

$$p(\sigma \mid D) = \frac{p(D \mid \sigma)p_0(\sigma)}{p(D)} \tag{16}$$

where $p(\sigma \mid D)$ is the probability density distribution for $\langle\sigma_z\rangle$ given the data, $D$, $p(D \mid \sigma)$ is the probability of obtaining the data,$D$, given $\langle\sigma_z\rangle$ has the value $\sigma$, and $p(D)$ is the probability of obtaining data,$D$, regardless of the value of $\langle\sigma_z\rangle$, which is given by

$$p(D) = \int_{-1}^{+1} p(D \mid \sigma)p_0(\sigma)\,\mathrm{d}\sigma \tag{17}$$

and which, further, ensures that $\int_{-1}^{+1} p(\sigma \mid D)\,\mathrm{d}\sigma = 1$, as it must.

If the expectation value of $\sigma_z$ is $\sigma$ , then the probability for spin up/down for a single count is $(1 \pm \sigma)/2$ and $p(D \mid \sigma)$ is given by the binomial distribution :

$$p(D \mid \sigma) = \binom{n}{n_\uparrow}(\frac{1+\sigma}{2})^{n_\uparrow}(\frac{1-\sigma}{2})^{n_\downarrow} \tag{18}$$

where $n = n_\uparrow + n_\downarrow$ . This gives us our formula

$$p(\sigma \mid D) = \frac{(1+\sigma)^{n_\uparrow}(1-\sigma)^{n_\downarrow}p_0(\sigma)}{\int_{-1}^{+1}(1+\sigma)^{n_\uparrow}(1-\sigma)^{n_\downarrow}p_0(\sigma)\,\mathrm{d}\sigma} \tag{19}$$

for updating our knowledge in the light of data obtained. In particular, this formula illustrates the epitome of the scientific process when we examine the limit of large number of data points. In that limit $(1+\sigma)^{n_\uparrow}(1-\sigma)^{n_\downarrow}$ becomes sharply peaked at $\sigma_p = (n_\uparrow - n_\downarrow)/(n_\uparrow + n_\downarrow)$ and $p(\sigma \mid D)$ becomes independent of $p_0(\sigma)$. This is an example of the more general principle, that sufficient data makes previous prejudices or errors irrelevant, which is discussed in Appendix C. However, there is one exception to this conclusion that occurs when $p_0(\sigma)$ is just a delta function i.e. there is absolute *a priori* conviction that $\sigma$ has a definite value. In that case, no amount of subsequent data from further experiments will change that conviction and one is outside the realm of science.

To obtain our effective density matrix as the result of collecting data, $D$, we need to consider $\langle O\rangle_D$ the expectation value of a general observable, $O$, using our current probability distribution, $p(\sigma \mid D)$, for $\sigma$. If the value of $\langle\sigma_z\rangle$ is taken as $\sigma$, then the expectation value of $O$ will be given by

$$\langle O\rangle_\sigma = \mathrm{Tr}[O\rho_\sigma] \tag{20}$$

where

$$\rho_\sigma = (1/2)\begin{pmatrix} 1+\sigma & 0 \\ 0 & 1-\sigma \end{pmatrix} \tag{21}$$

But all values of $\sigma$ are possible with probability density distribution $p(\sigma \mid D)$. So, the expectation value, $\langle O\rangle_D$, of $O$ given data $D$, is given by

$$\langle O\rangle_D = \int_{-1}^{+1} p(\sigma \mid D)\langle O\rangle_\sigma \,\mathrm{d}\sigma = \int_{-1}^{+1} p(\sigma \mid D)\,\mathrm{Tr}[O\rho_\sigma]\,\mathrm{d}\sigma \tag{22}$$

or

$$\langle O\rangle_D = \mathrm{Tr}[\int_{-1}^{+1} p(\sigma \mid D) O\rho_\sigma \,\mathrm{d}\sigma] = \mathrm{Tr}[O\rho_D] \tag{23}$$

where

$$\rho_D = \int_{-1}^{+1} p(\sigma \mid D)\rho_\sigma \,\mathrm{d}\sigma \tag{24}$$

Assuming we started from complete ignorance, so that, $p_0(\sigma)$, is just constant (and equal to $1/2$ for all $\sigma$), then it is straightforward to derive the effective density matrix

$$\rho_D = \frac{1}{(n+2)}\begin{pmatrix} n_\uparrow + 1 & 0 \\ 0 & n_\downarrow + 1 \end{pmatrix} \tag{25}$$

As a check, one notes that for no data at all, $n_\uparrow = n_\downarrow = 0$, we just obtain (15) representing complete ignorance, while in the limit of large data sets we obtain probabilities $n_\uparrow/n$ and $n_\downarrow/n$ respectively for spin up and spin down. The reader will also be able to verify the intuitively reasonable result that, if an observer collects some data, $D^{(1)} = (n_\uparrow^{(1)}, n_\downarrow^{(1)})$, and calculates a probability distribution, $p(\sigma \mid D^{(1)})$, and then a second observer takes this as prior information and updates it using their own data, $D^{(2)} = (n_\uparrow^{(2)}, n_\downarrow^{(2)})$, to obtain a probability distribution, $p(\sigma \mid (D^{(1)}\ \textit{then}\ D^{(2)}))$, then this will be the same as $p(\sigma \mid D^{(total)})$ where $D^{(total)} = (n_\uparrow^{(1)} + n_\uparrow^{(2)}, n_\downarrow^{(1)} + n_\downarrow^{(2)})$, that is, the probability distribution obtained if one just amalgamated the two data sets. This is the simplest case of two observers collaborating, simplest because they are both measuring the same observable. We now move on to the case of two or more observers measuring different observables and examine the meaning and significance of the Peierls Commutation and Product criteria.

# 6 The Peierls Commutation and Product Criteria

Peierls [7] asserts that the density matrices of different observers must commute, otherwise one may violate the uncertainty principle. It is not difficult to come up with what appears to be a simple counter example. Suppose two observers, $Z$ and $X$ have their separate, yet identical, Stern-Gerlach apparatus including identical sources of silver atoms. They each measure the number of spin up and spin down atoms along the $z$ and $x$ directions respectively. They summarize their results by each producing their own density matrices, $\rho^{(Z)}$ and $\rho^{(X)}$ given by

$$\rho^{(Z)} = (\frac{1}{2})(I + \langle\sigma_z\rangle\sigma_z) \tag{26}$$

and

$$\rho^{(X)} = (\frac{1}{2})(I + \langle\sigma_x\rangle\sigma_x) \tag{27}$$

where $Z$ and $X$ respectively calculate $\langle\sigma_z\rangle$ and $\langle\sigma_x\rangle$ from their own data. Observer $Z$ must set $\langle\sigma_x\rangle = \langle\sigma_y\rangle = 0$ in (9), by The Principle of Indifference [13], as they have no knowledge of the polarization in the $x$ and $y$ directions. Similarly $X$ must set $\langle\sigma_y\rangle = \langle\sigma_z\rangle = 0$ . The commutator of the observers' density matrices is readily found to be

$$[\rho^{(Z)}, \rho^{(X)}] = (i/4)\langle\sigma_z\rangle\langle\sigma_x\rangle\sigma_y \neq 0 \tag{28}$$

This commutator will only vanish if at least one of the expectation values vanishes which is not the case in general. So we seem to have a contradiction here. Peierls's commutation criterion seems to be saying that $Z$ and $X$ cannot be simultaneous observers even if they are using separate identical sources and separate pieces of equipment! This cannot be true and we must look for another interpretation. Suppose $Z$ and $X$ decide to combine their results and form a combined density matrix , $\rho^{(Z \text{ with } X)}$ using (9)

$$\rho^{(Z \text{ with } X)} = (\frac{1}{2})(I + \langle\sigma_x\rangle\sigma_x + \langle\sigma_z\rangle\sigma_z) \tag{29}$$

where $\langle\sigma_y\rangle = 0$ in (9), by The Principle of Indifference [13] since neither $Z$ nor $X$ has any knowledge of the polarization in the $y$ direction. It is readily verified that the eigenvalues of $\rho^{(Z\ with\ X)}$ are $(1 \pm \Gamma)/2$ where $\Gamma = +\sqrt{(\langle\sigma_x\rangle)^2 + (\langle\sigma_z\rangle)^2}$. We see that for these eigenvalues to be physically acceptable, then we must have $\Gamma \leq +1$ otherwise there must exist states with a negative occupation probability. Such probabilities violate the uncertainty principle; the author has not managed to find a proof of this in any textbook, so a proof is given in Appendix D, thereby establishing that the positivity of the density matrix and the Uncertainty Principle are completely equivalent statements, each implying the other, since the proof in Appendix D complements the standard text book proof that the positivity of the density matrix implies the Uncertainty Principle. So, it is impossible to find any physically realizable state for which $\Gamma > 1$ . However, if we construct the density matrix using data, then violations of the uncertainty principle can occur. Suppose the observers $Z$ and $X$ collect data $D^{(Z)} = (n_\uparrow^{(Z)}, n_\downarrow^{(Z)})$ and $D^{(X)} = (n_\uparrow^{(X)}, n_\downarrow^{(X)})$ respectively. Using the same procedure as in section 4, the effective density matrix on combining the two data sets is

$$\rho_D^{(Z\ with\ X)} = \begin{pmatrix} p^{(z,+1)} & \Delta_x \\ \Delta_x & p^{(z,-1)} \end{pmatrix} \tag{30}$$

with

$$p^{(z,+1)} = \frac{n_\uparrow^{(Z)} + 1}{n^{(Z)} + 2} \tag{31}$$

$$p^{(z,-1)} = \frac{n_\downarrow^{(Z)} + 1}{n^{(Z)} + 2} \tag{32}$$

and

$$\Delta_x = \frac{1}{2}(p^{(x,+1)} - p^{(x,-1)}) = \frac{1}{2}\frac{n_\uparrow^{(X)} - n_\downarrow^{(X)}}{n^{(X)} + 2} \tag{33}$$

Now it is readily verified that the eigenvalues of $\rho_D^{(Z\ with\ X)}$ lie outside the range $(0,1)$ and, in particular, $\rho_D^{(Z\ with\ X)}$ has a negative eigenvalue, if

$$(\Delta_x)^2 > p^{(z,+1)}p^{(z,-1)} \tag{34}$$

and, when this inequality is satisfied, there is then a violation of the uncertainty principle. So if, for instance, the spin down detector used by observer $X$ is very inefficient, then the value for $(\Delta_x)^2$ will be close to $1/4$ which happens to be the maximum possible value of $p^{(z,+1)}p^{(z,-1)}$. So, if there is any significant polarisation of the $z$ component of the spin, (complete polarisation will give $p^{(z,+1)}p^{(z,-1)}$ as zero), then there will be a violation of the uncertainty principle caused by the inefficiency of $X$'s spin down detector. In this case, $\rho_D^{(Z \textit{ with } X)}$ will have eigenvalues outside the range $(0, 1)$ and as mentioned in Appendix C no amount of fresh data, even good data, will correct the situation.

At this point we can see why the counter example to Peierls's commutation requirement put forward by Mermin [11] is not relevant. Mermin evaluated the density matrices of his two observers using theory alone and such density matrices can never violate the Uncertainty Principle (unless a mathematical error is made!) whether the density matrices commute or otherwise; using theory alone in our example above, we would always have $\Gamma \leq 1$ because $\langle \boldsymbol{\sigma} \rangle . \langle \boldsymbol{\sigma} \rangle = 1$ and hence a nonnegative density matrix is obtained and no possibility of violating the Uncertainty Principle arises. It is only when we consider density matrices formed using data that there is a possibility of violating the Uncertainty Principle, but only if the density matrices of the two observers do not commute. If they do commute, then no chance of violating the Uncertainty Principle is possible no matter how inaccurate the data.

In order to elucidate the Peierls commutation criterion further, suppose there is a third observer, $N$, who, like $Z$ and $X$, has set up an identical source of silver atoms, and has put them through a Stern-Gerlach apparatus to split them into two beams, but then puts one of the beams through a second Stern-Gerlach apparatus with a much more powerful inhomogeneous magnetic field so that it splits into two further beams according to whether the nuclear spin is up or down. While this second apparatus is impractical, the nuclear magnetic moment is about 4 orders of magnitude smaller than the electron magnetic moment, it certainly illustrates the point Peierls makes. We suppose that observer $N$ wants to amalgamate their data with that of an observer $E$ who has measured the polarization of the electron spin of an identical source of silver atoms; both $Z$ and $X$ are separately examples of observer $E$. Because the electron spin and the nuclear spin are commuting observables, we do not need to be specific about the directions of the inhomogeneous magnetic fields used by $E$ and $N$. We denote by $p_E^{(\pm 1)}$ the probabilities of electron spin up/down measured by $E$ and by $p_N^{(\pm 1)}$ the probabilities of nuclear spin up/down measured by $N$. The density matrices for $E/N$ are given by

$$\rho_D^{(E/N)} = \begin{pmatrix} p_{E/N}^{(+1)} & 0 \\ 0 & p_{E/N}^{(-1)} \end{pmatrix} \tag{35}$$

where the basis states are $(E/N, +1)$ corresponding to spin up and $(E/N, -1)$ corresponding to spin down. Now consider how the two observers modify their density matrices in the light of the knowledge of the data of the other. To start with $E$ has no knowledge of $N$'s data and must allocate equal probabilities of $1/2$ for nuclear spin up/down. So in the basis of, in a self evident notation, $[(E,+),(N,+)]$ , $[(E,+),(N,-)]$ , $[(E,-),(N,+)]$ , $[(E,-),(N,-)]$ , they must have a density matrix

$$\rho_{D,0}^{(E)} = \begin{pmatrix} p_E^{(+1)}/2 & 0 & 0 & 0 \\ 0 & p_E^{(+1)}/2 & 0 & 0 \\ 0 & 0 & p_E^{(-1)}/2 & 0 \\ 0 & 0 & 0 & p_E^{(-1)}/2 \end{pmatrix} \tag{36}$$

while observer $N$ having no knowledge of the electron spin polarization has a density matrix

$$\rho_{D,0}^{(N)} = \begin{pmatrix} p_N^{(+1)}/2 & 0 & 0 & 0 \\ 0 & p_N^{(-1)}/2 & 0 & 0 \\ 0 & 0 & p_N^{(+1)}/2 & 0 \\ 0 & 0 & 0 & p_N^{(-1)}/2 \end{pmatrix} \tag{37}$$

These two matrices evidently commute as they are both diagonal. When each observer becomes aware of the other's data they update their respective density matrices to arrive at a common density matrix

$$\rho_D^{(both)} = \begin{pmatrix} p_E^{(+1)}p_N^{(+1)} & 0 & 0 & 0 \\ 0 & p_E^{(+1)}p_N^{(-1)} & 0 & 0 \\ 0 & 0 & p_E^{(-1)}p_N^{(+1)} & 0 \\ 0 & 0 & 0 & p_E^{(-1)}p_N^{(-1)} \end{pmatrix} \tag{38}$$

Now, in sharp contrast to the common density matrix arrived at by observers $Z$ and $X$ , this matrix will never violate the uncertainty principle no matter how inaccurate the data. The counters will always provide data such that the quantities $p_{E/N}^{(\pm 1)}$ lie between zero and one and will obey $p_{E/N}^{(+1)} + p_{E/N}^{(-1)} = 1$ so that $\mathrm{Tr}[\rho_D^{(both)}] = 1$. The eigenvalues of $\rho_D^{(both)}$ will all be non-negative and no violation of the uncertainty principle is possible.

So the purpose of the Peierls commutation criterion is to ensure trouble free amalgamation of the data of two observers i.e. amalgamation without any danger of violation the uncertainty principle.

It is shown in appendix E that if the density matrices of two observers commute no matter how inaccurate the individual data, then the observables they are measuring must also commute, the same requirement from conventional quantum mechanics for two observables to be known precisely and simultaneously. This latter requirement is show to be necessary for imperfect measurements on a potentially mixed state system so that possible violations of the Uncertainty Principle can be avoided. In appendix F the immunity to uncertainty principle violations demonstrated above for electron and nuclear spin 1/2 systems is generalized to all systems with finite dimensional state spaces.

Peierls [10] also points out that the product of the density matrices of two observers should not be zero to avoid contradiction and the above analysis can illustrate this point. The product of the density matrices $\rho_{D,0}^{(E)}$ and $\rho_{D,0}^{(N)}$ of the observers $E$ and $N$, is given by

$$\rho_{D,0}^{(E)} \times \rho_{D,0}^{(N)} = \frac{1}{4} \begin{pmatrix} p_E^{(+1)} p_N^{(+1)} & 0 & 0 & 0 \\ 0 & p_E^{(+1)} p_N^{(-1)} & 0 & 0 \\ 0 & 0 & p_E^{(-1)} p_N^{(+1)} & 0 \\ 0 & 0 & 0 & p_E^{(-1)} p_N^{(-1)} \end{pmatrix} \tag{39}$$

If this product vanishes, each diagonal element must vanish and one finds that for that to be, at least , either both the probabilities $p_E^{(\pm 1)}$ or both the probabilities $p_N^{(\pm 1)}$ must vanish. This is clearly absurd and requiring that the product does not vanish is the minimum requirement to maintain coherence. The criterion that the product of the density matrices of two observers should be non zero is discussed more generally in Appendix G.

# 7 Peierls Revisited

In the light of the exposition in the previous sections, it is worthwhile revisiting Peierls's work, by quoting some of Peierls's final comments in his *More Surprises in Theoretical Physics* [10] and commenting on, and augmenting them. Firstly, on page 10, : *"It is possible for two observers to have some knowledge of the same system, and the knowledge possessed by one may differ from the other. For example, one may only have an incomplete or inaccurate view of the measuring instrument. In this situation the two observers will use different density matrices."* This emphasises the subjective nature of the density matrix of an observer to the extent that the two observers are viewing the same measuring instrument rather than carrying out the same experiment in separate laboratories. Our treatment of the Stern Gerlach experiment in section 5 is directly applicable to either situation, each observer constructing their density matrix according to the counts they each actually record. It is clear that Peierls is only describing the very first stage of the scientific process here. With one observer having "only an incomplete or inaccurate view of the measuring instrument", clearly the same and/or other observers will repeat the experiments : and provided only a small subset of the results are subject to systematic errors such as "an incomplete or inaccurate view of the measuring instrument", their influence will be systematically squeezed out. The analysis in section 5 gives a simple working model to illustrate the process.

Secondly, on pages 10 to 11: *"However, the information possessed by the two observers must not contradict the uncertainty principle. For example, if one observer knows the z component of an atomic spin [exactly], the other may not know [whether spin up or spin down is more likely for] the x component. This is because the measurement made by the first observer would have caused an uncontrollable interference with the x component. This limitation can be expressed concisely by saying that the density matrices appropriate to the two observers must commute with each other."*

The words inside the square brackets are the author's addition to aid clarity; and the $x$ and $z$ in the original have been interchanged for easier reference to the working in section 6 (and, indeed, throughout this paper) in which the basis states have been the spin up and spin down states in the $z$ direction .

To visualize what Peierls is saying we imagine the first observer passing an atomic spin through a Stern Gerlach apparatus with an inhomogeneous magnetic field in the $z$ direction. If that first observer has a detector in both the spin up and spin down exit channels, then they will be able to find out the polarization of the $z$ component of the spin the atom had. But, it is no longer possible to do any other experiment on that atom. Suppose, instead of trying to record the spin, the first observer allows a second observer to set up a Stern Gerlach apparatus with inhomogeneous magnetic field in the $x$ direction in the path along which the atom would emerge from the first observer's apparatus had it entered with spin up in the $z$ direction and for the second observer to set up detectors to measure the $x$ component of the atomic spin; so we are now considering sequential measurements on the same system rather than parallel experiments on identical systems. Then according to section 6, because all counts corresponding to spin down in the $z$ direction are automatically excluded from the observations so that we know that the atom is entering the second observer's apparatus with spin up, we obtain a density matrix $\rho^{(Z\ with\ X)}$ of the form $\begin{pmatrix} 1 & \Delta_x \\ \Delta_x & 0 \end{pmatrix}$. Now this matrix will always have one negative eigenvalue and another greater than unity and hence is not an acceptable density matrix; it violates the uncertainty principle. The only exception is for $\Delta_x = 0$ which corresponds to equal counts corresponding to spin up and spin down in the $x$ direction i.e. complete ignorance of the $x$

component of the spin. The precise knowledge of the $z$ component of the spin eliminates all possibility of knowing anything, except one's awareness of one's complete ignorance about the $x$ component.

If, however, the second observer measured the spin of the nucleus, then, according to our working in section 6, the two observers would arrive at the density matrix $\begin{pmatrix} p_N^{(+1)} & 0 \\ 0 & p_N^{(-1)} \end{pmatrix}$ (electron spin up assumed) where $p_N^{(+1)}$ and $p_N^{(-1)}$ are the probabilities deduced from the data for the $x$ component of the nuclear spin to be up and down respectively. As explained in section 6, these data dependent probabilities, can only lie between 0 and 1 no matter how inaccurate the data and the matrix automatically satisfies the uncertainty principle, a property that is a consequence of the commutation of the electronic and nuclear spin variables. Only when Peierls's criterion is satisfied can one be sure that violations of the uncertainty principle will be avoided. It is not difficult to see that the root of the problem in the solely electronic measurements is the fact that $\sigma_x$ has only off diagonal elements and this is related to the fact that it does not commute with $\sigma_z$ .

Thirdly, on page 11 of his *More Surprises in Theoretical Physics* , Peierls continues to provide a justification of the commutation criterion :

" *This can be seen by considering the representation in which one of these [density] matrices is diagonal. If the other does not commute with it, it must have off-diagonal elements in the representation. This means that it shows phase relations between the states which the first observer could have observed, and this would violate the uncertainty principle.* "

We have given a fuller proof of this result in appendix H, but it is important to emphasize that it is necessary to assume that all the information in the density matrix has, in principle, subjective origins to establish the result. If one does not assume this, but assumes that the density matrix is derived from theory and not experimental data, then it is easy to find putative counter examples.

One can deduce from section 4 and Appendices A and B that the "... phase relations between the states ..." are none other than the expectation values of the observable measured by the second observer or by subsequent measurements by the first observer of that same observable. And because the measurements of the first and second observables are subject to errors there can be no guarantee that violations of the uncertainty principle will not occur. And finally Peierls mentions his criterion to avoid a contradiction :

" *At the same time the two observers should not contradict each other. This means that the product of the two density matrices must not be zero.*

*Indeed, take a representation in which both are diagonal (which is possible if they commute); then there must be at least some states for which the probabilities assumed by both observers are non zero, and that means the product is non-zero.*"

We have shown, in section 6, explicitly for two observers measuring respectively the electronic and nuclear spin of an atom that, if the product of the density matrices of the two observers vanishes, then we end up with an absurd situation : all probabilities for the measurement outcomes for at least one of the variables must vanish. And we have also expanded on the proof of Peierls's theorem in Appendix G. In particular, we point out that, if Peierls's criterion is not satisfied, it will be impossible to reconcile the two observers' findings using further data and the Bayes Theorem.

# 8 Summary

We have provided an expansion and elucidation of Peierls's interpretation of quantum mechanics. Crucial to this has been the appreciation that the knowledge of an observer, encapsulated in their density matrix, can be expressed in terms of probabilities of observable outcomes even though those outcomes may be dependent on probability amplitudes and interference effects. As such, they can be updated as more data becomes available and eventually a consensus among the scientific community emerges. In particular, it has been shown that the density matrix in any representation with a finite basis can be written in terms of the expectation values of observable quantities and therefore in terms of probabilities of the outcome of the measurements being particular eigenvalues of those observables; the case for two dimensions [ spin 1/2 particles ] has been treated in detail and then generalized in appendices A and B. It has been shown how the density matrices can be constructed from experimental data for the Stern Gerlach experiment and how two observers performing such experiments can collaborate and amalgamate their data both for the case in which they are measuring the same observable or two different observables and how large data sets can minimize the influence of initial prejudices, the essence of the scientific method.

We have also shown that the criterion that the density matrices of two observers must commute is a generalisation of the usual quantum mechanical condition that two observables must commute if they are to be exactly known simultaneously without violating the Uncertainty Principle; but the Peierls criterion shows that the relevance of the condition can be extended into the realm of individual imperfect, even wildly inaccurate measurements; if the two observables do not commute, violations of the Uncertainty Principle may well occur. Perhaps it is helpful to state these results in just one theorem : if we want two observers/physicists to share data (including the results of, possibly inaccurate, calculations) concerning the state (either mixed or pure) of a quantum system constructed from arbitrary and unrestricted, but still possible belief, so that the Uncertainty Principle is never violated, then their observables must commute.

And, we have also illustrated how the criterion that the product of the density matrices of two observers must be non zero is necessary for there to be any hope of reconciling the results of two observers using further experiment.

In finishing this summary, the author would like to requote and reword Peierls's statement on pages 10 and 11 of his *More Surprises in Theoretical Physics*[10] about his commutation criterion, namely,

*"However, the information possessed by the two observers must not contradict the uncertainty principle. For example, if one observer knows the z component of an atomic spin , the other may not know the x component. This is because the measurement made by the first observer would have caused an uncontrollable interference with the x component. This limitation can be expressed concisely by saying that the density matrices appropriate to the two observers must commute with each other."* .

This statement has caused the author much puzzlement, and, on reflection, he offers his own version, in which extra words are included in square brackets : *"However, the information possessed by the two observers must not contradict the uncertainty principle. For example, if one observer knows the z component of an atomic spin [exactly], the other may not know [ whether spin up or spin down is more likely for] the x component. This is because the measurement made by the first observer would have caused an uncontrollable interference with the x component. This limitation can be [avoided and the condition necessary to do so can be] expressed concisely by saying that the density matrices appropriate to the two observers must commute with each other [no matter how inaccurate their data]."*

Finally, in parting, the author sincerely hopes that he has helped the reader appreciate further both the perspicacity and perspicuity that Peierls brought to this subject by spelling out two points of which Peierls must certainly have been aware, but constrained by circumstances to omit: firstly, that the density matrix can be expressed entirely in terms of probabilities and secondly, that these are subjective in the first instance, and so can be updated as further data and/or information becomes available to proceed towards greater objectivity by using the Bayes Theorem. I, for one, am grateful to Peierls for bringing Heisenberg's elegant idea to fruition and for providing the user of quantum mechanics, a nonspecialist in quantum foundations, with a coherent pragmatic approach.

# Appendix A : The Density Matrix in terms of the Expectation Values of Observables

We assume that the state space has finite dimension, $N$. We will show how the real and imaginary parts of each and every element of the density matrix/operator, $\rho$, can be expressed in terms of the expectation value, $\langle O\rangle = \text{Tr}[\rho O]$, of either some operator/dynamical variable, $O$, or a linear combination of two such. There is a distinction between operators/ dynamical variables and observables in general see e.g.[14] pp 26 and 28, in that the latter must have a complete set of eigenvectors. Fortunately, since we are working in a state space of finite dimension, all dynamical variables are observables: hermitian operators are an example of normal operators (for which $AA^\dagger = A^\dagger A$) which are well known to have a complete set of orthonormal eigenvectors, see e.g.[15] page 72, even in the presence of degeneracy. The reader will readily be able to confirm that all the operators/dynamical variables used here are indeed observables.

Introduce the $N$ observables $O(m,m)$ defined in some orthonormal basis (with vectors $|m\rangle, m = 1, 2, \ldots, N$ ) by

$$O(m,m) = |m\rangle\langle m| \tag{40}$$

for which it is easily verified that for the diagonal components of $\rho$

$$\langle m|\rho|m\rangle = \langle O(m,m)\rangle \tag{41}$$

Moving on to expressing the off-diagonal elements of $\rho$ in terms of expectation values of observables, introduce the $N(N-1)/2$ observables ( $1 \leq k < l \leq N$)

$$O(k,l) = a|k\rangle\langle l| + a^*|l\rangle\langle k| \tag{42}$$

we readily find

$$\langle O(k,l)\rangle = a\langle l|\rho|k\rangle + a^*\langle k|\rho|l\rangle \tag{43}$$

If we now define observables $O^{(x)}(k,l)$ and $O^{(y)}(k,l)$ by setting $a = 1/2$ and $a = -i/2$ in (42) respectively, then it is straightforward to find (for $1 \leq k < l \leq N$)

$$\langle k|\rho|l\rangle = \langle O^{(x)}(k,l)\rangle - i\langle O^{(y)}(k,l)\rangle \tag{44}$$

the elements $\langle l|\rho|k\rangle$ below the diagonal being just the complex conjugates.

# Appendix B : Density Matrix in terms of Probabilities of Observable Outcomes

We have already demonstrated in Appendix A how the density matrix can be specified in terms of the expectation values of various observables.

As a straightforward consequence, one can also write the elements of the density matrix, $\rho$, in an orthonormal basis, $|m\rangle$ $(m = 1, 2, \ldots, N)$, in terms of the probabilities of various observational results alone.

For the diagonal elements, we have

$$\langle m|\rho|m\rangle = \langle O(m,m)\rangle = p_m \tag{45}$$

where $p_m$ is the probability of finding the system in the state $|m\rangle$

For off-diagonal elements in the upper triangle ( $1 \leq k < l \leq N$)

$$\langle k|\rho|l\rangle = \langle O^{(x)}(k,l)\rangle - i\langle O^{(y)}(k,l)\rangle \tag{46}$$

Now the observables $O^{(x)}(k,l)$ and $O^{(y)}(k,l)$ each have two eigenstates each with non zero eigenvalue, $\sigma/2$ with $\sigma = \pm 1$ and $N-2$ eigenstates each with zero eigenvalue. Obviously, in evaluating the expectation value of either observable the eigenstates corresponding to zero eigenvalue will not contribute and we have using $\alpha = (x, y)$

$$\langle O^{(\alpha)}(k,l)\rangle = \sum_{\sigma} p_{k,l}^{\alpha,\sigma} \sigma/2 \tag{47}$$

or explicitly

$$\langle O^{(\alpha)}(k,l)\rangle = (p_{k,l}^{\alpha,+1} - p_{k,l}^{\alpha,-1})/2 \tag{48}$$

where $p_{k,l}^{\alpha,\sigma}$ is the probability of the system being in the eigenstate of $O^{(\alpha)}(k,l)$ with eigenvalue $\sigma/2$. Now introducing

$$\Delta p_{k,l}^{\alpha} = (p_{k,l}^{\alpha,+1} - p_{k,l}^{\alpha,-1}) \tag{49}$$

we can write for the elements above the diagonal,($1 \leq k < l \leq N$ ) ,

$$\langle k|\rho|l\rangle = (\Delta p_{k,l}^{(x)} - i\Delta p_{k,l}^{(y)})/2 \tag{50}$$

And finally the complete density matrix may be written formally as

$$\rho = \sum_m p_m O(m,m) + \sum_{k<l} [\Delta p_{k,l}^{(x)} O^{(x)}(k,l) + \Delta p_{k,l}^{(y)} O^{(y)}(k,l)] \tag{51}$$

which is clearly expressed in terms of observables and probabilities of measurement outcomes.

# Appendix C : the Bayes Theorem and the Scientific Method

We can use the Bayes theorem to evaluate the probability, $P[H \mid D]$, of a scientific hypothesis, $H$, being true, given that some data, $D$, has been collected, and also given that the a priori probability that $H$ is true, is $P[H]$. The theorem states that

$$P[H \mid D] = \frac{P[D \mid H]P[H]}{P[D]} \tag{52}$$

where $P[D \mid H]$ is the probability that data, $D$, would be collected if hypothesis, $H$, were true. $P[D]$ is the probability that the data, $D$, is collected regardless of whether $H$ is true or false :

$$P[D] = P[D \mid H]P[H] + P[D \mid \overline{H}]P[\overline{H}] \tag{53}$$

where $\overline{H}$ denotes the opposite hypothesis to $H$ so that $P[H] + P[\overline{H}] = 1$. One can now write the Bayes theorem as

$$P[H \mid D] = \frac{1}{1+A} \tag{54}$$

with

$$A = \frac{P[D \mid \overline{H}]}{P[D \mid H]}\frac{P[\overline{H}]}{P[H]} \tag{55}$$

Now suppose an experiment is set up to confirm the truth of hypothesis, $H$, such that

$$P[D \mid H] > P[D \mid \overline{H}] \tag{56}$$

If the experiment is repeated independently $n$ times, then the probability of this happening is $(P[D \mid H])^n$ and $A$ has to be replaced by

$$A_n = \left(\frac{P[D \mid \overline{H}]}{P[D \mid H]}\right)^n \frac{P[\overline{H}]}{P[H]} \tag{57}$$

As $n$ becomes sufficiently large, $A_n$ tends to zero, and $P[H \mid D]$ tends inexorably towards unity regardless of the value of $P[H]$, i.e. the original prejudices of the observer(s)/scientific community. This simple result neatly encapsulates the scientific method. One notes that it is essential for $P[H]$ to differ from either zero or unity. If $P[H]$ takes on either of these extreme values, then no amount of data collection will alter that belief.

In the light of the above demonstration of Bayesian updating, one wonders if, having got into a situation, such as described in Section 6, in which one has probabilities outside the range (0,1) whether further data can rectify the problem. Unfortunately, this is not the case as a relatively straightforward analysis shows. So great care has to be taken if one does not adhere to the Peierls Criterion and one tries to amalgamate data from observers using non-commuting observables.

# Appendix D : Proof that a negative eigenvalue of the density matrix can lead to a violation of the uncertainty principle.

We first demonstrate the result for the spin 1/2 case i.e. for 2 dimensions and then extend it to all finite dimensions. We work in a representation for which the density matrix is diagonal and write

$$\rho = \begin{pmatrix} \wp_1 & 0 \\ 0 & \wp_2 \end{pmatrix} \tag{58}$$

where $\wp_1$ and $\wp_2$ are the eigenvalues of the density matrix, $\rho$. Since one must have $\mathrm{Tr}[\rho] = 1$ , or $\wp_1 + \wp_2 = 1$, at least one of $\wp_1$ and $\wp_2$ must be positive. We need to show that there exists at least one pair of observables , $A$ and $B$ , for which the uncertainty principle is violated i.e. for which

$$\langle(\Delta A)^2\rangle\langle(\Delta B)^2\rangle < |\langle\Delta A \Delta B\rangle|^2 \tag{59}$$

whenever one of $\wp_1$ and $\wp_2$ is negative. Here $\Delta O = O - \langle O\rangle I$ for any observable $O$. For notational convenience, take the $z$ axis so that the basis states that diagonalize $\rho$ correspond to spin up and spin down with respect to that axis and consider the difference of the left and right hand sides of the above inequality i.e.

$$\langle(\Delta A)^2\rangle\langle(\Delta B)^2\rangle - |\langle\Delta A \Delta B\rangle|^2 \tag{60}$$

for the case $A = \sigma_x = \begin{pmatrix} 0 & 1 \\ 1 & 0 \end{pmatrix}$ and $B = \sigma_y = \begin{pmatrix} 0 & -i \\ +i & 0 \end{pmatrix}$ . It is readily verified that $\langle\sigma_\alpha\rangle = \mathrm{Tr}[\rho\sigma_\alpha] = 0$ for $\alpha = (x, y)$ so that $\Delta O = O$ for both $A$ and $B$ and hence

$$\langle(\Delta A)^2\rangle = \langle(\Delta\sigma_x)^2\rangle = \langle(\sigma_x)^2\rangle = \langle I\rangle = \wp_1 + \wp_2 \tag{61}$$

and similarly for $\langle(\Delta B)^2\rangle$. Therefore

$$\langle(\Delta A)^2\rangle\langle(\Delta B)^2\rangle = (\wp_1 + \wp_2)^2 \tag{62}$$

On the other hand,

$$|\langle \Delta A \Delta B \rangle|^2 = |\langle \sigma_x \sigma_y \rangle|^2 = |\langle \sigma_z \rangle|^2 = (\wp_1 - \wp_2)^2 \tag{63}$$

so that evaluation of (60) gives

$$\langle (\Delta A)^2 \rangle \langle (\Delta B)^2 \rangle - |\langle \Delta A \Delta B \rangle|^2 = 4\wp_1 \wp_2 < 0 \tag{64}$$

since $\wp_1$ and $\wp_2$ have opposite signs. So we have found a pair of observables for which the uncertainty principle is violated if $\rho$ has a negative definite eigenvalue. This is enough to establish the result that a negative eigenvalue for the density matrix can lead to a violation of the uncertainty principle for spin 1/2 systems.

To extend the result to an arbitrary number of dimensions, $N$ , is straightforward. One notes that $\mathrm{Tr}[\rho] = 1$ or $\wp_1 + \wp_2 + \ldots + \wp_N = 1$ still requires at least one of the $\wp_i$'s to be positive. Relabel the eigenstates of $\rho$ so that $\wp_1$ is positive and $\wp_2$ negative. For the observables $A$ and $B$ use $O^{(x)}(1,2)$ and $O^{(y)}(1,2)$ introduced in appendix $A$ which are equal to $\sigma_x/2$ and $\sigma_y/2$ in the space spanned by the eigenvectors of $\rho$ corresponding to the eigenvalues $\wp_1$ and $\wp_2$, but have all other elements zero. The proof of the violation of the uncertainty principle then proceeds as for the spin 1/2 case.

# Appendix E : Peierls Commutation Criterion and Conventional Quantum Mechanics

We demonstrate here that the Peierls Commutation Criterion, that the density matrices of two observers must commute for trouble free sharing of data, is a generalisation of the conventional quantum mechanics requirement that two observables must commute if they are to be known precisely and simultaneously, but now extended to imperfectly known mixed states rather than a single pure state. The crucial assumption here is that the density matrices commute in all circumstances no matter what knowledge, inaccurate or even fanciful, each of the observers may have. Now, the result would be a forgone conclusion if the density matrix could be an arbitrary function of the observable. In that case, we could just set $\rho^{(A)} = A$ and $\rho^{(B)} = B$ and the result follows. However, we have the constraints that $\rho$ is non-negative definite (it must have non-negative eigenvalues) and that $\mathrm{Tr}[\rho] = 1$. The subjective nature of $\rho$ means that there are no other constraints. So suppose we consider

$$\rho^{(A)} = \frac{A + \lambda_A I}{\mathrm{Tr}(A + \lambda_A I)} \tag{65}$$

where $\lambda_A$ is sufficiently large to make $\rho^{(A)}$ non-negative, i.e. larger than the modulus of any negative eigenvalue of $A$. We also have $\mathrm{Tr}[\rho^{(A)}] = 1$. With this assignment for $\rho^{(A)}$ and a similar one for $\rho^{(B)}$, the condition $[\rho^{(A)}, \rho^{(B)}] = 0$ leads to

$$\left[\frac{A + \lambda_A I}{\mathrm{Tr}(A + \lambda_A I)}, \frac{B + \lambda_B I}{\mathrm{Tr}(B + \lambda_B I)}\right] = 0$$

or $[A, B] = 0$ as in standard quantum mechanics. But rather than just ensuring that $A$ and $B$ may be known precisely and simultaneously, the condition $[A, B] = 0$ is now seen to be necessary to avoid violations of the Uncertainty Principle when potentially inaccurate data sets from observers $A$ and $B$ are amalgamated.

# Appendix F : Immunity from Uncertainty Principle Violations

We have demonstrated how an observer observing an electronic spin and a second observer observing a nuclear spin can combine their observations without fear of any violation of the uncertainty principle. We now want to demonstrate generally that two observers using commuting observables can do the same. Since, when the density matrices, $\rho^{(A)}$ and $\rho^{(B)}$, of two observers, A and B, commute, then their observables, $A$ and $B$, commute, we can construct a basis using the simultaneous eigenstates of $A$ and $B$. Observer A, in the absence of any knowledge of observer B's observations or knowledge will write the diagonal elements of their density matrix as

$$p^{(A)}_{(a,b)} = p^{(A)}_a \times (\frac{1}{N_B}) \tag{66}$$

where $N_B$ is the number of eigenvalues of $B$, since , while A from their own knowledge and observation can allocate a probability, $p^{(A)}_a$ , to a measurement of $A$ producing the result $a$, they are ignorant of any measurements by B and must assign the same probability, $1/N_B$, to the result of any measurement of $B$. Similarly, observer B in the absence of any knowledge of observer A's observations or knowledge will write the diagonal elements of their density matrix as

$$p^{(B)}_{(a,b)} = (\frac{1}{N_A}) \times p^{(B)}_b \tag{67}$$

where $N_A$ is the number of eigenvalues of $A$. If the two observers share their knowledge, they can each update their density matrix to

$$\rho^{(A\ with\ B)}_{(a,b),(a',b')} = p^{(A)}_a \times p^{(B)}_b \delta_{(a,b),(a',b')} \tag{68}$$

Because both $p^{(A)}_a$ and $p^{(B)}_b$ lie between zero and unity, this matrix can never have negative eigenvalues and hence the uncertainty principle cannot be violated even for wildly inaccurate probability assignments.

# Appendix G : Peierls Product Criterion

In section 6 we demonstrated, for an example involving electronic and nuclear spins, how Peierls's product criterion must be obeyed to avoid absurd results. We now extend the discussion to higher dimensions.

The application of the product criterion assumes that the density matrices, $\rho^{(A)}$ and $\rho^{(B)}$, of two observers, A and B, commute, so that we can use the result of the previous appendix, appendix F, to write the product of the two density matrices as

$$\rho^{(A)}_{(a,b)} \times \rho^{(B)}_{(a,b)} = \frac{p^{(A)}_a \times p^{(B)}_b}{N_A N_B} \tag{69}$$

Peierls states [10] that the product of $\rho^{(A)}$ and $\rho^{(B)}$ cannot be zero in order to avoid contradiction and the above expression helps clarify this point. If the product of $\rho^{(A)}$ and $\rho^{(B)}$ is zero, then each and every product $p^{(A)}_a p^{(B)}_b$ must be zero and consequently $\rho^{(A\ with\ B)}_{(a,b),(a',b')} = 0$ as can be seen from the expression for $\rho^{(A\ with\ B)}$ given in appendix F. But this is impossible and, even worse, a situation which one cannot rectify by further experiment. To appreciate this, note that for the product $p^{(A)}_a p^{(B)}_b$ to be zero at least one of $p^{(A)}_a$ or $p^{(B)}_b$ must be zero. Suppose we assume $p^{(A)}_a \neq 0$, then $p^{(B)}_b = 0$ and, indeed, all $p^{(B)}_b = 0$ and $\rho^{(B)}$ is just the null matrix, an absurd situation. The only course of action is to discard whatever knowledge led to B's probablility assignments of zero as they cannot be used as *a priori* probabilities in any application of the Bayes theorem to update B's probablility assignments in the light of new data, as pointed out in Appendix C. So, the product of the density matrices of two observers must not be zero.

# Appendix H : Violations of the Uncertainty Principle resulting from Non-commutivity of Density Matrices

It was demonstrated in section 6 how violations of the uncertainty principle come about when using inaccurate data for a spin 1/2 system. The cause of the violations was the combination of inaccurate data from two different observers measuring non-commuting observables. We will now demonstrate this result for a system in the general case i.e. with a state space of any finite dimension.

Let's call two observers A and B after the observables $A$ and $B$ they are individually measuring and denote by $\rho^{(A)}$ and $\rho^{(B)}$ their individual density matrices. In a basis using the eigenstates of $A$, $\rho^{(A)}$ is diagonal with elements $p_a^{(A)}$. The zeros in the off-diagonal elements represent A's ignorance and it is tempting for A to remove some of their ignorance by using part of B's data. Suppose A becomes aware that B assigns a value $\rho_{a,a'}^{(B)} \neq 0$ to the element $(a, a')$. Then A may well be tempted to incorporate that knowledge and modify their diagonal matrix to form

$$\begin{pmatrix} \cdots & \cdots & \cdots & \cdots & \cdots \\ \cdots & p_a^{(A)} & \cdots & \rho_{a,a'}^{(B)} & \cdots \\ \cdots & \cdots & \cdots & \cdots & \cdots \\ \cdots & \cdots & \cdots & \cdots & \cdots \\ \cdots & \rho_{a',a}^{(B)} & \cdots & p_{a'}^{(A)} & \cdots \\ \cdots & \cdots & \cdots & \cdots & \cdots \end{pmatrix} \tag{70}$$

Now it is well known, see e.g.[16] problem 7 Chapter VII page 293, that a matrix of this form must have

$$p_a^{(A)} \times p_{a'}^{(A)} \geq |\rho_{a,a'}^{(B)}|^2 \tag{71}$$

if the eigenvalues are to all be non-negative and a violation of the uncertainty principle is to be avoided. But the problem is that the off-diagonal element $\rho_{a,a'}^{(B)}$, coming from B's observations, is independent of $p_a^{(A)}$ and $p_{a'}^{(A)}$ which come from A's observations, and there is no reason why the inequality should hold. Hence A's modification of their density matrix using B's data is in danger of violating the uncertainty principle.

# Acknowledgements

Much of the material here originally appeared as my arxiv posting ('The Peierls Interpretation of Quantum Mechanics' 1805.11162) and it is appropriate that the acknowledgements there be repeated here :

From the start of this work I have greatly benefited from detailed and illuminating discussions and correspondence with Drs D T Cornwell and B A Foreman. I am also indebted to Professors S J Clarke and T Lancaster for an important and enlightening discussion early on. I am also very grateful to Professors R B Stinchcombe and J T Chalker for their efforts , help and encouragement . And finally I would like to thank Professors S J Clarke and R A Abram for their general help and support with publication and helpful remarks on the manuscript.

To those acknowledgements I need to add further thanks to David Cornwell and Brad Foreman for continuing discussion and correspondence. I have also enjoyed continuing support and interest from Stewart Clarke at Durham. And our editor, Tuck Choy, has provided much interesting and stimulating feedback and encouragement.

Finally, I would like to dedicate this chapter to the memory of Professor Sir Rudolf Peierls FRS who died 30 years ago in 1995. I had the privilege of studying for my DPhil. at Oxford with him as my supervisor (1968 - 1971). My thesis topics were 'The Momentum of a Light Wave in a Refracting Medium' and 'The Surface Specific Heat of a Crystal Lattice'. So, little quantum mechanics, and certainly no interpretations thereof! And my career was largely involved with the application of semiconductor physics theory to night vision, optical communications and nanocrystals. It was only a decade or so ago, long after his passing, that I started to study his writings on the knowledge interpretation of quantum mechanics. While his writings were the epitome of perspicacity and perspicuity, and a delight to read, I found, on reflection, that there were certain points that were not clear to me. From my interaction with Professor Peierls as a student and beyond (he always answered letters), it was clear to me that my ignorance was the problem. I greatly regretted that he was no longer available to consult! Anyway, I set to work with the result you see in this chapter. I anticipated that it would be a rewarding experience and I have found it to be a rich one indeed. I hope the reader will experience the same; and will come to appreciate that the knowledge interpretation is yet another great legacy left to us by Heisenberg.

## References

[1] Werner Heisenberg. *The Physical Principles of the Quantum Theory.* Mineola, NY 11501: Dover, 1949. ISBN: 486-60113-7.

[2] P.C.W. Davies and J.R. Brown. *The Ghost in the Atom.* Cambridge, UK: Cambridge University Press, 1986. ISBN: 987-0-45728-6.

[3] L D Landau and E M Lifshitz. *Quantum Mechanics.* Oxford, UK: Pergamon, 1965.

[4] Rudolf Peierls. *Bird of Passage.* Princeton, NJ: Princeton University Press, 1985. ISBN: 0-691-08390-8.

[5] R H Dalitz and R B Stinchcombe. *A Breadth of Physics.* Singapore: World Scientific, 1988.

[6] Lahti and Mittelstaedt. *Symposium on the Foundations of Modern Physics pp187-196.* Singapore: World Scientific, 1985.

[7] Rudolf Peierls. " In defence of 'measurement' ". In: *Physics World* January (1991), pp. 19–20.

[8] J S Bell. " Against 'measurement' ". In: *Physics World* August (1990), pp. 33–40.

[9] Rudolf Peierls. *Surprises in Theoretical Physics.* Princeton, NJ: Princeton University Press, 1979.

[10] Rudolf Peierls. *More Surprises in Theoretical Physics.* Princeton, NJ: Princeton University Press, 1991.

[11] N David Mermin. "Compatibility of state assignments". In: *Journal of Mathematical Physics* 43 (9) (2002), pp. 4560–4566.

[12] N G van Kampen. "Het Postmoderne Obscurantisme". In: *Nederlandse Tijdschrift voor Natuurkunde* A56(2) (1990), pp. 13–17.

[13] E T Jaynes. *Probability Theory.* Cambridge, UK: Cambridge University Press, 2003.

[14] P A M Dirac. *The Principles of Quantum Mechanics 4th Edition.* Oxford, UK: Oxford University Press, 1958.

[15] M A Nielsen and I L Chang. *Quantum Computation and Quantum Information.* Cambridge, UK: Cambridge University Press, 2000.

[16] A Messiah. *Quantum Mechanics.* Mineola, NY 11501: Dover, 1999. ISBN: 486-60113-7.

# Chapter 12

## Field emission tunnelling as a window onto fundamental issues in quantum mechanics

Richard G. Forbes[1,*]

[1]*Quantum Sciences Group, School of Mathematics and Physics, University of Surrey, Guildford, Surrey GU2 7XH, UK.*

The physical processes of field electron emission (FE) and electrostatic field ionization (ESFI) both involve quantum-mechanical (QM) tunnelling. They are two of the (perhaps) six historical paradigm examples of QM tunnelling. A modern relevance of FE and ESFI is that each forms part of the theory of various modern technologies and scientific techniques. However, there are fundamental quantum aspects of both FE theory and ESFI theory that are not fully understood. This chapter aims to identify some of these fundamental difficulties and to offer some ideas for wider discussion.

Keywords: field electron emission, electrostatic field ionization, wave-mechanical tunnelling, quantum fundamentals, utilisations of quantum mathematics, "seeing electrons", the arrow of time.

CONTENTS



* Permanent alias for author to whom correspondence should be addressed: r.forbes@trinity.cantab.net; current institutional e-mail: r.forbes@surrey.ac.uk .

### I. INTRODUCTION

#### A. Background

This chapter relates to the theory and problems of two field-induced processes that can take place near the surface of a metal or other material, particularly when the material has the form of a sharp, needle-like solid "emitter" with a rounded apex, typically of radius of order 100 nm or less. (This shape facilitates electrostatic field enhancement at and around the emitter apex.)

Field electron emission (FE) occurs when an electron is extracted from a negatively charged emitter by the presence of a strong local surface electrostatic (ES) field, of magnitude typically of order a few V/nm, directed towards the surface. This field can be described as a "negative conventional (textbook) surface ES field", and can be denoted by the negative parameter $E$. The extraction mechanism is electron tunnelling. The simplest theoretical case (of interest here) is tunnelling from a metal surface into vacuum, with the metal modelled by a Sommerfeld-type free-electron model.

By longstanding convention, the field-like parameters appearing in FE-related equations, such as eq.(6) below, are made positive. Hence, for use in equations, it is convenient to define a positive parameter $F$ as equal to $-E$.

Electrostatic field ionization (ESFI) is the process whereby a gas atom or molecule is ionized by a very strong local surface ES field close to a positively charged "emitter". In this case the field is directed away from the emitting surface, and can be described as a "positive conventional surface ES field". Its value is typically in the range 20 V/nm to 60 V/nm for the noble gases (but depending on the ionization energy of the species involved).

When the gas atom or molecule is in space well above the emitter, the electron tunnels into free space, towards the emitter, and ends up in the emitter. The resulting positive ion is then repelled from the emitter surface.

When the gas atom or molecule is sufficiently close to the emitter then the electron can tunnel directly into a metal electron state above the emitter Fermi level. (The exclusion principle inhibits tunnelling into states below the Fermi level.) In scientific instruments based on ESFI, such as the field ion microscope, the dynamics of the operating gas prior to ESFI can be of importance, but this is of limited relevance here.

A distinction needs to be drawn between ESFI and so-called "strong field ionization", where the ionization is induced by the oscillating electric component of an electromagnetic wave. Although related, these processes are best treated as physically different, with our interest here being primarily in ESFI.

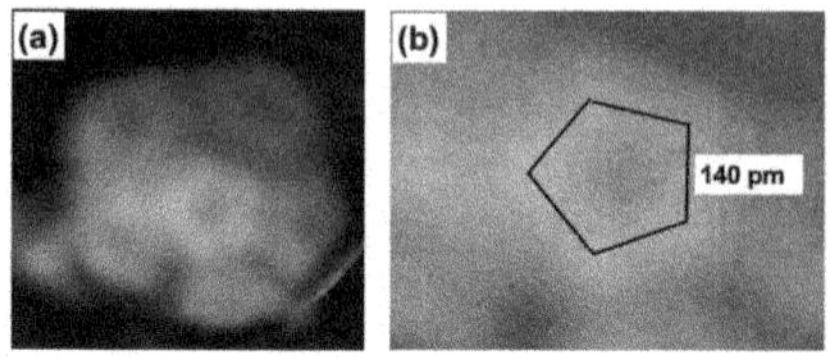


FIG. 1. (a) Field electron micrograph showing electron bonds associated with a five-membered carbon ring at the apex of a carbon nanotube. (b) Enlarged version of the central ring, with carbon nucleus separation marked. (Adapted from Fig. 2(a) in Ref. [1].) Micrographs were taken at room temperature. Approximate magnification is of order $10^8$. Copyright (2000) The Japan Society of Applied Physics.

### B. Motivation

FE and ESFI are of scientific and technological significance for three main reasons.

(a) They are two of the (perhaps) seven historical paradigm examples of quantum tunnelling.

(b) They are part of the theory of various modern technologies, both implemented and proposed (speculatively in some cases), and of many of the "machines of nanotechnology". A relevant list is given in Appendix A.

(c) FE may be part of the cause of electrical breakdown in vacuum or in air or in other gases. For example, electrical breakdown might cause failure of the X-ray sources in medical scanners, or limit the performance of high-voltage particle accelerators, or be part of the physics of lightning strikes or lightning-strike protection. The detailed physical mechanisms of electrical breakdown are not yet fully understood in all cases.

There are also two specific reasons for future interest in FE theory. The field electron (emission) microscope can now image individual electron bonds at the apex of a carbon nanotube [1] (see Fig.1). Thus, opportunities for careful fine-scale sub-atomic experiments and comparisons with theory may exist in due course (though no experiments of this type have yet been carried out). Second, it is now established [2] that DNA mutation is facilitated by the quantum tunnelling of a hydrogen-related entity. Any future improvements in FE tunnelling theory, as discussed below, might conceivably have wider implications in due course, including in quantum biology and in the future subject of quantum botany.

Quantum-mechanical theories of FE and ESFI have been under development for around 100 years but are still far from fully in place. The late President of the UK Institute of Physics (Marshall Stoneham) considered that developing an accurate theory of field electron emission was amongst the most difficult unsolved problems of theoretical physics [3]. The present author is not in a position to agree or disagree with this view, but has worked in FE and ESFI theory for over 50 years and considers that deep fundamental problems do exist in both subjects, and that these problems may link to wider issues in discussing and applying quantum mechanics.

This chapter aims to describe some of these perceived problems, and to offer some preliminary suggestions about how to discuss them. The aim is to initiate a wider discussion about specific points of interest in the context of FE and ESFI theory and the related technologies. E-mail responses from readers with more-specialized quantum-mechanical knowledge than the author has would be welcome.

Part of the thinking behind this chapter is as follows. If a quantum-mechanical problem has remained incompletely solved for nearly 100 years, then perhaps it is timely to discuss ideas that may be unorthodox from the point of view of conventional quantum mechanics. Another part of thinking is that, before developing the detailed mathematical models that will eventually be needed, it will be useful to first discuss qualitative issues and then, where possible, discuss simple mathematical models, even if these are necessarily inaccurate.

## II. ALTERNATIVE UTILISATIONS OF QUANTUM MATHEMATICS

For convenience, the term "quantum mathematics" is used here to describe the formalisms and related mathematics of quantum mechanics. Also, to make discussion more definite, focus is on describing the behaviour of electrons and (in order to avoid mathematical and conceptual complications) on considering only the simplest situations that involve the behaviour of an electron in space.

The author's perception is that, in the high-field-magnitude physics of interest here, quantum mathematics is applied to electron behaviour in two distinctively different ways. In what might be called the "matter-distribution utilisation", quantum mathematics is used to describe the distribution in space of electron matter (or, in some special cases, such as electrons travelling in free space, the time-dependence of this matter distribution). For example, if we wish to calculate the ES field at some point in space close to a strongly polarized atom, then the atom's electron-wave-function-structure provides necessary information about the "source" dipole-like charge distribution to which classical electrostatics is applied. At a more basic level, simple cartoons of atomic orbitals are widely used in teaching quantum mechanics and in qualitative discussions in chemistry and materials science.

It can be argued that in this utilisation there is a sense in which the electron wave-function describes the "real" behaviour of the electron in the context of interest. This utilisation is close to Schrödinger's original interpretation of the wave-function (see eq. (18) in [4]), subsequently abandoned in favour of what became called the "Copenhagen interpretation". (Also see further comments below.)

Alternatively, what might be termed the "pathway-choice utilisation" is used in situations where there is a choice of paths that an electron in a "non-stationary" state might take. In this case, quantum mathematics is used to evaluate the probabilities that different paths will be followed. It is presumed that, in some contexts, different pathways to the same final state may be coherent: in such cases probability amplitudes (rather than probabilities) would need to be summed.

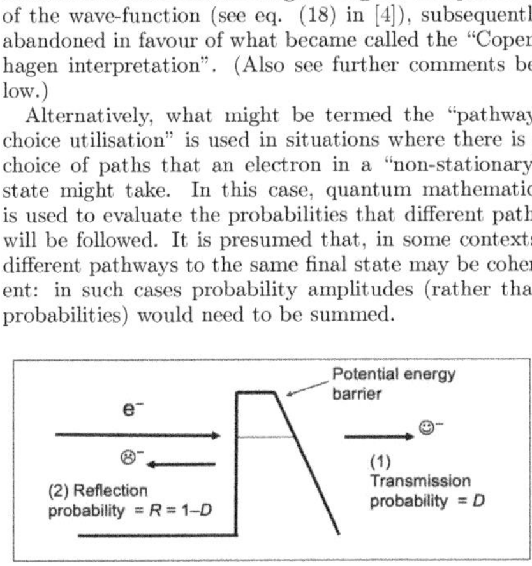


FIG. 2. To illustrate schematically that an electron approaching a potential-energy barrier has a choice of two pathways: (1) transmission; or (2) reflection.

A specific example of the pathway-choice utilisation is the tunnelling process in field electron emission, illustrated schematically in Fig. 2. Here, one first creates a travelling wave-function that represents an electron approaching the barrier. Applying quantum mathematics generates a situation where the total output of the mathematics includes both a travelling wave-function representing an electron transmitted through the barrier and a travelling wave-function representing an electron reflected from the barrier. The barrier transmission probability is found by comparing the squared moduli associated with the incoming and transmitted electron wave-functions.

Since it must presumably be a fundamental principle of physics that "real electrons" cannot split into parts, it follows that the "pathway-choice utilisation" of quantum mathematics is not a full description of the behaviour of "real electrons".

This thought leads on to an apparent fundamental difficulty with FE tunnelling theory. It is well established that the escaping electron certainly has an interaction with the emitter electrons that is in the nature of a correlation interaction (or probably an exchange-and correlation (E&C) interaction). However, due to the extreme theoretical difficulty of doing otherwise, this interaction is usually modelled as the classical image potential-energy associated with a classical point electron situated at a point in space outside but close to the emitter surface. A justification for doing this can be given [5, 6]. In the context of the presumed wave-like (and hence "distributed") behaviour of real electrons, this point in space can presumably be taken to represent the centre-of-charge associated with the wave-function of the escaping real electron. The outcome (when the Sommerfeld model is used to model a planar metal-like emitter) is that the tunnelling potential-energy (PE) barrier is modelled as the so-called Schottky-Nordheim (SN) barrier (see Appendix B). Quantum mathematical methods for evaluating transmission probabilities for the SN barrier, at least approximately, are well established (see Appendix B).

But what if one wants a more accurate calculation of transmission probability? Presumably we would need to know "as exactly as possible" how the correlation (or exchange-and-correlation (E&C)) contribution to the PE barrier varies with the position of an escaping electron's centre of charge, as the escaping electron moves through the barrier. If there is any established method of doing this reliably within the framework of existing quantum mechanics, then the author is not aware of it. It would seem that any suitable theory would, in some way, have to disregard the possibility of wave-reflection effects during escape. Thus, the pathway-choice utilisation of quantum mathematics would presumably not work at this detailed level.

It is also not obvious how one would formulate even a qualitative picture of what is "really" happening to an escaping electron. The author has a (very) rough picture of a tunnelling electron as behaving something like an octopus squeezing through a crack. A travelling electron, extended in space in the direction of travel, but of finite length, comes up to the barrier. When the front of the electron encounters the barrier then the whole electron slows down, and "amplitude" gradually transfers through the barrier. On reaching the further side of the barrier, the transmitted part of the electron is presumably subject to the applied electrostatic field, which presumably tends to pull it away from the barrier. On the face of it, there appears to be a question (not commonly asked) of "precisely why doesn't the electron break apart in such circumstances – what keeps the electron together?"

In this discussion of alternative utilisations, there is also a question of what is meant by a "quantum-mechanical (QM) measurement". The author's short answer is that QM measurement is *observed* pathway choice, in an experimental context established by a human experimentalist. The author's thinking is that the observation can be made either by a human observer or by a machine that ideally is able to record the time and results of its observation process.

## III. WAVE-FUNCTION INTERPRETATION AND THE LINGUISTICS OF QUANTUM MECHANICS

### A. Alternative interpretation for a "material wave-function"

An issue in the history of quantum mechanics has been the "interpretation of the wave-function". The author's view is that the different utilisations of quantum mathematics require different interpretations for the squared modulus of the wave function. I also have the view that clear understanding is often impeded by the language used to discuss these issues, which sometimes seems to unnecessarily involve the classical concept of a classical point entity (in the present discussion a classical point electron).

For simplicity, consider the case of a hydrogen atom. Let the related conventional wave-function be $\psi(r)$, where $r$ is the electron-matter position vector, measured relative to the position of the atomic nucleus. A widely encountered interpretation of the wave-function (e.g., see comments on Born rule [7]) is that its squared modulus $\psi(r)\psi^*(r)$ yields the probability density for measuring the electron as being at position $r$. As discussed below, I have difficulties with this form of interpretation.

In the context of the "matter distribution" utilisation of quantum mathematics, the following alternative is proposed, for discussion purposes at this stage. In the context of the International System of Quantities (ISQ) [8], let $n_1$ denote the amount-of-substance of a system that contains one specified entity, in this case an electron. The parameter $n_1$ is a fundamental atom-level constant that plays the same role with respect to the ISQ quantity amount-of-substance as the elementary charge $e$ does for charge and the unified atomic constant mass constant $m_u$ does for mass.

If we further introduce the name "entity" as the (provisional) name of an atomic-level unit of amount-of-substance, one can write

$$n_1 = 1 \text{ entity}. \tag{1}$$

This unit, the "entity", plays the same role, for amount-of-substance, as the atomic-level unit the Dalton does for mass.

If we introduce the further rule that (when no confusion would occur) the word "entity" can be replaced by the name of the entity of interest, then in the present case we have

$$n_1 = 1 \text{ electron}. \tag{2}$$

Using $n_1$, one can define a new wave-like quantity $\Psi_n(r)$ and its complex conjugate $\Psi_n{}^*(r)$ by the equations

$$\Psi_n(r) = n_1{}^{1/2}\psi_n(r);\ \Psi_n{}^*(r) = n_1{}^{1/2}\psi_n{}^*(r). \tag{3}$$

where $\psi_n{}^*(r)$ is the complex conjugate of $\psi_n(r)$. From which it follows that:

$$\Psi_n(r)\,\Psi_n{}^*(r) = n_1\psi(r)\psi^*(r); \tag{4}$$

$$\int \Psi_n(r)\,\Psi_n{}^*(r)\mathrm{d}V = n_1 = 1 \text{ electron}. \tag{5}$$

where $\mathrm{d}V$ is the element of volume. In the context of these equations, $\Psi_n(r)\,\Psi_n{}^*(r)$ can be interpreted as the (amount-of-substance) concentration of electron matter at position $r$.

Similar procedures, but using $e$ or $m_e$ instead of $n_1$, would lead to squared moduli that can be interpreted as the electron charge-density magnitude or the electron mass density at position $r$. For example, the charge-density interpretation of the wave-function is widely used to illustrate the outcomes of modern density functional theory (DFT) calculations – see [9] for a FE example.

Two advantages of the form of wave-function interpretation embodied in eq. (4) are as follows. First, it avoids the linguistic implication, implicit in the common statement of wave-function interpretation, that somehow the electron is a point-like entity capable of being found at a definite point in space. Second, it avoids use of the terms "probability" or "probability density", and the resulting awkward question "probability of what?"

Obviously, probabilities are essential in the pathway-choice utilisation of quantum mathematics, but they seem a conceptual hindrance when discussing the matter-distribution utilisation of quantum mathematics. It would perhaps be possible to interpret the conventional squared modulus $\psi(r)\psi^*(r)$ as the "probability density for finding electron matter at position $r$", but it seems a more precise statement to let $\Psi_n(r)\,\Psi_n{}^*(r)$ define the concentration of electron matter at position $r$.

More generally, one can interpret what is being proposed here as the extension, to the wave-function level, of the concept of "amount-of-substance". This was introduced (as a "seventh SI base-quantity") as part of the 1970s reforms to the international system of measurement, and has subsequently been applied to "count atoms" and to describe the rates of observable atomic-level processes (e.g., see: [10], Section 3.2.4; or [11]).

### B. "Point electrons" and the presumed impossibility of negative kinetic energy

As already noted, the commonly stated interpretation of $\psi(r)\psi^*(r)$ tends to carry the implication that the electron "really" is a point entity capable of being found at some point in space, as opposed to "really" being a distributed entity that (in space-based formulations) can only be found-by-measurement in some distributed state defined by the measurement apparatus. For this commonly stated "point-electron" interpretation, a particular difficulty arises when the point of interest is inside a tunnelling barrier. At such a point, a point electron would have negative kinetic energy. So this would raise the wider question of whether a real electron could have negative kinetic energy (ignoring issues related to the Dirac equation and positrons).

It is clear that a *classical* point electron cannot have negative kinetic energy. But the issue of whether a real electron could have negative kinetic energy is, philosophically, a different question. My answer to this question is "certainly not", but it is not completely obvious whether this is an axiom about the nature of the "real world" or an experimental fact. (If it is an experimental fact, rather than an axiom, what precisely are the experiments that show this?)

More generally, if a distributed (in space) electron traversing a barrier is to have positive kinetic energy overall then the related matter-distribution wave-function must "stick out of the barrier" at one or both ends. Thus, if it is true that a "real electron" (or other entity) cannot have negative kinetic energy, it follows that – in the models of reality commonly used – a real electron (or other entity) must be a distributed entity.

There are models of reality in which it is possible that this conclusion could not be drawn, for example Bohm's "pilot wave" model, in which – according to some commentators (e.g. Norsen [12]) – a point-like entity can cross a tunnelling PE barrier. But it seems that the conclusion certainly can be drawn in the commonly used models.

Obviously, behind the present discussion there lies the wider issue of whether physics should concern itself only with the results of measurements (see Wikipedia [13] for a useful introductory summary and further references). Those who adopt this view presumably might regard my "matter distribution" interpretation of quantum mathematics as philosophically invalid. However, the widespread use of cartoons depicting electron orbitals and also the fact that we can use field electron microscopy (FEM) to "see electron bonds" suggests that this interpretation is in fact empirically useful, even if currently we do not fully understand the physics of the FEM imaging process (see Section 5). There could also be the philosophical question of whether Fig. 1 could be considered to provide a rough measurement of electron behaviour.

As already has been noted, diagrams representing electron charge density are very widely used in papers describing the results of modern density functional theory.

## IV. MORE LINGUISTIC AND RELATED ISSUES

Some other linguistic and philosophical issues seem to arise when real electrons are assumed to be distributed in space. In particular, it seems a meaningful question to ask: "How long is a real electron, in the context under discussion?" In high-energy physics an electron might be tightly localised and thus appear "nearly point-like". However, in solid-state physics the electron length presumably might be determined by its coherence length, and presumably might be 20 nm or more. In ESFI, if the picture is valid that a real electron tunnels out of the atom over a period of time, then this would seem to imply that in some contexts the length of an electron might be a function of time.

There also seem to be linguistic issues relating to the way that the Heisenberg Uncertainty Principle is sometimes formulated. For a discussion in one spatial dimension $z$, a common form of statement is that the uncertainty $\Delta z$ in measuring the electron position and the uncertainty $\Delta p$ in simultaneously measuring the electron momentum $p$ are related by $\Delta z \Delta p \geq \hbar$, where $\hbar$ is the reduced Planck constant. But, again, what does the term "electron position" mean if/when a real electron is 20 nm long?

An alternative might be to reformulate the uncertainty principle as a principle that relates minimum electron (etc.) kinetic energy to electron (etc.) length. The basic idea, as applied in one dimension, would be to consider an electron (or other entity) that is effectively confined to a region of space of finite length $L$. The smaller this length, then the higher the minimum electron kinetic energy $E_{\text{min}}$.

The simplest illustration is an electron in a one-dimensional Sommerfeld well with infinitely high walls. The lowest energy state (corresponding to a half-cycle of a sinusoidal wave) has circular wave-number $k = \pi/L$, and energy $E_{\text{min}} = h_{\text{P}}^2/8m_{\text{e}}L^2$ where $h_{\text{P}}$ is Planck's constant and $m_{\text{e}}$ the electron mass. Obviously, an approach of this kind avoids using the concept of the position of a point electron.

For a numerical example in the above Sommerfeld-well case, consider a one-dimensional well of length 10 nm. The kinetic energy of the lowest-lying electron state is found (to 2 sig. fig.) to be 3.8 meV.

It may also be arguable that the "particle" aspects of the principle of wave-particle duality could be formulated in a clearer way. For some people, "wave-particle duality" (as applied to matter) seems to mean that matter is not infinitely divisible, but comes in the discrete chunks that we now call "particles" (but might be better called "material entities" or by the old-fashioned name "corpuscles"). (The extensive use of the term "particle" by Newton, in his discussion of classical mechanics [14, 15], may tend to associate point-like behaviour with the term "particle".) With this interpretation of "wave-particle duality", so-called particles are always (all the time) discrete material entities and are always (all the time) wave-like distributed entities.

For others, "wave-particle duality" (as applied to matter) seems to mean that sometimes it is necessary to describe entity behaviour by quantum mathematics, but that sometimes it is more convenient to describe entity behaviour by the classical mathematics of point objects. A particular example is the Rutherford alpha-particle scattering experiments. Rutherford's original paper [16] used classical mechanics to analyse his results. But modern discussions might prefer to use the quantum mathematics of scattering theory. It is arguable that the usual statements of wave-particle duality could usefully be replaced or supplemented by a statement about "fundamental physical principles of existence and behaviour" (see Appendix C).

## V. THE CONCEPTUAL PROBLEMS OF "SEEING ELECTRONS" IN THE FIELD ELECTRON MICROSCOPE

As already indicated, it is now possible to use the field electron microscope to image the bonds in a five-atom carbon ring at the apex of a carbon nanotube. The image in Fig. 1 probably had a diameter of order around 1 cm on the fluorescent phosphor screen. The screen image is a highly magnified and (probably) slightly blurred representation of the across-surface variations in the local electron emission current density emitted from the outside of the tunnelling PE barrier that holds the electron into the bond.

It can probably be assumed (in accordance with conventional thinking) that: (a) the observed images, as recorded by some form of electronic "photographic" detector/recorder, are the product of the arrival at the screen of many individual electrons, one after another; and that (b) each electron causes light to be emitted from a phosphor grain (or a group of atoms) in an area of the screen that is very very much smaller than the area of the screen image. Thus, over time, from the time when the emitter is turned on by application of a step-like increase in applied voltage, the observed image changes from being very "spotty" to having smoothly varying optical-intensity variations in the plane of the screen.

There is an apparently unsolved fundamental question as to precisely how this happens, given that the electron behaviour inside the bond must presumably be wave-like. For an individual electron, what is incident on the inside of the PE barrier is presumably an "electron wave" that is laterally coherent across the whole area of the bond. Mathematically, under the influence of the ES-field configuration between the emitter and screen, this wave presumably spreads out to cover the whole area of the screen image. But it would seem that the effect of the impact of the electron wave is localised to a very small screen area. Alternative conceptual possibilities (each with their own significant conceptual difficulties) would seem to be as follows.

If lateral localisation occurs at the microscope screen, then there has be some mechanism whereby all the charge and kinetic energy associated with the impinging widespread wave-front very quickly "drains laterally" into one particular small location at the phosphor screen. To the present author, this does not seem plausible.

Alternatively, one might hypothesise that some mechanism acting when the wave emerges from the barrier or during its travel from emitter to screen causes the wave to become "laterally highly localised". But the author is aware of no obvious mechanism to achieve this.

A thought is that at present we treat the interaction between an electron and the electrostatic field classically, which also means the field acts continuously. Possibly what is needed is a deeper physical understanding of this interaction (or, maybe, of the interaction with the sources of the field), and of how to treat these things "in a proper quantum-mechanical fashion" (whatever that is). Feynman's approach to electron-electron interactions (e.g., [17], in which electrons interact by exchanging virtual photons, might be one starting point for further thought.

It might also be worthwhile to attempt an experimental demonstration that would confirm the "conventional thinking" described above.

This problem is not fully understood and is unsolved.

## VI. FIELD EMISSION TUNNELLING AND THE ARROW OF TIME

There is continuing interest in the question of how the arrow of time emerges from a quantum mechanical discussion of the behaviour of the world. (For a brief introduction and further references, see Wikipedia [18].) It is usually considered that quantum mathematics is time-reversible, both in the wave-mechanics and in the matrix-mechanics implementations. Thus, it is sometimes argued, it is difficult to immediately see how a unique arrow of time can emerge from quantum mathematics. For example, a recent paper [19] suggests that two opposing arrows of time can emerge from a detailed study of dynamics.

However, the author's perception is that the quantum mathematics used in the pathway-choice discussion of FE tunnelling is **not** time-reversible as regards its statistical implications: rather it obeys/gives rise to a different fundamental principle of physics, namely transmission reciprocity. (This is a special case of the more general principle of scattering reciprocity, and can also be seen as an implication of the principle of detailed balance.)

For simplicity, consider the Schottky-Nordheim barrier discussed in Appendix B. For a given set of barrier parameters let the transmission probability for an electron approaching the barrier from the left be $D_{\mathrm{L}\to\mathrm{R}}$, with $D_{\mathrm{L}\to\mathrm{R}} < 1$. Transmission reciprocity means that the transmission probability $D_{\mathrm{R}\to\mathrm{L}}$ for an electron approaching the same barrier from the right is equal to $D_{\mathrm{L}\to\mathrm{R}}$, with $D_{R \to L} < 1$. It is easy to see that this is necessarily so within the JWKB approach to evaluating transmission probability: the probability depends on the JWKB integral which is the same whether you evaluate it from left to right or right to left. Hence, if you reverse time (and hence velocity) on a transmitted electron moving from left to right, and make it approach the barrier from the right, then it is not 100% certain that the electron will go back to the left of the barrier. Hence the original change in electron state, from the left of the barrier to the right of the barrier, is "not always time-reversible".

Clearly, if you simply reverse time in the thinking about the Schrödinger equation, you could formally recombine two "partial electrons" travelling in opposing directions into a single electron travelling to the left, but this formal mathematics does not (in my view) correspond to "real" electron behaviour.

Even when an electron approaches the barrier with energy sufficient to cross the barrier above its peak energy level, significant wave-mechanical reflection can occur, and similar arguments to those above apply.

In principle, it looks as if this type of argument could be extended further, to discuss the time-dependent behaviour (including the occupancy of different system parts) of a large system containing many individual non-time-reversible pathway choices. It is to be presumed that the statistical behaviour of such a system would not be time-reversible.

In general terms, the mechanisms just discussed seem to constitute one possible way to cause the entropy of a system to increase with time, leading to the existence of an arrow of time. This is not assumed to be the only possible causal mechanism for the arrow of time, others being related to changes in coherence and/or entanglement [20, 21].

## VII. THE CONCEPTUAL PROBLEM OF NEAR-SURFACE ELECTROSTATIC FIELD IONIZATION

Most field electron emission theory is currently based on Sommerfeld-type models that assume that emitters have classically smooth surfaces. An urgent problem of FE theory is to "get atoms into the theory". The work of Oppenheimer in 1928 [22, 23] showed that, in principle, the transmission process in FE could be thought of as equivalent to ESFI of surface atoms. Thus, it is thought that a possible route into atomic-level FE theory is to look again at the theory of the ESFI of a hydrogenic ion in its ground state. (A hydrogenic ion has a single electron surrounding a nucleus with charge $Ze$, where $e$ is the elementary (positive) charge, as before, and $Z$ is the nuclear charge number. It is helpful to discuss the ESFI of a hydrogenic ion, rather than the ESFI of a hydrogen atom, because it is desirable to get the charge-number $Z$ into the mathematical treatment.)

ESFI theory is also relevant to the theory of gas field ion sources and of the field ion microscope, and to the process of field post-ionization (FPI) that occurs in the context of atom probe tomography (APT). In APT, FPI is used both to explain aspects of emitted ion dynamics and to estimate near-surface ES-field values.

A good starting point for any fresh look at the theory of ESFI would be an ISQ version of the treatment given by Landau and Lifschitz (LL) in their well-known textbook on quantum mechanics, first published in English in 1958 [24]. The present author formulated (but did not publish) an ISQ treatment of this kind some years ago [25]; this treatment now needs updating, but the mathematical results are still thought valid. The third English edition [26] of the LL textbook, published in 1977, is the most useful because it gives a Gaussian-system version of the formula for ESFI rate-constant, as well as the atomic-units proof given in the 1958 edition. This enables the LL Gaussian-system formula to be compared with ISQ-system formulae.

To discuss planar FE theory in a slightly improved way, Forbes and Deane have introduced a new special mathematical function (SMF) $\mathrm{v_{FD}}(x)$, where $x$ represents the independent variable in the Gauss Hypergeometric Differential Equation and $\mathrm{v_{FD}}(x)$ is a special solution of this equation (for example, see [27]).

This function $\mathrm{v_{FD}}(x)$ can also be applied to line-integrals used in ESFI theory and it can be argued that the resulting mathematics is superior to that used by Landau and Lifschitz. LL note that their treatment is expected to be valid only in the limit of low electrostatic (ES) fields. For a hydrogen atom in free space, both treatments generate the same result in the limit of low ES fields, but the new treatment is expected to be slightly more accurate at higher fields.

Part of the reason seems to be as follows. The new SMF $\mathrm{v_{FD}}(x)$ is one of small group of SMFs that require TWO infinite series in order to provide a series definition. For such functions the process of Taylor expansion does not work well. The LL treatment of ESFI involves Taylor expansion and thus might be expected to not work well in mathematical detail.

However, aside from purely mathematical issues, there seem to be fundamental physical issues. The combination of the applied linear ES field and the Coulombic field due to the ionic nucleus generates a set of curved ES field lines that can usefully be described in parabolic coordinates. Close to the ionic nucleus there is a region where a classical point electron would have positive kinetic energy and "downfield" of the atom there is also a region where the point electron would have positive kinetic energy.

In between these regions there is a PE barrier region. The ES-field distribution in this region might be expected, in some sense, to guide a real electron through the barrier. The requirement for conservation of electron angular momentum about the ionic nucleus would in reality complicate accurate physical and mathematical discussion; for present purposes we can disregard this complication. Because of the large difference in mass between the electron and the ionic nucleus, we also disregard issues relating to the need to use reduced mass.

The central issue is how to set up the physics for calculating the electron transmission probability through the barrier region. LL use parabolic coordinates to define an infinite number of possible "paths" through the barrier. LL then consider each path and (using a JWKB-type theory) calculate relatively how much contribution to the final result is made via this path. An integration over a downfield plane normal to the applied field, taking into account what the classical velocity (in this plane) of a point electron would be, yields the total probability of ESFI per unit time, i.e. the ESFI transmission rate-constant.

There could be questions about the detailed validity of this approach, but the procedure may work more-or-less adequately when the hydrogenic entity is in free space. However, there is an issue of procedure when the entity is close to a metal surface and there is an exchange-and-correlation interaction (modelled as an image interaction) between the tunnelling electron and the surface. The issue is where to place the image charge. If one has a conceptual theoretical model in which quantum mechanics represents the outcome of an ensemble of alternative electron paths, each of which is traversed by a point-like "full electron", then the simple position for the image charge is opposite the instantaneous position of the nucleus.

On the other hand, LL seem to be implementing a model of reality in which an electron is divided into infinitesimally small elements, each of which traverses a different path through the barrier and has its own path-related transmission behaviour. If one assumes, for the purposes of generating an E&C interaction, that these elements act coherently, the logical position for the image charge would seem to be opposite the ion nucleus. The question of where to put the image charge is currently unresolved and seems to deserve wider discussion.

## VIII. TUNNELLING-INTEGRAL VERSUS OVERLAP-INTEGRAL FORMULATIONS OF QM TRANSMISSION THEORY

In the above discussion, quantum-mathematical aspects of transmission theory have, where needed (as in Appendix B), been discussed using a tunnelling-integral formulation of transmission theory. Physically, this approach derives ultimately from the 1928 work of Fowler and Nordheim (FN) [28] on field electron emission (but note the detailed theory in this FN paper is seriously flawed – for example see [29]). Virtually all later theoretical work on FE, including the important 1956 reformulation of FE theory by Murphy and Good [30] (in order to take E&C effects into account), adopts some variant of this underlying approach.

Tunnelling-integral-type approaches have also been widely used in ESFI theory, in particular: (a) in LL's theoretical discussion noted above; (b) as one approach in the useful review of hydrogen atom ESFI by Yamabe et al. [31]; (c) in discussions of how field ion microscopes work (for example, by Gomer [32]); and (d) in discussions of FPI theory (for example, by Kingham [33] and by Lam and Needs [34]).

An alternative basic approach is the overlap-integral type of formulation that derives ultimately from the 1928 papers by Oppenheimer [22, 23]. In this approach, separate Hamiltonians and wave-functions are established to describe electron motion in initial and final states in regions of space separated by the barrier. The transmission probability then depends on the degree of overlap of the exponentially decaying wave-function tails inside the barrier.

In the field ion microscope, ESFI takes place near and outside the so-called critical surface which (for a helium atom) is the locus of He-nuclear positions such that the energy level of the topmost electron orbital in the helium atom aligns with the emitter Fermi level. When discussing the imaging of an emitter surface atom, in circumstances where operating-gas distribution effects are not significant, the tunnelling-integral formulation suggests that the electron transmission rate-constant (ETR) primarily depends on the magnitude of the ES field between the critical surface and the emitter surface, whereas the overlap-integral formulation suggests that this ETR primarily depends on the distance between the critical surface and the nucleus of the imaged emitter surface atom. Thus, there is an opportunity to investigate how well the alternative formulations can explain the experimental details of field ion images.

A convenient discussion situation (used historically) is the helium ion imaging of a relatively small tungsten (111) crystal facet, as illustrated in Fig. 3. Two image features require explanation. The first is the fact that atoms in this facet can be resolved, which requires that the ETR for a gas-atom-nucleus position in the critical surface that is "over a substrate atom" be sufficiently greater than the ETR for a critical-surface position that is "over a point midway between atoms". The second is that the corner atoms in the facet image more brightly than the edge atoms between them.

By 1985, work by Homeier and Kingham [36] and by the present author [37] had eventually shown that, in the tunnelling-integral formulation, the ES-field variations across the emitter surface (between the critical surface and the emitter surface) were sufficient to explain atomic resolution.

Earlier, in the 1970s, it had been found [38–40] that overlap-integral formulations apparently could not provide a satisfactory explanation of field ion imaging. In the view of Goldenfeld et al. ([40], p.338) this was because "the behaviour of the (wave-)function in the complete potential(-energy) barrier...is essential in defining the value of the full probability flux through the barrier". An extensive theoretical re-investigation/review of the ESFI of the hydrogen atom in free space, carried

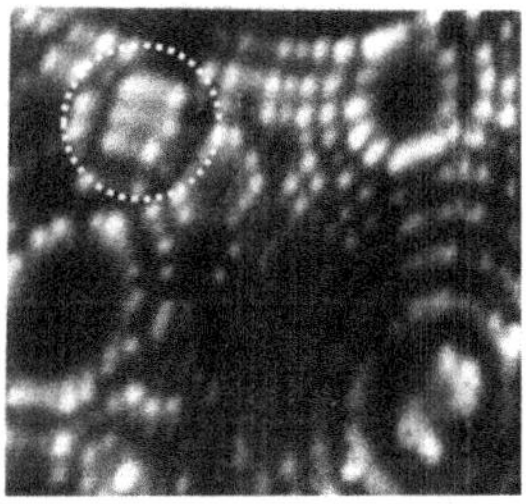

FIG. 3. Part of helium-ion image of tungsten emitter, taken near 80 K. A small W(111) crystal facet visible in the figure is circled. In the bottom atom-row, the corner atoms are imaged more brightly than the edge atom or the interior atom. (Adapted from Fig. A:31c in Vol. II of Ref. [35].)

out by Yamabe, Tachibana and Silverstone in 1977 [31], reached a similar conclusion: namely that "the barrier region dominates practical calculations, which are surprisingly sensitive to the accuracy of the wave-function there".

Returning to Fig. 3, the underlying physical reason why the corner atoms image more brightly is that (in accordance with Gauss' theorem) these atoms carry a higher effective positive charge than edge atoms or interior atoms (because there is less electron charge around a protruding emitter surface atom). In consequence, for the corner atoms: (a) the critical surface is further from the emitter nucleus than for edge and interior atoms; and (b) the field in the related barrier region is expected to be higher.

This effect is shown schematically in Fig. 4 below. Thus, qualitative expectation is that the tunnelling-integral formulation will be able to explain corner-atom brightness, but the overlap-integral formulation will not. Obviously, this conclusion is qualitative. It would be helpful to have it re-investigated using modern density-functional-theory techniques, but this has not yet been done.

Thus, the provisional conclusions from the above analysis of FIM image formation are that, in tunnelling contexts, tunnelling-integral formulations seem likely to be more reliable than overlap-integral formulations, and that (in some contexts) overlap-integral formulations may not even be able to predict results that "go in the right direction".

Another significant difference between the two formulations discussed above is the following. In the formula for the electron transfer rate-constant (ETR), a parameter relating to the effective density of final electron states (in the emitter) currently appears in the overlap-integral formulation but not in the tunnelling-integral formulation. Conceptually, if there are more states available for the electron to tunnel into than free-electron-type theory would predict, then the ETR is expected to be higher. And vice-versa.

This argument has recently been used [41] to make predictions about "chemical contrast" in FIM images, i.e. the imaging of "surface impurity" atoms in FIM images, albeit in the context of an overlap-integral formulation. (See this paper for one method of incorporating density-of-states effects into overlap-integral-type theory.) Thus, there is reason to think that effects of this general kind should probably be incorporated into tunnelling-integral theories. A brief summary of the essential merits of the two approaches is given in Table 1 below.

This problem with tunnelling-integral formulations is more obvious in the situation where two solids with different non-free-electron-like band structures are separated by a penetrable asymmetric PE barrier that can be treated as planar. This is adequately the case with so-called "Schottky-barrier devices", as discussed by the present author in a recent conference presentation [42].

If the whole system of electrons (both sides of the barrier) is in joint thermodynamic equilibrium, then the principle of detailed balance must apply. In the case of states on either side of the barrier that have equal total-energy and equal normal-energy components, this seems to require the following: that (for these states) the expressions for the electron-transfer-rates from left to right and from right to left must *both* contain density-of-states-related parameters corresponding to the different band-structures on *both* sides of the barrier. The existing theory of Schottky-barrier devices appears not to meet this requirement [42].

The conclusion drawn by the author from this Section's discussion is that neither tunnelling-integral formulations nor overlap-integral formulations are complete, correct and accurate in their present forms. It seems, as claimed by Stoneham [3], that the problem of developing a fully correct theory of quantum tunnelling (and, more generally, of quantum transmission) is not yet fully solved.

## IX. CONCLUSIONS

This chapter has attempted a survey of what appear to be fundamental theoretical problems arising in the context of the theories of field electron and field ion emission, and has offered some suggestions about how to discuss them. The survey is not quite complete, since it does not cover fundamental problems relating to the process of field evaporation (FEV), in which surface atoms are removed from the emitter surface as ions, under the influence of a very high electrostatic field. The fundamental issue for FEV is precisely how the process of surface-atom ionization takes place at a highly charged surface. However, the author's view is that there are preliminary scientific issues (see [11] for a recent discussion) that need to be resolved before a discussion of more fundamental issues becomes timely.

The proposals that emerge from this survey can be

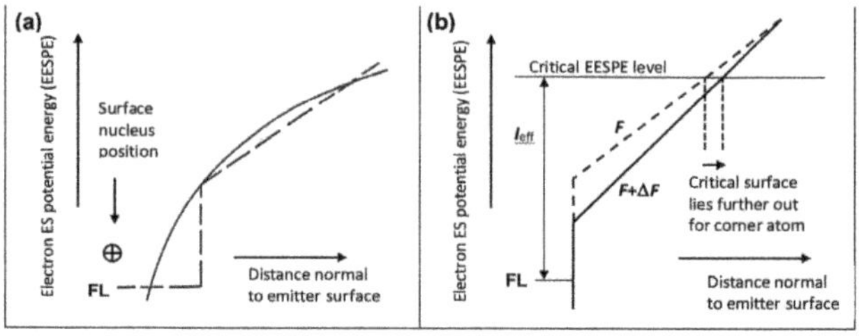


FIG. 4. Schematic diagram illustrating the modelling of electron electrostatic potential energy (EESPE) variations at a charged surface. (a) Fitting of a straight-line equivalent-barrier to the EESPE section through a surface-atom nucleus. (b) Straight-line equivalent barriers for an edge atom (broken line) and for a corner atom (full line). The extra positive charge (actually reduced electron charge) at the corner site means that the ES field is higher there and the critical surface is further out.

TABLE I. Comparison of different theoretical approaches to evaluation of electron transfer rate-constants in the context of field ion imaging.

| *Theoretical issue* | *Tunnelling-integral theory* | *Overlap-integral theory* |
|---|---|---|
| Can the theory qualitatively explain FIM images? | Yes | Apparently not |
| Does the theory currently contain a term relating to final-density-of-states effects for non-free-electron metals? | No | Yes |

classified in three groups, as follows.

*Definite proposals*

(1) The issue of whether a "real" material entity can have negative kinetic energy is an important preliminary fundamental question about models of reality and hence about quantum mechanics. The author's answer is "certainly not", but there is a question of whether this answer is an axiom about the nature of physical reality or an experimental fact.

(2) When the answer to (1) is "no", this implies that (specifically in the context of electron tunnelling, and presumably more generally) electrons – and presumably all material entities – are "really" distributed entities rather than point entities.

(3) Given the above, it would probably be helpful if the language we use to discuss quantum mechanics (or the utilisation of quantum mathematics) were revised to remove all reference to "the position of a (point) electron or other material entity". (Except where absolutely essential, as in Bohm-type theories.)

(4) As regards the tunnelling of material entities, neither of the main approaches (using tunnelling integrals or using overlap integrals) is currently complete, correct and accurate. There is an unfulfilled need to develop "more accurate" theory (and there is almost certainly a need to develop more precise methods of comparing the outcomes of tunnelling experiments with theory). A "more accurate" theory could likely be useful in various technological and industrial contexts. A "highly accurate" theory of quantum tunnelling seems likely to eventually be needed as part of the theory of DNA mutation.

*Other suggestions*

(5) It may be helpful to talk, not about "alternative interpretations of quantum mechanics", but about "alternative utilisations of quantum mathematics".

(6) It has been argued that, although the matter-distribution utilisation of quantum mathematics may sometimes be time-reversible, the pathway-choice utilisation is not. Hence, in a large system, the arrow of time seems not to be a mystery but could emerge from many individual pathway-choice events conceptually analogous to FE tunnelling. (But also in other ways, as well.)

(7) Where convenient, it might be clearer in principle to talk about the behaviour of "material entities" (or the behaviour of photons) rather than about the behaviour of "particles".

(8) There could be merit in replacing or supplementing the uncertainty principle by a principle that relates the minimum kinetic energy of a material entity to the length or volume of space to which it is restricted.

(9) There could be merit in replacing or supplementing the concept of wave-particle duality by a more comprehensive set of principles about the behaviour of matter and radiation, for example as suggested in Appendix C.

*Unresolved fundamental issues*

(10) In the situation where a tunnelling electron has a correlation or exchange-and-correlation interaction with other electrons in the "emitter", it appears unclear (even in principle) how to carry out an accurate calculation of the time-dependence of this interaction, and hence an accurate estimation of transmission probability.

(11) It is unclear how to explain precisely how the field electron microscope images the carbon bonds in a carbon nanotube.

(12) In the Landau and Lifschitz approach to the theory of electrostatic field ionization (ESFI), it is unclear how to deal correctly with image-charge effects when gas-atom ESFI takes place close to a conducting surface.

(13) There seems a fundamental question, not commonly asked, as to precisely why the front part of an electron emerging from a tunnelling barrier, and subject to the pull of the applied electrostatic field, does not "pull apart" from the rest of the electron.

As noted in the introduction, the primary purpose of this chapter has been to offer suggestions for scientific discussion, with the specific aims of improving theories of field electron and field ion emission, but also in the hope they might be useful more generally. Responses, including informed criticisms, will be welcomed.

As regards procedure, for the most part all that can be done in the near future is to raise these suggestions in other forums. However, there are four proposals on which scientific work could reasonably begin as soon as time permits. I perceive the order of priority to be (most plausible first): (6), (9), (4) and (12). However, the issues I would most like to see solved are (10) and (11).

## APPENDIX A: TECHNOLOGIES POTENTIALLY UTILISING FIELD ELECTRON OR FIELD ION EMISSION

***Established technologies***

*Electron microscopy, in its various forms
*Electron beam lithography
*Gas field ion sources & related scanning ion machines
*Liquid metal ion sources
*Focused ion beam (FIB) machines
*Field electron microscopy & related techniques
*Field ion microscopy & related techniques
*Atom-probe tomography & related techniques
*Field desorption mass spectrometry
*Electron sources for new "fast" forms of medical X-ray "tomographic" machines
*Scanning probe lithography (in the field electron emission regime)

***Technologies under development***

*Electron sources for microwave generators (US military research)
*Many different emitting materials for large-area field electron sources
*Field desorption ion sources (which have been applied to fingerprint recognition)
*Field (ion) emission electric propulsion of spacecraft
*Space-vehicle neutralisers based on field electron emission
*New forms of imaging atom-probe
*Miniaturised mass spectrometers (e.g. for space applications)
*Miniaturised electron microscopes
*Ion implantation systems based on liquid metal ion sources
*Lightning-strike protectors
*Amorphous metal non-linear resistors (AMNR devices)

***Technologies under exploration***

*Nanoscale vacuum-channel and air-channel transistors
*The surface technique of "Near Field Emission Scanning Electron Microscopy"
*Technologies based on microgas discharges
*Neutralisers for Hall thrusters for spacecraft propulsion
*New battery forms based on electrostatic energy storage (probably unlikely to happen)
*Silent drones driven by electrostatic propulsion

## APPENDIX B: THE SCHOTTKY-NORDHEIM BARRIER

The Schottky-Nordheim (SN) barrier is the potential-energy (PE) barrier above a smooth planar metal surface when the joint effects of an applied negative ES field of magnitude $F$ and classical image effects are taken into account. For a SN barrier with zero-field height equal to the local work function $\phi$, the difference $M(\phi, F, z)$ between the electron potential energy and the electron kinetic-energy component normal to the surface is

$$M(\phi, F, z) = \phi - eFz - e^2/16\pi\epsilon_0 z \qquad (6)$$

where $z$ is distance outwards from the metal surface, $e$ is the elementary (positive) charge and $\epsilon_0$ is the electric constant (aka the vacuum permittivity). In the simplest tunnelling-integral formalism (the so-called first-order JWKB formalism), the electron transmission probability $D$ for escape from the emitter is approximately given by

$$D \approx \exp[-(8m_e/\hbar^2)^{1/2} \int_{z_1}^{z_2} M^{1/2}(\phi, F, z)\mathrm{d}z \qquad (7)$$

where $m_e$ is the electron mass in free space, $\hbar$ is the reduced Planck constant, and the integral is taken between the zeros $z_1, z_2$ of $M^{1/2}(\phi, F, z)$.

## APPENDIX C: SUGGESTED "FUNDAMENTAL PHYSICAL PRINCIPLES OF EXISTENCE AND BEHAVIOUR"

It is suggested that the principle of wave-particle duality could be replaced or supplemented by a set of "fundamental physical principles of existence and behaviour". A tentative first draft for such a set is as follows. Emphasis is on content rather than precise wording, and ordering is provisional.

1) Travelling electromagnetic (EM) radiation is not infinitely divisible: all EM radiation consists of travelling EM entities now called photons.

2) Travelling photons are distributed in space and their motion is always wave-like.

2a) A travelling photon has energy $h_P\nu$, where $h_P$ is Planck's constant and $\nu$ is the frequency of the EM radiation.

2b) When a photon is absorbed by matter then normally the whole energy of the photon is transferred to the matter.

3) Matter is not infinitely divisible: all forms of matter are composed of discrete entities that can be called "material entities".

3a) Material entities are always distributed in space and their behaviour is always wave-like.

3b) However, in some circumstances it may be convenient to model the behaviour of a material entity as the motion of a mathematical point placed at its centre of mass or charge.

4) In some circumstances, when a large number of material entities are interacting with each other, it may be convenient to represent the collective behaviour of some part of the assembly of entities as a "smeared out" continuous distribution of matter or charge.

4a) Examples of this are classical space-charge models and the use of "jellium" in surface-physics models.

Due to the possible ambiguity in the interpretation of the word "particle", its use is avoided in the above statement of principles.

A difference from the usual statement of wave-particle duality is the emphasis that photons and material entities are always (*all the time*) discrete entities and are always (*all the time*) distributed in space, and that their behaviour is always (*all the time*) wave-like.

[1] Y. Saito, K. Hata, and T. Murata, Field emission patterns originating from pentagons at the tip of a carbon nanotube, Jpn. J. Appl. Phys. **29**, L271 (2000).

[2] L. Slocombe, M. Sacchi, and J. Al-Khalili, An open quantum systems approach to proton tunnelling in DNA, Comm. Phys. **5**, 109 (2022).

[3] A. M. Stoneham, Private communication to the present author during a research visit to Professor Stoneham at University College London, in October 2001.

[4] E. Schrödinger, Second lecture, in *Collected Papers on Wave Mechanics (Third augmented English edition)* (AMS Chelsea Publishing, 1982) pp. 168–180.

[5] J. Bardeen, The image and Van der Waals forces at a metallic surface, Phys. Rev. **58**, 727 (1940).

[6] R. G. Forbes, Comments on the continuing widespread and unnecessary use of a defective equation in field emission related literature, J. Appl. Phys. **126**, 210901 (2019).

[7] See Wikipedia entry on "Born rule" (checked 20 October 2025).

[8] *International Standard ISO 80000-1:2022, Quantities and Units, Part 1: General* (International Organization for Standardization, Geneva, 2022).

[9] C. P. de Castro, T. A. de Assis, R. Rivelino, F. de B. Mota, C. M. C. de Castilho, and R. G. Forbes, Restoring observed classical behavior of the carbon nanotube field emission enhancement factor from the electronic structure, J.Phys. Chem. C **123**, 5144 (2019).

[10] M. K. Miller and R. G. Forbes, *Atom Probe Tomography: The Local Electrode Atom Probe* (Springer, New York, 2014).

[11] R. G. Forbes, On the scientific status of the theory of field desorption and field evaporation, 2024 Workshop on Frontiers of APT Physics, Data Processing, Analysis and Reconstruction, Arlington, Virginia, USA, August 2024; doi: 10.13140/RG.2.2.25625.53602.

[12] T. Norsen, The pilot-wave perspective on quantum scattering and tunnelling, Amer. J. Phys. **81**, 210901 (2013).

[13] See Wikipedia article on "The Copenhagen Interpretation". (Checked: 20 October 2025).

[14] I. Newton, *Philosophia Naturalis Principia Mathematica (in Latin)* (Royal Society, London, 1676).

[15] I. Newton, *The Principia (English translation of Ref. 14)* (Snowball Publishing, 2010).

[16] E. Rutherford, The scattering of $\alpha$ and $\beta$ particles by matter and the structure of the atom, Phil. Mag., Ser. 6 **21**, 669 (1911).

[17] R. P. Feynman, *QED* (Penguin, London, 1985).

[18] See Wikipedia article on "Arrow of time". (Checked: 20 October 2025).

[19] T. Guff, C. U. Shastry, and A. Rocco, Emergence of opposing arrows of time in open quantum systems, Sci. Repts. **16**, 3658 (2025).

[20] W. H. Zurek, *Decoherence and Quantum Darwinism* (Cambridge Univ. Press, 2025).

[21] J. Al-Khalili and E. K. Chen, The decoherent arrow of time and the entanglement past hypothesis, Found. Phys. **54**, 49 (2024).

[22] J. R. Oppenheimer, Three notes on the quantum theory of aperiodic effects, Phys. Rev **13**, 66 (1928).

[23] J. R. Oppenheimer, On the quantum theory of the autoelectric field currents, Proc. Nat. Acad. Sci. **14**, 363 (1928).

[24] L. D. Landau and E. M. Lifschitz, *Quantum Mechanics* (1st English edition, Pergamon, Oxford, 1958).

[25] R. G. Forbes, ISQ derivation ("derivation in SI units") of a formula for the electrostatic field ionization rate-constant for a hydrogenic atom in its ground electronic state (2014), arXiv:1412.1821v4.

[26] L. D. Landau and E. M. Lifschitz, *Quantum Mechanics* (3rd revised English edition, Buterworth-Heinemann, Oxford, 1977).

[27] R. G. Forbes, Renewing the mainstream theory of field and thermal electron emission, in *Modern Developments in Vacuum Electron Sources*, edited by G. Gaertner, W. Knapp, and R. G. Forbes (Springer Nature, Switzerland, 2020) Chap. 9, pp. 387–447.

[28] R. H. Fowler and L. Nordheim, Electron emission in intense electric fields, Proc. R. Soc. Lond. A **119**, 173 (1928).

[29] R. G. Forbes and J. H. B. Deane, Transmission coefficients for the exact triangular barrier: an exact general analytical theory that can replace Fowler and Nordheim's 1928 theory, Proc. R. Soc. Lond. A **467**, 2927 (2011).

[30] E. L. Murphy and R. H. Good, Thermionic emission, field emission, and the transition region, Phys. Rev. **102**, 1464 (1956).

[31] T. Yamabe, A. Tachibana, and H. J. Silverstone, Theory of the ionization of the hydrogen atom by an external electric field, Phys. Rev. A **16**, 877 (1977).

[32] R. Gomer, *Field Emission and Field Ionization* (Harvard Univ. Press, Cambridge, Mass., 1961).

[33] D. R. Kingham, The post-ionization of field evaporated ions: A theoretical explanation of multiple change states, Surf. Sci **116**, 273 (1982).

[34] S. C. Lam and R. J. Needs, Model-potential calculations of tunnelling rate-constants for the field ion microscope, J. Phys. Condens. Matter **5**, 1195 (1993).

[35] R. G. Forbes, *Field Ion Microscopy at Very Low Temperatures*, Ph.D. thesis, University of Cambridge (1970).

[36] H. H. H. Homeier and D. R. Kingham, Effects of local field variations on the contrast of a field-ion microscope, J. Phys. D: Appl. Phys. **16**, L115 (1985).

[37] R. G. Forbes, Seeing atoms: The origins of local contrast in field-ion images, J. Phys. D: Appl. Phys. **18**, 973 (1985).

[38] S. P. Sharma and G. K. Schrenk, Effect of periodic surface-potential variation on high-field tunneling in field-ionization processes, Phys. Rev. B **2**, 598 (1970).

[39] H. Iwasaki and S. Nakamura, Field ionization processes in field ion microscopy, Surf. Sci. **33**, 525 (1972).

[40] I. V. Goldenfeld, E. N. Korol, and V. A. Pokrovsky, Field ionization theory, Int. J. Mass Spectrom. Ion Phys. **5**, 337 (1970).

[41] S. Bhatt, S. Katnagallu, J. Neugebauer, and C. Freysoldt, Accurate computation of chemical contrast in field ion microscopy, Phys. Rev. B **107**, 335413 (2023).

[42] R. G. Forbes, Apparent conceptual oversight in the statistical mechanics of the reverse-biased Schottky junction, and equivalent devices (2022), 6th Symposium on Schottky Barrier MOS Devices (SSBMOS), Giessen, Germany, September 2022; doi: 10.13140/RG.2.2.34438.04163.

# Chapter 13

**From Faraday and Maxwell to Quantum Physics**
**The early story of the Electromagnetic Vector Potential**

**Basil Mahon**

**Abstract**. Before its appearance in quantum physics, the vector potential, also known as the electrotonic state, played a part in the development of the theory of electromagnetism. Two of the main creators of that theory gave it the prime role but a third regarded it as a mere mathematical contrivance and banished it from the stage. However, the advent of the quantum has brought about its revival and the vector potential is now essential to quantum mechanics and quantum field theory.

## Introduction

J. B. S. Haldane was an outstanding British scientist who made contributions to physiology, genetics, biochemistry, statistics, and philosophy. But he is probably best known for voicing his suspicion that the universe is not only queerer than we suppose but queerer than we *can* suppose.[i]

Indeed! One need only think of the quantum. Before 1900, who could possibly have supposed that energy came in lumps? Planck himself mistrusted his monstrous creation and for five years went on looking for a more plausible alternative. Only after young Einstein showed how beautifully the quantum explained the photoelectric effect did the world begin to accept that the universe really is that queer.

Yet along with the queerness brought in by new discoveries we have some old concepts that survive and even flourish in their new environment. I'd like to take you through the story of one staunch survivor of the quantum revolution – the vector potential. Essential to quantum mechanics and quantum field theory, it can be traced back, through the work of James Clerk Maxwell, to that of Michael Faraday.

I'll be taking us up to around 1905. Developments since then are beyond my range of competence but Tuck Choy and Miguel Ortuno will be completing the story for you.

## A blacksmith's son

I'll start in 1820. Faraday was 29, set for an honourable if unspectacular career as an analytic chemist at the Royal Institution in London, where he was assistant to the famous Humphry Davy.

Simply getting to this position had been a stupendous achievement. A blacksmith's son, he had only rudimentary schooling before being apprenticed to a bookbinder at 14. The books were a godsend for Faraday: all kinds came into the shop and he became an avid reader. The world began to open up for him but the more he read the more conscious he became of his own ignorance.

Then he came across a book that could have been written for him alone – *The Improvement of the Mind*, by Isaac Watts. He threw himself with all the vigour of youth into Watts' programme of self-improvement. He told friends to put him right when he made grammatical errors in speech and began to keep what was then called a commonplace book, setting down facts, especially scientific ones, for future reference. He was careful to use precise language and, most of all, to be guided by *observed* fact. And he took to heart Watts' advice to be wary of drawing general conclusions from a few particular observations. Here were the seeds of the scientific method that Faraday made his own: observe; record; think, taking imagination to its limits; but draw no conclusions without solid experimental proof.

By a happy combination of determination, dedication and luck he eventually managed to get a job at the Royal Institution. It was the lowest post, called "fag and scrub", but before long he was helping Davy with experiments, writing up the results, and even publishing some experiments of his own.

There was no outward sign that 29-year-old Faraday would become a great discoverer, but this was about to change. His first big test came when momentous news arrived from Denmark. Hans Christian Oersted had placed a compass near a current-carrying wire and seen its needle move.

So electric currents produced magnetism. But how? Faraday was ready for the challenge. In some ways his lack of formal education, even his total ignorance of mathematics, was an advantage, because it freed his imagination. He was able to think outside the box because he didn't know where the box was!

It was someone else who led the way. André Marie Ampère blazed a trail in Paris. Within a few months he had produced a mathematical theory of electromagnetism that won almost universal acceptance. Faraday had met Ampère and they got on well. He couldn't follow any of Ampère's mathematics but nevertheless managed to find something in his friend's theory that didn't seem right. Ampère had assumed that the motion of Oersted's magnetic needle was the result of repulsions and attractions that acted in straight lines between it and the wire, but to

Faraday this seemed wrong. What happened, he felt, was that the wire induced a *circular* force around itself and that everything else followed from that. To prove it, he did an experiment that was so simple and so elegant that it takes one's breath away. Mine, anyway.

He fixed a bar magnet in an upright position to the bottom of a basin with sealing wax and filled the basin with mercury so the only the top of the magnet showed above the surface. Then he dangled a short length of stiff wire from an insulated stand so that its bottom end dipped into the mercury and connected one terminal of a battery to the top end of the wire and the other terminal to the mercury. The wire and the mercury now formed part of a circuit that would remain unbroken even if the bottom end of the wire moved. And it did move – in rapid circles around the magnet. Faraday had not only shown that the new force acted in a circle; he had made the world's first electric motor.

By 1831, everyone knew how to make magnetism from electric currents. Surely it should be possible to make electricity from magnetism. Many had tried but all had failed. Faraday's attempts had fared no better than anyone else's. Then he thought of winding two separate wire coils on opposite sides of an iron ring, passing a current through one coil to magnetise the ring, and watching to see whether the magnetism would induce a current in the other coil. To get the strongest possible magnetic effect, he had a 6-inch diameter wrought-iron ring specially made and wound as many turns of wire as would fit into each coil. He connected one coil to a battery and a switch and the other to a galvanometer, so forming a primary circuit to provide the magnetism, and a secondary circuit, in which he hoped a current would flow.

When he closed the switch the galvanometer needle gave a brief twitch and then returned to its resting position. Now a steady current was flowing in the primary circuit but absolutely nothing was happening in the secondary one. But when he opened the switch, shutting off the primary current, the galvanometer needle twitched again, this time in the opposite direction.

No wonder it had taken so long to find a way to generate electricity from magnetism. Nothing happened except when the magnetic influence was *changed.* Now he soon found a much easier way to make electricity from magnetism. When he connected a coil of many turns to a galvanometer, took an ordinary bar magnet, pushed it into the coil cavity and pulled it out again, the galvanometer needle swung vigorously, first one way then the other. He'd made the world's first electric generator!

His report on the experiment was plain and factual but he allowed himself a brief venture into theory, describing how he believed the current was induced in the secondary circuit of the iron ring experiment. He wrote:

> Whilst the wire is subject to either volta-electric or magneto-electric induction, it appears to be in a peculiar state: for it resists the formation of an electric current in it, whereas, if in its common condition, such a current would be produced; and when left uninfluenced it has the power of originating a current, a power which the wire does not possess under common circumstances. This electrical condition of matter has not been recognized, but it probably exerts a very important influence in many if not most phenomena produced by currents of electricity. I have, after advising with several learned friends, ventured to designate it as the electro-tonic state.[ii]

Faraday's revelation of this hitherto unrecognised state of matter failed to generate much interest among his contemporaries. Nevertheless, here was one of the great moments in science. With simple apparatus and almost supernatural prescience Faraday had not only discovered electromagnetic induction, he had identified the electrotonic state, which later came to be called the magnetic vector potential.

**Faraday ploughs a lone furrow**

Before and after every experiment Faraday had been thinking deeply along lines different from everyone else's. In his mind's eye, every electric current created a circular force around itself and every magnet created a pattern of forces that acted along curved lines through and around it. Though invisible, these lines needed only a sprinkling of iron filings on a piece of paper held over a magnet to reveal themselves. They were physically present in space. He first called them magnetic curves but later changed the name to magnetic lines of force.

Now he could explain induction. In his experiments with the coil of many turns a current flowed while he was pushing the magnet into the coil or pulling it out, so making the wire of the coil intersect the magnet's lines of force. It was the act of intersection, or cutting, that made the current flow. From this he developed Faraday's law of induction, an early version of which stated that the "quantity of electricity thrown into the circuit" was "directly as the amount of curves intersected".[iii]

What about his electrotonic state? Hard as he tried, he failed to detect any direct physical sign of it, but it wouldn't leave his thoughts. He wrote "again and again the idea of an electro-tonic state has been forced on my mind".[iv] He felt in his bones that it somehow played a part in the induction process, possibly the central role.

Faraday spent most of the rest of his career working on electricity and magnetism. By studying electrolysis he came up with the idea of electric lines of force, to complement the magnetic ones. His many discoveries included the laws of electrolysis and the eponymous Faraday effect, by which the plane of polarisation of light is rotated when it passes through a magnetic field. He also showed, by demonstration with his "floating needle" experiment, that there are no magnetic poles – the supposed poles of a permanent magnet are simply the entry and exit points of the magnetic lines of force, which curve round to form continuous loops.

Scientists elsewhere in Europe tried to explain Faraday's experimental results by erecting theories of ever-increasing mathematical complexity. These were all based on the concept of instantaneous action at a distance along an imaginary straight line connecting two material objects, by analogy with Newton's law of gravity. Faraday was universally acknowledged to be a superb experimenter, but almost no one gave him any credit as a theorist. His ideas about forces in space were regarded as flights of fancy, not to be taken seriously because they were expressed in words, not in mathematics. This is how the British Astronomer Royal, Sir George Airy expressed the prevailing view.

> I can hardly imagine anyone who practically and numerically knows the agreement [between observation and calculations based on action at a distance] to hesitate an instant between this simple and precise action on the one hand and anything so vague and varying as lines of force on the other.[v]

Since Davy's time Faraday had worked alone. He had no close colleagues and never took on a pupil. A big question hung in the air. Who, if anyone, would be willing and able to translate his radical vision into mathematics so that others could share it? There seemed to be only one candidate – William Thomson, better known today as Lord Kelvin. He was a prodigy who had enrolled at Glasgow University the age of ten, taken top prizes in all subjects, gone on to Cambridge, and then been appointed professor of natural philosophy at Glasgow University at the age of 22.

While in his mid-teens he had found an equation for Faraday's electric lines of force, by analogy with heat flow. A little later he immersed himself in the little-studied mathematics of vectors, hoping to find more general ways to express Faraday's ideas. He rescued George Green's work from obscurity by finding three old copies of his self-published book[vi] gathering dust on a shelf in a Cambridge tutor's office, seeing at once that it was pure gold, and having it republished, Thomson was almost certainly the true author of Stokes' theorem, and he introduced the mathematical form of what we now call the vector potential.

This was brilliant work but the mercurial Thomson rarely worked on one topic for long. He had many interests and during the 1850s became deeply engaged in the heroic Atlantic telegraph cable project. It seemed as though Faraday's radical ideas might wither on the vine, but early in 1857 he received in his post a copy of a paper called "On Faraday's Lines of Force". The author was a young professor of natural philosophy at Marischal College, Aberdeen, called James Clerk Maxwell. The paper began "The present state of electrical science seems particularly unfavourable to speculation". Faraday must have feared for what was to come, but this was just Maxwell's way of introducing his use of an unusual analogy. After reading and re-reading the paper, Faraday wrote back:

> I received your paper, and thank you very much for it. I do not venture to thank you for what you have said about "Lines of Force" because I know you have done it for the interests of philosophical truth; but you must suppose it is work grateful to me, and gives me much encouragement to think on. I was at first almost frightened when I saw the mathematical force made to bear on the subject and wondered to see that the subject stood it so well.[vii]

Faraday had found his successor.

**What's the go o' that?**

Maxwell was born in 1831, 40 years after Faraday. He was brought up in Galloway in southern Scotland, where his father owned a small estate. Even as a toddler, Maxwell was fascinated by the world and determined to find out how it worked. He was always asking

"What's the go o' that?", and if he wasn't satisfied with the answer he'd say "Yes, but what's the *particular* go of it?" He had a wonderfully happy early childhood but his mother died when he was 8 and he had a disastrous time under a private tutor who used rote-learning methods. James absolutely refused to learn by rote – he saw no sense in it. After a year he rebelled and was sent away to school in Edinburgh.

He had a rough time at first. His classmates made fun of his country clothes and accent, and there was more rote teaching which he hated. He seemed to be slow in class and acquired the nickname "Dafty". But he was a tough little fellow; he stuck it out, made friends, began to do well and was soon in the top group in all subjects. At 14 he surprised everyone by investigating and writing about the kind of oval curves you can draw using pins, string and a pencil. His father showed this to one of the professors at Edinburgh University and it became Maxwell's first published paper.

At 16 he left school for Edinburgh University. The plan at this stage was for him to spend four years there before training to become a lawyer, as his father had been. As things turned out, happily for us, he left Edinburgh for Cambridge after three years and became one of the world's great scientists. But, paradoxically, the three years at Edinburgh did much to make him the kind of scientist he was. I'll offer three reasons,

First, he studied philosophy and brought the results of this to his scientific work. This enabled him to give free rein to his imagination yet, at the same time, to apply a strict professional scepticism. He never got carried away by his own ideas, or got bogged down by them. He never got stuck in a rut. Uniquely among scientists, he could return to a subject and take it to new heights with an entirely new approach. He did this twice with electromagnetism.

Second, he read a vast amount on all kinds of topics. And he didn't just read, he analysed, appraised, and remembered. So he always had a vast store of knowledge on tap.

And third, he spent many hours making his own experiments at home, using a motley collection of improvised equipment and materials. In one set of experiments, using pieces of old glass cut to shape and heated in the kitchen stove, he demonstrated the technique of photo-elasticity that became a great help to designers of structures like dams and bridges. If they made a scale model of the structure from a transparent substance, put it under stress, and shone polarised light though it, the strain patterns would show up as coloured lines, revealing any weak points in the structure. All this home experimenting gave him a commanding insight into nature's processes. He had an uncanny intuition: his guesses were nearly always right but even when they were wrong they had a way of leading to good outcomes.

Shortly before Maxwell left Edinburgh for Cambridge, one of the professors wrote to the master of Trinity College, telling him what to expect.

> He is not a little uncouth in manners, but is withal one of the most original young men I have met ... He is a singular lad, and shy [but] very clever and persevering … I am aware of his exceeding uncouthness, as well mathematical as in other respects ... I thought the Society and Drill of Cambridge the only chance of taming him and advised his going … I should think he might be a discoverer.[viii]

**Maxwell in Cambridge and Aberdeen**

Society and drill! Maxwell had plenty of both at Cambridge. The lively company of his fellow students brought him out of his shell. He revelled in the camaraderie and badinage, and joined in discussions about almost anything. He was even elected to the Apostles, an elite discussion group who kept their number at twelve by electing a new member whenever somebody left.

Maxwell's response to the drill of Cambridge was nicely summed up by one of his friends.

> … he brought to Cambridge a mass of knowledge which was really immense for so young a man but in a state of order appalling to his methodical private tutor. Though the tutor was William Hopkins, the pupil to great extent took his own way …[ix]

Nevertheless, he came second in the Tripos and was equal first in the even harder exam for the Smith's Prize.

While still at Cambridge, first as a "bachelor scholar" and then as a fellow of Trinity College, he did two groundbreaking pieces of work. The first was to show that you can make any colour by mixing red, blue and green light in the right proportions – exactly the principle used in colour TV today. He did it by simply by taking a disc marked like a pie chart with different colours and spinning it fast so the eye sees only a single *new* colour. The triangular diagram he drew to explain things differs only in detail from today's chromaticity diagrams. Giving a talk at the

Royal Institution a few years later he wanted to demonstrate the principle with some sort of display that could be easily seen from the back seats, so he made the world's first colour photograph and projected it on to a screen.

The second piece of work was his first venture into electricity and magnetism. Luckily, he knew just who to ask for advice on what to read and in what order. His cousin Jemima had married Hugh Blackburn, who was a friend and colleague of Willam Thomson at Glasgow University. There's no record of Thomson's reply but he probably said "read Faraday first" because that's what Maxwell did. He was struck at once by Faraday's openness and integrity and, as he read on, he marvelled not only at the precision of the great man's experiments but also at the power and subtlety of the reasoning that followed. To Maxwell, Faraday's ideas rang true. He had found a kindred spirit and a new source of inspiration. He wrote to a friend: "I hope to see my way through the subject and arrive at something intelligible in the way of a theory".

After much reading he decided on a limited first objective – to find a way to represent the *geometry* of Faraday's lines of force mathematically – and he chose an analogy well-suited to this purpose – the steady flow of an imaginary weightless and incompressible fluid through a porous medium. The streamlines of flow represented electric or magnetic lines of force and the *speed of flow* at any point represented the intensity of the electric or magnetic force there. He called the paper "On Faraday's Lines of Force" and in Part1 he used only words and a few simple equations to describe in detail the geometry of static electric and magnetic fields in space, including how surfaces of equal electric potential intersect the electric lines of force at right angles. In the course of this he introduced the distinction between what he called "intensities", like our field forces **E** and **H**, that are summed along a line, and "quantities", like our flux densities **D** and **B**, that are summed over an area.

What of Faraday's electrotonic state? Maxwell ended Part 1 by saying, more or less, that he didn't yet understand it well enough to write anything useful, but that in Part 2 he hoped, in his words, "to discover a method of forming a mechanical conception of this electro-tonic state adapted to general reasoning". Like Thomson, he immersed himself in vectors, and tried everything he could think of. It was hard work because nobody had yet devised a way of using a single symbol to represent a vector. Everything was in Cartesian form, so each vector needed three symbols and most vector relationships needed a trio of equations.

He achieved only partial success in his rather enigmatic aim. He found out how to express the electrotonic state mathematically but was unable to assign it a mechanical role. But he did make an important connection that's central to our story. I'll translate what he did into our usual vector language and take the liberty of paraphrasing his reasoning very freely. First, he defined the electrotonic state vector **A** by saying that its time derivative gave the induced electric field force (with a minus sign). Then he saw that, in order for Faraday's law of induction to be true, the curl of **A** had to be equal to the magnetic flux density **B**. This meant that Faraday's electrotonic state was also the vector potential that Thomson had identified earlier. As Faraday had suspected, it had a prime role, its time derivative giving us electricity and its space derivative giving us magnetism.

This was a huge conceptual breakthrough but Maxwell confined himself to a modest and cautious statement. "… [These] expressions … may possibly be the symbolic representative of a real law of nature. Such a law was suspected by Faraday from the first. …"[x]

There was something else. Maxwell knew from his study of vectors that the curl of the gradient of any scalar function was zero. This meant that one could add the gradient of any scalar function to the vector potential without affecting the result. He didn't develop this point but it turned out to have important consequences later when the subject of gauge freedom came into play. I'll leave it to my more knowledgeable colleagues to explain this when they continue the story.

By the time the paper was published Maxwell had taken up his first job, as Professor of Natural Philosophy at Marischal College in Aberdeen. He had a warm welcome from his new colleagues – they were probably glad to have somebody young to talk to – but he missed the lively buzz of Cambridge and wrote to a friend: "No jokes of any kind are understood here. I have not made a joke for 2 months and if I feel one coming on I shall bite my tongue."

There were classes to take, including evening lectures to working men, and exams to set and mark but he took it in his stride and continued his work on colour vision, but the most remarkable work he did was on two new topics. He proved mathematically that Saturn's rings were composed of many solid objects orbiting independently, and, in his first venture into kinetic theory, derived the first-ever statistical law in physics – the distribution of molecular speeds in a gas. He published nothing on electricity and magnetism during his 4 years in Aberdeen but that was Maxwell's way. He'd leave a topic for months, or years, and then return to it with totally new ideas.

In 1860 Marischal College merged with another with another college in Aberdeen and they needed only one professor of natural philosophy. Extraordinary though it seems to us, they chose Maxwell's rival for the post. Few people then had any idea of the value of Maxwell's research work and none of those who did lived in Aberdeen! Happily, King's College, London needed a professor of natural philosophy, Maxwell applied, and got the job.

**Maxwell's mechanical model**

The time had come for another go at electricity and magnetism. This time he needed a *physical* model – one that would go beyond geometry and yield *dynamical* equations. He felt certain by now that something rotational was going on in a magnetic field and imagined all space filled with a fluid in which vortices could exist, and a region in which adjacent vortices were all rotating the same way with their axes parallel. There would be a tension along the spin axes and a sideways pressure, an exact analogy with magnetic lines of force,

As his thoughts developed, Maxwell replaced the vortices with solid, tiny, close-packed spherical cells that could spin. The combined effect of many cells spinning was the same as that of the vortices: rows of them with their spin axes aligned represented magnetic lines of force. Then he put even tinier particles *between* the cells to act like the idle wheels that engineers put between two gears that need to rotate in the same direction. They represented particles of electricity. In the presence of an electromotive force, they would move bodily between the cells, constituting an electric current. At least, this would happen if the cells were in a part of space occupied by a conducting material. If the space was occupied by an insulating material, the cells would hold on to their little particles, so preventing a current from flowing.

With this rather bizarre apparatus, Maxwell managed to explain magnetic forces and electromagnetic induction. And here, at last, was a mechanical interpretation of Faraday's electrotonic state, alias the vector potential. It was the electromagnetic momentum of the magnetic field at any specified point. In his words;

> It is what the electromotive force at the point would be if the currents, etc to which the lines of force are due, instead of arriving at their current state by degrees, had started instantaneously from rest with their actual values. It corresponds to the impulse which would act on the axle of a wheel in a machine if the actual velocity were suddenly given to the

> driving wheel, the machine being previously at rest.[xi]

What about electrostatic forces? The cells in a region occupied by a dielectric subjected to an electromotive force would hold on to their little particles, so preventing a current from flowing. But if the cells were elastic they would distort under stress like springs, so allowing the particles to move a short distance. Now everything fell into place. The spring-back force exerted by the stressed cells represented electrostatic force, and the movement of the little particles constituted a displacement of electricity that Maxwell described in these words

> This displacement does not amount to a current, because when it has attained a certain value it remains constant but it is the commencement of a current and its variants constitute currents in the positive or negative directions, according as the displacement is increasing or diminishing.[xii]

He called this new type of current the displacement current and added it to the familiar conduction current to give the total current, thereby completing the linkage between electricity and magnetism. In the scholarly words of Daniel Siegel, the displacement current "perfected the enduring foundation for modern electromagnetic theory".[xiii] It was also the key that opened the door to vast new regions of science and technology.

But if you look up displacement current in a textbook you won't see any reference to vortices or spinning cells. Instead, you'll probably find a statement that the displacement current was brought in as a necessary correction to Ampère's law that would ensure the continuity of charge. Such a correction was indeed necessary, and Maxwell's displacement current provided it, but this was not the way he arrived at the displacement current. Discovery doesn't always take the logical route: the origin of the displacement current was Maxwell's notion that the cells in his construction kit model could be springy.

This wasn't all. A medium that has both inertia and elasticity should be able to transmit waves, and his now had both. He calculated the speed of the waves and it turned out to be within 2 per cent of the speed at which light had been measured. He had united electricity, magnetism and light.

Here was one of the greatest moments in science, but his paper, called "On Physical Lines of

Force", caused barely a ripple. The reaction of one of his friends was typical. He said:

> The coincidence between the observed velocity of light and the calculated velocity of a transverse vibration in your medium seems a brilliant result. But I think a few such results are needed before you can get people to believe that every time an electric current is produced a little file of particles is squeezed along between two rows of wheels.[xiv]

People couldn't understand that the model was intended merely as an analogy that would assist in finding the relevant mathematical relationships. They saw it as an ingenious but flawed attempt to portray the true mechanism, for which the search would continue. Even the most enlightened of Maxwell's contemporaries thought that the next step would be to refine the model. But he decided instead to put the model on one side and build the whole theory from scratch, using only the laws of dynamics.

**Maxwell's dynamical theory**

He called his new paper "A Dynamical Theory of the Electromagnetic Field" and, after setting the scene, began with a bold statement:

> … there is an aethereal medium filling space and permeating bodies, capable of being set in motion and of transmitting that motion from one part to another, and of communicating that motion to gross matter so as to heat it and affect it in other ways,[xv]

He then described and worked through all the phenomena that needed to be accounted for. In this he used mechanical analogies – for example electric displacement was "a kind of elastic yielding" and the effect of the magnetic field surrounding a current was to "endow it with a kind of momentum". He emphasised that such references were "illustrative, not explanatory", but that when he spoke of the energy of the field he wished to be understood literally, because "all energy is mechanical energy".[xvi]

Analogies were useful in another way. They enabled him to use Lagrange's formulation of the laws of dynamics, which employed so-called generalised coordinates and so avoided the need to specify a particular mechanism for his medium. In the Lagrangian analysis, electromotive force could play the role of a mechanical force, current could play the role of a velocity and so on. It worked because Maxwell's formulae for electric and magnetic energy translated directly into the corresponding ones for mechanical energy. When he produced his Equations of the Electromagnetic Field it was all bona fide but one can see how it must have seemed to many of his contemporaries like a conjuring trick.

All the equations were still in Cartesian form. There were six vector triples and two single equations making 20 in all. The vector potential had a prominent place, where it was represented by the letters F, G, H. There was clearly redundancy in this set of equations; Maxwell could easily have reduced their number but chose not to do so. He later gave his reasons, saying "to eliminate a quantity which expresses a useful idea would be a loss rather than a gain at this stage of our enquiry".[xvii] As usual, he was right.

When Maxwell presented his paper to the Royal Society in 1864 the audience was bewildered; they simply did not know what to make of it. A theory based on a bizarre model was bad enough but one with no model at all was beyond comprehension. Among the doubters was William Thomson, who said that Maxwell had "lapsed into mysticism".[xviii] He and others were rooted in their Newtonian world and had failed to see that Maxwell was opening the way to new regions of physics, in which some phenomena not only do not need a mechanical model but *cannot* be represented by a mechanical model. When Einstein said "One epoch ended and a new one started with James Clerk Maxwell" he may have been thinking of this moment. But nobody noticed it at the time.

After five years in London Maxwell spent another six working from home at his Scottish estate before answering a call from Cambridge to become founding Director of the Cavendish Laboratory in 1870. His work during this time covered far too big a range to be covered today but it included a book, *The Theory of Heat*, in which he introduced his thought experiment "Maxwell's Demon", which perplexed the world's best physicists for 60 years; reciprocal diagrams in the theory of structures; and the first-ever paper on automatic control systems,

In 1873 Maxwell published his monumental *Treatise on Electricity and Magnetism*. It was the first proper book on the subject and covered everything that he knew. His own theory was there but you had to hunt for it – the subject of electromagnetism doesn't come in until Article 475 in Volume 2. Here at last, the Equations of the Electromagnetic Field appeared with single letters for vectors. This wasn't yet the system of vector analysis that we use today – we'll come to that shortly. Maxwell had heard from his friend P. G. Tait about quaternions, invented by William Rowan Hamilton – he of the Hamiltonian function. They were strange things – each quaternion was a combination of a 3-dimensional vector and a scalar but it had a single-letter symbol. Maxwell didn't need the full panoply of quaternions but he used some of Hamilton's notation to give his vectors single-letter symbols and write each of his vector equations in one line.

Knowing that this format would be new and strange to readers, he decided to include the original 20 Cartesian equations as well. And to make his vector symbols distinctive he employed German Gothic script, with the result that his vector equations had a strange Teutonic aura. Leading them was the equation for the vector potential, its symbol a very Gothic-looking A.

At the time Maxwell died, in 1879, almost nobody understood his theory, and so things remained for a while after. It sat like an exhibit in a glass case, admired by some but out of reach. By using single symbols for vectors he had reduced the number of equations from 20 to eight but there still seemed to be a maze of symbols with little coherence. The person who simplified the theory and made it accessible was an unemployed former telegraph operator who had never been to university called Oliver Heaviside.

**Heaviside recasts Maxwell's theory**

Heaviside was born in 1850, the youngest of four sons in a respectable but rather poor family. Scarlet fever at the age of 6 left him partially deaf and he found himself excluded from street games because he couldn't hear what the other boys were saying. He was thrown on his own resources and began to build a defence against the barbs of life. A stubborn independence took hold of him, almost against his will, and held its grip until the day he died. He did well at school but university was beyond the family's resources so he spent 2 years studying at home, reading everything he could find on scientific topics. This privilege hadn't been extended to his brothers, who were working and contributing to the family income. His uncle by marriage, the famous inventor Charles Wheatstone, had probably interceded on his behalf.

At 18 he began work as a telegraph operator on the newly opened Anglo-Danish link. These were still early days for undersea telegraphy and the operators were free to improvise and experiment with their equipment – indeed they had to, to keep the traffic flowing. Heaviside became a star trouble shooter, but for him each problem was not merely something to be fixed but a means of probing the strange ways of electricity, which often baffled even the most experienced of his colleagues.

Young men were drawn to the work by good wages and the excitement of being at the forefront of technology. They learnt their craft on the job. Nobody really knew how electricity worked: people thought it was something you developed a feel for, like riding a horse. And the prevailing attitude to mathematics was neatly summed up by the weekly trade journal *The Electrician*.

> In electricity there is seldom any need for mathematical or other abstractions… they may for all practical purposes be dispensed with.[xix]

Heaviside was coming to quite a different view. He had already begun to see that the only way to understand electricity was by harnessing the power of mathematics. And the moment he saw Maxwell's *Treatise* in the Newcastle public library he was captivated. His mission in life was now clear. With Maxwell as his guiding star, he set out to find out all he could about electricity and spread the word. After 6 years as a telegrapher, he left his job and began his career as an unpaid researcher into electricity.

In the course of this career, he more or less invented the subject of electrical engineering science, and this included presenting Maxwell's theory to the world in a new and compact form that people could understand. Heaviside didn't like the prominence Maxwell had given to potentials, which he regarded as "metaphysical" because they couldn't be measured, and he took the momentous decision to concentrate only on the electric and magnetic field forces. The result was the tight and clear set of four equations that everybody now calls "Maxwell's equations". The vector potential had been banished, along with the electric scalar potential, and nobody seemed to miss them.

Heaviside also invented the system of vector analysis that became the standard language for describing any vector field. He had eagerly investigated quaternions but found them of little use, so he dropped the scalar parts, leaving plain vectors together with some simple algebraic rules. Remarkably, Josiah Willard Gibbs in America went through the same sequence at about the same time and came up with the same system.

Until 1888, Maxwell's theory of electromagnetism was only one of several competing theories but all changed when Heinrich Hertz succeeded in producing and detecting electromagnetic waves in his laboratory in Karlsruhe. All theories based on action at a distance dropped out, leaving Maxwell's in sole command. Driven by the march of technology over the following decades, electromagnetism became a recognised subject for inclusion in educational courses and books. And at the heart of the subject was Heaviside's version of Maxwell's theory. The way of the world when it comes to learning and applying Maxwell's theory was, and still is, Heaviside's way.

One can see why. Heaviside had recast Maxwell's theory into a form that ordinary physicists and engineers could use in their daily work. He had cleared the fog and helped them to visualise the forces and fluxes, and the flow of energy in space.

How would Maxwell have reacted to his favoured vector potential being so rudely dismissed? The answer probably lies in one of the essays he'd written for the Apostles discussion group at Cambridge.

> The dimmed outlines of phenomenal things all merge into one another unless we put on the focusing glass of theory and screw it up sometimes to one pitch of definition and sometimes to another, so as to see down to different depths through the great millstone of the world.[xx]

Heaviside had screwed the focusing glass to reveal the two field forces and the results were splendid, but one should not forget that at a deeper pitch of definition there lies the vector potential, whose variation in space gives us one of the field forces whose variation in time, when added to the gradient of the electric scalar potential, gives us the other and whose variation in time, when added to the gradient of the electric scalar potential, gives the other.[xxi]

**Three great shocks**

Around the turn of the century, three huge shocks followed in quick succession: J. J. Thomson's discovery of the electron, Planck's discovery of the quantum, and Einstein's special theory of relativity. The scientific world had been violently shoved from what we might still call the Newtonian era into the modern era. How would Maxwell's equations cope? The answer is astonishingly well.

Take special relativity. Maxwell's aetherial medium needed to operate in absolute space and time so when Einstein abolished those it no longer had a home. But everything turned out well because the mathematics of special relativity stems directly from Maxwell's equations. Given Einstein's two axioms, Maxwell's equations actually *require* space and time to operate together as Minkowski's spacetime. And spacetime itself is a sufficient medium for the electromagnetic field. So, Maxwell's medium was not needed because his equations had supplied their own. You could say that it had an honourable retirement.

The discovery of the electron had a profound effect, not on the equations themselves (though some modification was needed for microphysics) but on what they represented and how they were interpreted. The process was complicated – at least one scholarly book has been written about it – but the main change was that electric charge was no longer just an artifact of the field, as Faraday and Maxwell had thought. It existed in its own right, as a property of matter. For the first time, the true relationship between the field and matter was revealed as a kind of partnership: charges reside in matter but their effects are transmitted by the surrounding field.

Even these two shocks were as nothing compared with that of the quantum. It posed, and still poses, fundamental questions about reality. It performs an essential service to our universe – stabilising everything by enabling a balance to be struck between the amount of radiation bodies emit and the amount they absorb. In a sense, the quantum saved Maxwell by curbing high-frequency radiation – without this constraint on his equations, matter would have radiated all its kinetic energy away long ago in the so-called ultraviolet catastrophe. With this constraint, his equations remain valid at all scales large enough for the minute graininess of energy to be ignored, and even at smaller scales they act as a kind of underpinning for quantum-based theories.

Maxwell never claimed or believed that his theory represented ultimate reality. To him it was work in progress and his assumptions about the primacy of the field and the necessity of an aethereal medium were, in effect, working assumptions, to be discarded immediately if disproved. The wonder is that despite being wrong in these matters he managed to produce equations that absorbed the shocks and still form a key component of our understanding of how the universe works. How was he able to do this? Perhaps even he couldn't explain it. An editorial in the Times Educational Supplement in 1925 said that Maxwell was "to physicists, easily the most magical figure of the nineteenth century".

There's a much-used quote from Richard Feynman on a similar theme. He said:

> From a long view of the history of mankind – seen from, say, ten thousand years from now – there can be little doubt that the most significant event of the nineteenth century will be judged as Maxwell's discovery of the laws of electrodynamics.

He wasn't thinking here just of the four famous equations that you see on T shirts. He also said:

> In the general theory of quantum electrodynamics, one takes the vector and scalar potentials as the fundamental quantities. **E** and **B** are slowly disappearing from the modern expression of physical laws: they are being replaced by **A** and ϕ.[xxii]

The wisdom of Maxwell's decision to keep all options open has been handsomely borne out. Heaviside had done the world a great service by making Maxwell's theory accessible but his four equations were not enough to cope with the queerness of the quantum. Feynman and others needed to bring in the vector and scalar potentials, which had been dropped by Heaviside on the grounds that they were mere mathematical constructs with no physical existence. Perhaps the term "physical existence" has a different meaning in the quantum world from its everyday meaning, but I understand that most experts now believe that the vector potential does truly exist physically, a fact demonstrated by the Aharonov-Bohm effect. (Similar attempts to prove the physical existence of the scalar potential have so far failed.)

**Finale**

Something not in doubt is that the vector potential plays a fundamental role in quantum mechanics and quantum field theory. Sadly, I'm not in the least qualified to talk to you on this subject so my contribution is almost done. Let's take a final look at our two main characters.

Faraday and Maxwell are popularly depicted as a partnership of genius, with Faraday as the practical-minded experimenter and Maxwell as the mathematical theoretician. This portrayal grossly undervalues Faraday. He was not only a great experimental scientist but also a great theoretical scientist. He introduced the concept of the field, and that alone is enough to put him in this category. The popular portrayal also wrongly implies that Maxwell was a mathematical genius. He wasn't. His genius lay in *using* mathematics to describe and explore the physical world. James Jeans wrote this eloquent description of his method.

> … every mathematical step in the analysis is full of physical significance; every symbol is alive, and represents a clear physical concept in the mind, even when the symbol does not represent anything in nature. An obvious instance is the magnetic vector potential, which Maxwell had pictured as the measure of the intensity in Faraday's electrotonic state.[xxiii]

Jeans, of course, wrote this before Aharonov and Bohm had set to work on the vector potential.

Now is the time for me to hand over the baton my colleagues, who will complete the story by bringing us up to the present day. It's been a pleasure and an honour to talk to people who are at the cutting edge of science today. I hope you have a happy and productive conference.

**Acknowledgements**

I thank Tuck Choy and his colleagues for inviting me to contribute to this conference. As a late-starting writer I am exceedingly lucky to have had generous help and encouragement from some fine people, including past and present Trustees of the James Clerk Maxwell Foundation and fellow-writers Francis Everitt, Bruce Hunt, and my co-author of *Faraday, Maxwell, and the Electromagnetic Field*, the late Nancy Forbes.

**References**

[i] Haldane, C. B. S. *Possible Worlds and Other Essays*. London: Chatto and Windus, 1927
[ii] Faraday, M. *Experimental Researches in Electricity*, series 1, paragraph 60
[iii] Faraday, M. *Experimental Researches in Electricity*, series 28, paragraph 3,115
[iv] Faraday, M. *Experimental Researches in Electricity*, series 29, paragraph 3,269
[v] Bence Jones, H. *The Life and Letters of Faraday*, vol 2, page 353, Philadelphia: Lippincott, 1870
[vi] Green, G. *An Essay on the Application of Mathematical Analysis to the Theories of Electricity and Magnetism*. Nottingham: Printed for the author by T. Wheelhouse, 1828. Re-published in several parts in Crelle's Journal, 1850-54
[vii] Campbell, L. and Garnett, W. *The Life of James Clerk Maxwell*, 1st ed. London: Macmillan, 1882. Sonnet Software version, page 252
[viii] Letter from James Forbes to William Whewell in Harman, P. M. *The Scientific Letters and Papers of James Clerk Maxwell*, 3 vols, vol.1. Cambridge: Cambridge University Press, 1990, 1995, 2002
[ix] Tait, P, G. "James Clerk Maxwell", *Proceedings of the Royal Society of Edinburgh* 10 (1878=80) 331-339

[x] The quotation and the paraphrased reasoning are from a lecture titled "On Faraday's electrotonic state" that Maxwell gave to the Cambridge Philosophical Society prior to the publication of his paper "On Faraday's Lines of Force". See Doncel, M. G. and de Lorenzo, J. A. "The electrotonic state, a metaphysical device for Maxwell, too?", European Journal of Physics 17 (1996) 6-10
[xi] Simpson, T. K. *Maxwell on the Magnetic Field,* page 165. New Brunswick: Rutgers University Press, 1997
[xii] Simpson, T. K. *Maxwell on the Magnetic Field,* pages 171-2. New Brunswick: Rutgers University Press, 1997
[xiii] Siegel, D. M. *Innovation in Maxwell's Electromagnetic Theory*. Page 85. Cambridge: Cambridge University Press, 1991. Siegel was representing how the displacement current is generally taught to students.
[xiv] Campbell, L. and Garnett, W. *The Life of James Clerk Maxwell*, 1st ed. London: Macmillan, 1882. Sonnet Software version, page 163
[xv] Maxwell, J. C. (Torrance, T. F. ed,) *A Dynamical Theory of the Electromagnetic Field*, pages 34-5. Edinburgh: Scottish Academic Press, 1982
[xvi] Maxwell, J. C. (Torrance, T. F. ed,) *A Dynamical Theory of the Electromagnetic Field*, page 70. Edinburgh: Scottish Academic Press, 1982
[xvii] Maxwell, J. C. *Treatise on Electricity and Magnetism*, vol 2, page 254. Oxford: Clarendon Press, 1891

[xviii] Reported in Templeton, J. M. and Herrmann, R. L. *The God Who Would Be Known: Revelations of Divine Contemporary Science*, page 161. Templeton Press, 1998
[xix] From an editorial in *The Electrician* on 30 November 1861. This is ten years before the time of the story but things had changed little.
[xx] Maxwell's essay "Are There Real Analogies in Nature" in Harman, P. M. *The Scientific Letters and Papers of James Clerk Maxwell*, 3 vols., vol.1, page 377. Cambridge: Cambridge University Press, 1990, 1995. 2002
[xxi] The relationship is not quite direct. Curl **A** gives **B**, which is related to the field force **H** by $\mathbf{B} = \mu\mathbf{H}$.

[xxii] Feynman, R. P., Leighton, R., and Sands, M. *The Feynman Lectures on Physics*, vol,2, chapter 1, page 3. New York: Basic Books, 2010
[xxiii] Jeans' essay "Clerk Maxwell's Method" in *James Clerk Maxwell: A Commemoration Volume,* page 96. Cambridge: Cambridge University Press, 1931

# Chapter 14

## From Faraday and Maxwell to Quantum Physics. The later story of the Electromagnetic Vector Potential.

Tuck Choy and Miguel Ortuño*

*Departmento de Fisica-CIOyN, Universidad de Murcia, Murcia 30071, Spain.*

## Abstract

With the advent of quantum mechanics by Heisenberg in 1925 exactly a century ago, the quantization of the electromagnetic field became an important goal for our founding fathers, whom we are here to celebrate. It was realized very soon that a consistent picture of quantum electrodynamics (QED) requires the quantization of not just the electromagnetic field $\mathbf{A}(\mathbf{r}, t)$, but the electron field $\psi(\mathbf{r}, t)$ as well. The electron field is now a Dirac spinor field and it becomes a major partner for the vector potential. Together these two fields form a duet which yearns for a unification which is not often emphasized in text books, but no one knows how to do this yet. The underlying structure appears to be a super-symmetric one where bosons and fermions live comfortably as a single field that could by some yet unknown process produce QED. This area is still at the forefront of current research with names like Super-symmetric Quantum Electrodynamics (SQED), strings and others. Ironically for a variety of reasons, the number of fields in SQED has to be increased and not decreased, defying the objective of an economical unification. Furthermore it is now known that Maxwell's theory is incomplete and that there are situations due to boundaries or topology that demand the vector potential must itself be gauge invariant. This is another topic of current research which we can only briefly survey in this paper. To keep one's feet to the ground, we will show some novel toy models such as those proposed by Jennison (1979) which will be used to illustrate how one might "create" an electron from just pure light. While crude, these models should be looked at in the spirit of Maxwell's original "mechanical models" for light which despite their crudeness, were able to guide Maxwell to propose the displacement current, central to his electromagnetic theory, see Basil Mahon's paper [1]. We will also show that it is unexpectedly easy to operate in the QED regime, (no big particle accelerators needed) with something as common as an everyday mobile phone. Physics students may now do experiments in QED even from their home kitchen table! We shall recall attention to some of Heisenberg, Dirac and Wigner's insightful remarks in 1972 on the occasion of Dirac's 70th birthday symposium in Trieste, and end with some speculations about gravity waves and quantum mechanics.

(This paper is dedicated to the celebration of 100 years of quantum mechanics, on the anniversary of Heisenberg's founding paper on the subject in July 1925, delivered by Miguel Ortuño at the IQSA 2025 Conference at Tropea in Calabria, Italy from 30 June to 4th July 2025).

---

* tuckvk3cca@gmail.com

"All these fifty years of conscious brooding have brought me no nearer to the answer to the question, 'What are light quanta?' Nowadays every Tom, Dick and Harry thinks he knows it, but he is mistaken Albert Einstein [2]."

## I. INTRODUCTION

Let us begin by reminding the reader of the above quote from the Einstein who first discovered the 'photon' in 1905. Einstein was known to have gone to Westminster Abbey in June 1921 for the first time to lay a wreath on the tomb stone of Isaac Newton. Reporters surrounding him asked if he thought he was the successor to Newton. Einstein replied that "That statement is not quite right, I stood on Maxwell's shoulders". Most reporters at that time we would guess did not know who James Clerk Maxwell was. Maxwell would have pointed out that he in turn had Michael Faraday's shoulders to stand on. Now we must remember that in the 1870s, the great electromagnetic wave theory of Maxwell had yet to be experimentally verified. Maxwell died young at 48 in 1879 and it is likely that had he lived longer, being then Director of the Cavendish Laboratory at Cambridge, he would have initiated research into electromagnetic waves. This task was left to a brilliant young PhD student of Hermann von Helmholtz by the name of Heinrich Hertz who published in 1887 the first demonstration of the existence of 'invisible' electromagnetic waves as predicted by Maxwell, using the Hertz spark oscillator, now commonly known as the spark gap transmitter. Hertz's studies were extensive and indeed profound, in fact his published works include studies of reflection and diffraction of radio waves (at about 70 cm wavelengths or 0.5 GHz), including their transmission through material media, for example through bricks! An astute student of science may be able to find a hint of an early demonstration of quantum mechanical tunnelling in his reports [3]. We shall have more to say about this later. Tragically Hertz also died young, at just 37, his calibre as a theoretical and experimental physicist was a hard act to follow, except perhaps by Enrico Fermi. However he did not think that these 'electric waves' would have any practical applications. It was left to the dedication and imagination of one man, who was born in Bologna of mixed Irish and Italian heritage that gave the world modern radio technology that we have so taken for granted these days. As a matter of fact, it was exactly 124 years ago, that Guglielmo Marconi succeeded towards the end of 1901, against all criticisms from his scientific colleagues, in demonstrating that he could send radio waves across the Atlantic. When news of this broke out in early 1902 the share price of the transatlantic cable companies on the London stock exchange fell [4]. This was the exciting history of wireless. Marconi used a 10kW transmitter for his transatlantic experiments costing 50,000 pounds sterling in 1901 or about 7.5 million in today's money. Nowadays radio amateurs can send messages across the Atlantic with a garden dipole and a $10 piece of electronics running merely 100mW of RF power. In fact that is a lot of power; transmissions from Texas at a mere $1mW$ to Western Australia is shown in Fig. 2. That's a distance of over 16,000 kms, nearly three times that from Newfoundland to the UK with 10 million times less power than Marconi used in 1901! Marconi would be pleased. Unfortunately this is still way too high compared to the quantum limit at this frequency; more on this later.

Figure 1: Milliwatt WSPR transceiver built using a Raspeberry Pi Zero. Image courtesy of Greg Lawler K6FTP, https://wspr-zero.com/gallery/

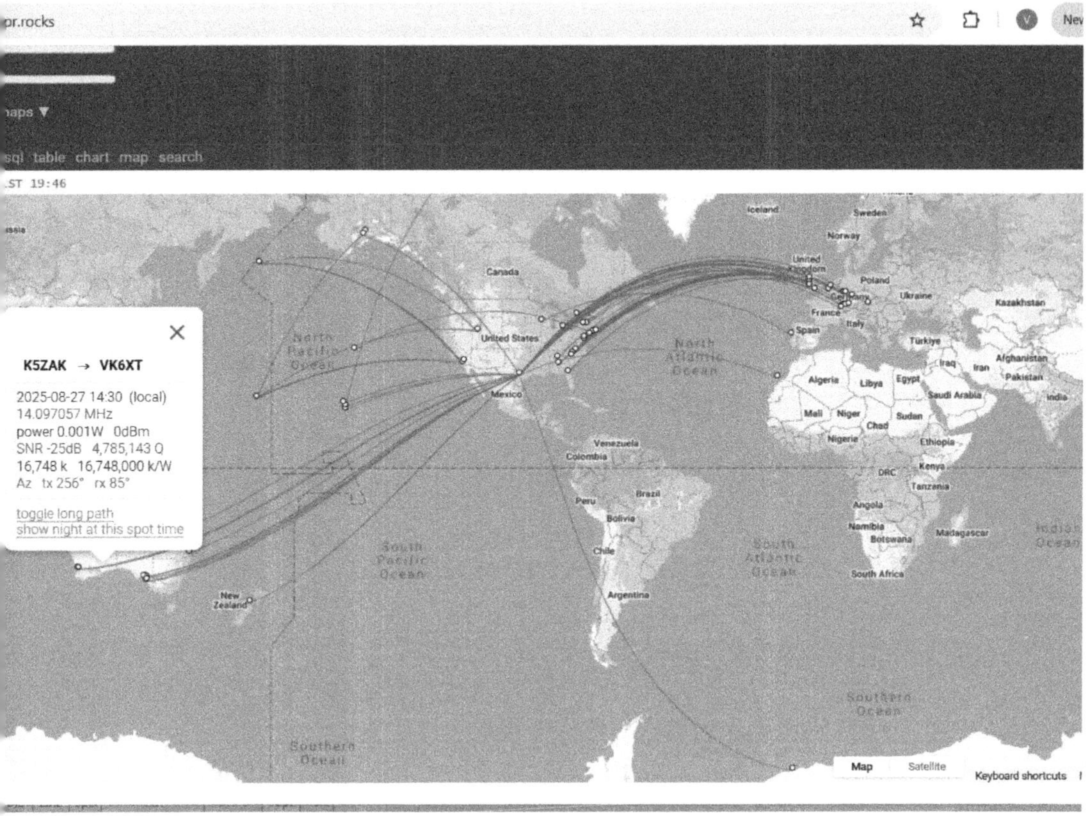


Figure 2: WSPR 1mW transmissions, highlighted is transmission from Texas to Western Australia.

Back to 1901 for quietly behind in the background, Max Planck revolutionized modern physics by reluctantly introducing the light quantum hypothesis to explain the black body radiation spectrum. Then from 1911 to 1925, starting from the Bohr-Rutherford model of the atom; the old quantum theory took shape and was developed by Bohr, Sommerfeld and Wilson to explain atomic spectra for more complicated atomic spectra. Einstein also played a key role by introducing the photon concept for the photo-electric effect and spontaneous radiation coefficients to explain atomic radiative transitions, see quote above, but there were discrepancies which these theories could not explain. Then the chain of events rapidly changed when Werner Heisenberg discovered his matrix mechanics and within months of this discovery in 1926, in their famous three-men paper, Born, Jordan and Heisenberg made their first attempt to apply matrix mechanics to the quantization of the electromagnetic field according to Maxwell's theory [5]. Their preliminary attempt was premature and courageous. It took several more years before a quantum theory of light was finally adopted with the notable names of Heisenberg, Pauli, Dirac, Peierls, Landau and Fermi as key players. All of these players had to involve the vector potential $\mathbf{A}(\mathbf{r}, t)$ in one way or another. Dave Delphenich has given a good account of Heisenberg and Pauli's early work at this conference [6]. I shall briefly fill in the gap on the others. Yet would Einstein agree that we have now answered his question 'What are light quanta?'. I doubt it. In this article we shall carry on the task that Basil Mahon set up in his previous paper [1] by first bringing you up to date on the development of what is now widely accepted as the quantum electromagnetic vector potential $\hat{\mathbf{A}}(\mathbf{r}, t)$, with the hat on $\hat{\mathbf{A}}$ implying it is a quantum operator. However let's first go back to Maxwell.

At the end of Part 1 of "On Faraday's Lines of Force" [7] Maxwell said that he hoped, in Part 2, "to discover a method of forming a mechanical conception of this electro-tonic state adapted to general reasoning". Earlier in the same paragraph he had explained that "The idea of the electro-tonic state, however, has not yet presented itself to my mind in such a form that its nature and properties may be clearly explained without reference to mere symbols...", But in Part 2 he failed to get beyond "mere symbols". Some of the symbols, though, did provide a significant conceptual breakthrough ($\mathbf{B} = curl\mathbf{A}$ and $\mathbf{E} = -d\mathbf{A}/dt$). So that was the best he could do at that time. In retrospect it's hard to see how Maxwell hoped to find a mechanical analogy when the model he was working with was based on the flow of a massless fluid. To get a mechanical (or physical) conception he had to have mass in his model and this came six years later in his paper "On Physical Lines of Force", where the vortices , or spinning cells, had mass and hence momentum. It's a great pity that Faraday had lost his mental powers by the time Maxwell's "On Physical Lines of Force" came out. Would Faraday have been satisfied by Maxwell's representation of the electro-tonic state? We'll never know for sure but Faraday had given a favourable hint when he said: "Again and again the idea of an electro-tonic state has been forced on my mind; such a state would then constitute the *physical lines of magnetic force*." (Our italics.) Maxwell on the other hand could not have been satisfied with his mechanical model, which students are rarely taught these days, powerful though it was in predicting the 'displacement current'. He wanted more and three years later he published his 'Dynamical theory of the Electromagnetic Field [8]. However note the prophetic words of Faraday who is known to have chosen his words carefully. The electro-tonic 'state' must be one of the most powerful and conceptually significant concepts in the history of science discovered by an experimentalist with hardly any mathematical training. Maxwell, though, was a mathematically trained experimentalist. Few people now see him as an experimentalist but he was, and in fact some of his experiments were done in his own home in London [9]. He took great pains to design electromagnetic standards for measurement, he developed the theory and did experiments on the viscosity of gases and opto-ellasticity and later became the first director of the now esteemed Cavendish laboratoy at Cambridge. Finally in Part III of the'Dynamical Theory' [8] of 1864, he put his finger on Faraday's electro-tonic state as the 'electromagnetic momentum' of the electromagnetic field. That was the best he could do to go beyond "mere symbols". Unfortunately many later generation students are taught that the vector potential is not directly observable, a great error that did not come from Maxwell. With hindsight we can see how impressive both men were, because we had to wait until Niels Bohr who in 1911 spectacularly introduced the 'stationary state' and with it the famous 'quantum jumps' that greatly disturbed Schrödinger. The quantum state then became common usage after Dirac's book in 1931. All these preoccupation with quantum states must surely come from Michael Faraday's obsession with his electro-tonic state. Subsequently Dirac identified that the Maxwell vector potential is the quantum electromagnetic potential $\hat{\mathbf{A}}(\mathbf{r}, t)$, with the hat on A that actually creates or destroys these 'photon' states as we are now taught. Maxwell observed that his vector potential has a gauge degree of freedom. He did not make much use of it. We now know that through the work of C.N. Yang and Robert Mills, Weinberg, Salam and Glashow, the gauge freedom can be generalised and with it and the idea of spontaneous symmetry breaking, the Standard Model with gauge symmetry group $SU(3) \times SU(2) \times U(1)$ became the pillar of modern particle physics see Yang 2014 [10]. Unfortunately, however, these advances do not bring us any closer to answer Einstein's fundamental question. In fact they complicate the situation somewhat. The model requires the Higg's mechanism to provide mass, and for that a massive Higgs boson of some $125GeV/c^2$ had to be postulated and fortunately subsequently detected. One of the main discrepancies currently is the cosmological constant problem or vacuum energy catastrophe.

Extensive microwave background measurements of the universe currently place an upper bound of $2.5\times10^{-47}GeV^4$ in so called geometrized units ($c = 1, G_m$(gravitational constant) $=$ 1) [11]. Unfortunately the best current theoretical estimate based on quantum field theory is $10^8 GeV^4$, a discrepancy of some 55 orders of magnitude. Zero point energy is of fundamental significance, it prevents all atoms from collapsing into the nucleus in the ground state and is ultimately responsible for the stability of our very existence [12]. The vacuum energy catastrophe is as serious as and maybe even more serious than the ultraviolet catastrophe that had plagued classical black-body radiation theory given by the Rayleigh-Jeans law. We do not yet know the answer to such questions. Lev Landau for one never accepted quantum field theory as a logically consistent theory because of the divergent nature of the zero point energy both of the vacuum and in higher order perturbation calculations. This is not to undermine the remarkable success of the renormalization programme, as current research has shown that the Standard model is both renormalizable and mathematically consistent. However significant difficulties arise at higher energies where the supposedly graviton energy or where curved space-time issues come into play as there is as yet no unified field theory that includes gravitation. These issues become naturally important when attempting to answer cosmological questions such as the Big Bang and the origin of the universe, space-time and matter. Elsewhere one of us has alluded to the fundamental question [13] as to whether it is the quantum laws that dictate cosmology or vice-versa. If the latter, then any hope to further understand the secrets of the photon in the way Einstein desired would be a very formidable challenge indeed.

## II. UNIFIED QUANTUM ELECTRODYNAMICS

The early pioneers of quantum mechanics were faced with a number of problems in trying to extend the finite particle quantum mechanics of Heisenberg and Schrödinger to a field theory with an infinite degree of freedom. Recall that in classical mechanics, we approach this by a process of coarse graining as in continuum mechanics which leads to the well known classical theories of elasticity and hydrodynamics, which are also extendable to mesoscopic systems [14]. How do we do this for a quantum system, such as a quantum fluid or for the quantization of the classical electromagnetic field theory of Maxwell? We note that at the same time as others were struggling with these problems, Paul Dirac had been trying to find the relativistic wave equation for the electron, which because of the superposition principle required a first order differential equation in time. The breakthrough came when he proposed the Dirac equation in 1928 which required a new algebra, his $\gamma$ matrices later identified as a Clifford algebra and also that the wave function must be extended to a four component object, later identified as a four component spinor. The spinor concept was first introduced by the French mathematician Élie Cartan in 1913 but it was rediscovered independently by Dirac in 1928. It is now well known that the triumph of this theory is the prediction of the anti-particle of the electron, namely the positron with its opposite positive charge and identical mass. However Dirac was embarrassed by the existence of negative energy states. The effort to quantize the electromagnetic field ran into the same problem. Initially there were attempts by Landau and Peierls to discover the Schrödinger equation for light [15]. At the same time Heisenberg and Pauli [6] came up with commutation relations between the $E$ and $B$ fields which did not focus on the vector potentials. It became clear later, through the work of Paul Dirac and Pascual Jordan that the vector potential is a very important object for the quantization of the electromagnetic field. In his book of 1931 Dirac had spelled out the beautiful number operator representation of the harmonic oscillator which he could readily apply to an infinite set of non-interacting harmonic oscillators [16]. As it was already well known from the classic textbook on the theory of sound by Lord Rayleigh [17], who wrote down the theory of normal modes, Dirac immediately saw how to use normal modes to quantize Maxwell's theory using the results he had obtained for the harmonic oscillators [16]. Dirac realised that the theory has to be relativistic, so canonical quantization must be extended to a covariant form. Commutators between the four components have to be worked out which lead to the well known boson commutators for photons by identifying the Fourier coefficients of the normal modes as creation and annihilation operators. However there is a problem because quantization rules, in this case the commutators, are subjectable to the choice of gauge. Indeed there could be logical inconsistencies without further conditions which Dirac called supplementary conditions. The first one was introduced by Enrico Fermi in (1932), which requires the choice of the Coulomb gauge (see later section III) which is now adopted in most textbooks. Dirac however realised that this was a major inconvenience and clearly he also saw something deeper here. He generalised Fermi's approach to any arbitrary gauge and developed the whole method of constraint quantization to deal with it [18]. To go into it will take us too far afield but in so doing he opened the door to constraint quantization which later in the skillful hands of Ludvig Dmitrievich Faddeev of the then USSR in the 1960s, enabled the path integral quantization of non-Abelian gauge field theories started by Yang and Mills as mentioned earlier and today also the Standard model. In his lectures of 1964 it became clear that Dirac was in fact tackling the Hamiltonian quantization of the Einstein theory of General Relativity through his method of constraint quantization. He died without succeeding but the path he laid was so rich that it is still a major area of research; the results he left behind on weak gravity quantization for example were ground breaking. Still the infinite zero point energy of the vacuum remains, and later the infinities in QED, which required a renormalization or other schemes to handle which we will not go into here. Instead, in this talk, we will present symbolically the first two equations of the Maxwell-Dirac electrodynamics which describe a fundamental feature crying out for a solution, namely the unification of the electron and photon fields.

$$\frac{1}{c^2}\frac{\partial^2 A}{\partial t^2} - \nabla^2 A = J = \Lambda(\psi)A, \tag{1}$$

$$(\gamma^\mu \partial_\mu - mc)\psi = -A\psi. \tag{2}$$

The LHS of the first equation, written in the Lorentz gauge gives the wave equation for the electromagnetic potential $A(r)$ in the presence of the electron current source $J(r) = e\psi^\dagger\gamma^\mu\psi$ which could also include an applied current if there is one on the RHS. In the second equation we have the same for the Dirac electron field $\psi(r)$ which is as mentioned earlier a spinor, while the RHS source is now the electromagnetic $A(r)$ field which may be viewed as a source. With some skillful manipulations the first equation can also be written to look like the second with the RHS a $\Lambda$ tensor operator acting on the $A$ field. This is quite remarkable, for details see Ruei[19]. These equations published almost thirty years ago, gave the first author a feeling of deja vu. One should recall it was as early as 1897 that the British mathematician and physicist Henry Cabourn Pocklington first studied the theory of the current driven wire antenna [20] that in general has a current that must be self-consistent with the electromagnetic wave it generates and vice-versa. He solved this through a series of approximations which leads to the now famous Pocklington integral equation. It was a masterpiece in classical electrodynamics. Here we can see the same self-consistent phenomena at a more fundamental level, that of Compton scattering. Carrying this further, we can now see that the EM field and the Dirac field must be essentially two sides of the same coin. It is therefore compelling to seek a unification of both into a super symmetric field (SQED). Exactly how to do this is still an open question. It is also not obvious that this can be done on this $(A, \psi)$ pair alone without involving the larger family of particles in the Standard model i.e. at much higher energy than the Compton wavelengths of the electron, about $10^{-11}$ cm or its classical Thomas radius $10^{-13}$ cm and at much higher energy scales than $MeV$. However one experimental fact needs to be taken into consideration. This is because between the low 1 MeV energy level for electron-positron annihilation with photon production or its inverse and a correspondingly very low $10^{-4}$ probability for neutrino-antineutrino production, it seems that no other leptons are found until the $\mu^-$ muon (at 105.7 MeV, lifetime 2.2 $\mu s$ and the $\tau^-$ tauon (at 1.8 GeV, lifetime 0.3 ps) with their anti-particles and corresponding neutrinos. Then there is a large empty territory for leptons all the way even until some $80 - 90GeV$ for electro-weak W and Z boson production, a puzzle? Such a unification should allow us to see what the process of electron-positron annihilation is all about. This remains a hope for the future. We can only quote Dirac who considered the electron to be the most basic, most fundamental of all the elementary particles apart from the photon. By great good fortune we inherit a universe in which the first and most basic, perhaps the most important lepton is stable with such a low energy scale of 0.5 MeV (lifetime estimate $\sim 10^{28}$ years), without which there will be no electronics, no radios, no televisions, no mobile phones, no computers; not to mention no chemistry, no biology and therefore we wouldn't be here!

## III. EXTENDED MAXWELL ELECTRODYNAMICS

It is well known that the Maxwell-Faraday electrodynamics is locally gauge invariant. This means that all of Maxwell's equations are unchanged upon the substitution for the vector potential

$$\mathbf{A}(\mathbf{r}, t) \rightarrowtail \mathbf{A}'(\mathbf{r}, t) = \mathbf{A}(\mathbf{r}, t) + \nabla\chi(\mathbf{r}, t), \tag{3}$$

which incidentally also requires a simultaneous change in the scalar potential $\phi(\mathbf{r}, t)$ so that:

$$\phi(\mathbf{r}, t) \rightarrowtail \phi'(\mathbf{r}, t) = \phi(\mathbf{r}, t) - \frac{1}{c}\frac{\partial\chi\mathbf{r}, t)}{\partial t}; \tag{4}$$

where $\chi(\mathbf{r}, t)$ is an arbitrary function that defines the gauge. The two common ones being the Coulomb gauge $\nabla\cdot \mathbf{A}(\mathbf{r}, t) = 0$ and the radiation or transverse gauge where $\nabla\cdot \mathbf{A}_\perp(\mathbf{r}, t) = 0$. This is known as gauge fixing, where in the last equation $\mathbf{A}_\perp$ is the transverse part while the longitudinal part $\mathbf{A}_{\|}$ is taken to be always zero. In recent years extensions of Maxwell's theory have been put forward for certain systems which may also require the **A** field itself to be gauge invariant. For this we require a stronger condition: that is, in addition to the above gauge transformations we require the potentials to also acquire a compensating gauge field that leaves the potentials unchanged. For example in the Coulomb or radiation gauge we further demand that upon a gauge transformation:

$$\mathbf{A}(\mathbf{r},t) \rightarrowtail \mathbf{A}'(\mathbf{r},t) = \mathbf{A}(\mathbf{r},t) + \nabla\chi(\mathbf{r},t) - \nabla\chi'(\mathbf{r},t) = \mathbf{A}(\mathbf{r},t). \tag{5}$$

The compensating field is usually topological since $\chi$ and $\chi'$ cannot be equal but only their gradients must cancel i.e. $\nabla\chi = \nabla\chi'$. Interestingly since gauge invariance is well known to be associated with charge conservation, this entails that there must also be topological charges associated with this extension of Maxwell's theory. This issue is intricate and we won't be able to go into it here [21]. We mention this here because other extensions of Maxwell electrodynamics have been around for quite a while, for example since Born-Infeld [22] (1934) and later Dirac [27] (1962). These were introduced initially with the aim to remove divergences of the self energy but are gaining a revival of interest due to string theory and may be the key to a SQED unification as mentioned.

## IV. A POOR MAN'S QED EXPERIMENTS

Before moving on, let us now talk about the quantum limits and where the poor man can experiment with QED. Here I will give you a beautiful formula from quantum electrodynamics. it comes from Landau and Lifshitz (1979) [28] and you can check it yourself. By equating the electromagnetic energy per $m^3$ i.e. $E^2/(4\pi)$ (in Gaussian units) of an EM wave, with the Planck energy quanta density $\frac{h\nu}{\lambda^3}$ then the field strength of $E$ (in order of magnitude) to get into the quantum regime is given approximately by:

$$E \leq \sqrt{2\hbar c}\,\frac{\nu^2}{c^2} \approx 0.25F^2(\mu V/m), \tag{6}$$

where $\nu$ is the frequency in Hz while $F$ is the (dimensionless) frequency in GHz i.e. $\nu = F \times 10^9$ Hz. A static field can never be quantum [28], but as it turns out, at the frequency of 1.2 GHz, your mobile phone frequency, the E field strength works out as about $0.25\mu$ V/m which is the typical sensitivity of its receivers, (or -119 dBm in rf engineer's units, assuming a 1 m antenna with 50 ohms load). This will give you about one photon per $m^3$ sensitivity, slightly better than the human eye. So a keen hacker can easily do photon counting by modifying their mobile phones and writing a bit of software; many of these, including their valuable GPS receivers, are being scrapped at present. We refer you back to our question earlier and ask our German colleagues to check if Heinrich Hertz in 1889 got anywhere close to this number in his experiments? Photon counting should yield a Poisson distribution for such an experiment confirming the quantum nature of light. You may wonder where is a good source? I would recommend the hydrogen 21 cm line at about 1.42 GHz , first predicted by the Dutch astronomer Frederick van de Hulst from Leiden in 1944. It is an ideal source with very stable frequency other than Doppler shifts from the Galactic spirals but you will need to pick up a satellite dish from EBay.

Note that the quantum limit given by eqn(6) is by no means absolute. Consider Voyager 1 now coming up to 24.6 billion km or about one light day away from us, with its still operational 23 W transmitter. The number of photons arriving works out with Voyager 1's 3.7m dish antenna with a gain of 47 dBi, as approximately 60 nano photon per $m^3$. You can still receive the signals, but you will need a lot of real estate and cash for NASA's Deep Space communication systems use antennas of up to 70m diameter dishes (with about 73 dBi gain) , liquid helium cooled low noise amplifiers, very clever signal processing software and powerful computers to do the crunching.

Now you may also wonder if your smart phone's camera can also be used for photon counting. The answer is not quite, but there are reported experiments suitable for home experimenters for low intensity bioluminescence detection. The sensitivity is about $10^7$ photons over 180 seconds of integration time, or about $10^5$ photons per second on average [29]. One of the us (TCC) owns an Astro T7 CCD camera for astrophotography work which is an improvement but it costs about the same price as the mobile phone. Suitable photon counting cameras for astronomy will costs you above €2000, hence the GHz receiver is still a better choice if you wish to experiment.

## V. INSTRUCTIVE TOY MODELS

Now it is time to take a look at some toy models for the electron. Some toy models we think are necessary even though crude, remember Maxwell's mechanical models which must have played some role in guiding him towards the correct EM theory [1]. Lack of toy models is what might have led to many attacks on string theory. Some of the toy models we shall describe may be crude but if extended into higher dimensions may unify both the beautiful mathematics and the underlying physics of string theory. Firstly, research into the classical inner structure of the electron started from none other than the esteemed Hendrik Lorentz (1853-1928) famous for the Lorentz transformation, the Lorentz force , the Lorentz oscillator model for dielectric dispersion, Lorentz theory of metals, the theory of the Zeeman effect etc for which he shared the Nobel prize with his student Pieter Zeeman. We are sure our colleague and friend Bernard Nienhaus can tell you a lot more about Lorentz and the Institute now named after him than we can. Lorentz and Max Abraham (1875-1922) from Göttingen were competing for a theory for the inner structure of the electron using classical electrodynamics. Note that this was before Einstein's special theory of relativity and it was during these investigations that Lorentz discovered the Lorentz transformations - the power of toy models. Both Lorentz and Abraham considered the model of a spinning electron with Thomas classical radius $r_s \sim 10^{-13}$ cm and uniform charge density $e$. Later on Paul Dirac too joined in but much later and took on the problem in 1934, after he had already won the Nobel prize in 1933 for his quantum mechanical theory of the relativistic electron. They are now famous for the Abraham-Lorentz-Dirac force or the radiation reaction force, radiation damping or self-force for the electron, a problem with a long history [23] with many players including Pauli, Planck, Poincare, Landau etc, so we need no excuse for bringing it up again. Unfortunately, again we do not have time to discuss everything in great detail, but for the interested reader we recommend the excellent book by Rohrlich [24] as well. Suffice to say, Abraham-Lorentz-Dirac gave a theory for the inertial mass of the electron considered as a spinning body of uniform charge $e$ and of finite dimension $r_s$. As we are here to celebrate Heisenberg, let us not forget to remind you too of his great contributions in S Matrix theory as a new concept for elementary particles, his positron theory in which he re-interpreted Dirac's equation for the electron as a 'classical' field equation for a spin 1/2 particle subject to field quantization conditions. He also made ground breaking work on high energy meson production in cosmic ray showers. Cosmic rays are a poor man's ultra-high energy particle source, if you are prepared to climb mountains or fly balloons for the reward of detecting 1 TeV cosmic rays and their meson cascades, nowadays well within the reach of a dedicated physics undergraduate. With the explosion of activities in string theory in the 80s we now wish to discuss some toy models to help you see where we can go with some crude models. We shall introduce Professor Roger Jennison, who held the chair of physical electronics and also radio astronomy at Canterbury during the 1960s and 70s until his retirement. He founded the Electronics Laboratories there and as part of his work he had dedicated some time to researching building, theorizing and experimenting on microwave cavity models for the structure of the electron [25, 26]. The main idea is to capture some light and this is easy to do with microwaves e.g. from the 3K cosmic background radiation of about 3mm wavelength, see Fig 3 as originally suggested by Jennison [26]. It is admittedly not so easy at the $\gamma$ ray wavelength of $10^{-11}$ cm for the real electron experimentally, so here we will have to use our imagination. If one could trap an EM travelling wave say propagating in the clockwise direction, around a circular transmission line. Now by spinning the circular cavity in the anti-clockwise direction at the speed of light $c$ in the medium, the wave is now effectively stationary in the laboratory frame. Note that it is a trapped travelling wave and not a standing wave. For all intents and purpose, the trapped wave is now identical to a static E field and a static B field. If the trapped light is exactly one wavelength, then the model will have a B field that could represent an electron with a magnetic dipole moment. In fact Jennison argued that the g factor will be exactly 2 in such a model [25].

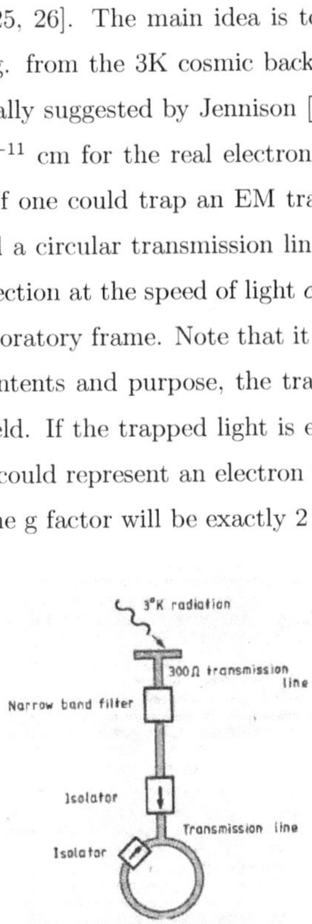


*Fig. 2. More practical arrangement for putting the wave into the circular line.*

Figure 3: Reproduced from Electronics World http://www.electronicsworld.co.uk with permission[26].

Although he did not spell it out in detail, we will quantify it as follows. The magnetic moment of such an object is given by $\mu = \frac{e\hbar}{2mc}s$ with $s$ the spin quantum number. Now the magnetic energy $\frac{\mu_0 B^2}{8\pi}$ is only half the total energy $\mathcal{E} = mc^2$ hence this is equivalent to reducing $m$ by half . Therefore we have $\mu = \frac{e\hbar}{mc}s$ i.e. a $g$ factor of two. In this toy model the elementary charge is no longer fundamental but is given by the radius $r_e$ from equating the electrostatic energy $\frac{e^2}{r_e}$ to $mc^2 = 0.5(\text{MeV})$. Now the radius of such an "electron" $r_e$ is in turn given by $\frac{r_e}{\lambda_c} = 2\pi\alpha$ in terms of the Compton radius $\lambda_c$ with $\alpha \approx \frac{1}{137}$, the fine structure constant. This is about the same as the Thomas radius. Unfortunately for this model the corresponding E field is that of a charged electric dipole not quite the monopole charge of the electron, so more creative thinking is needed here. Since the model is classical it is worth noting that upon quantization, topological constraints come into play as in Dirac's famous monopole argument [31]. Also we are dealing with rotating frames, not quite the inertial frames of special relativity, so this picture may be over simplified. We offer this as food for thought especially for those string theorists who might like to make contact with the low energy physics laboratory. Jennison an engineer and astronomer had also offered an LED rotating construction that demonstrates his model now published but not so well known. It is one that is easy to construct, and, more importantly, easy to experiment with. This is shown in Fig 4 taken from Jennsion[26].

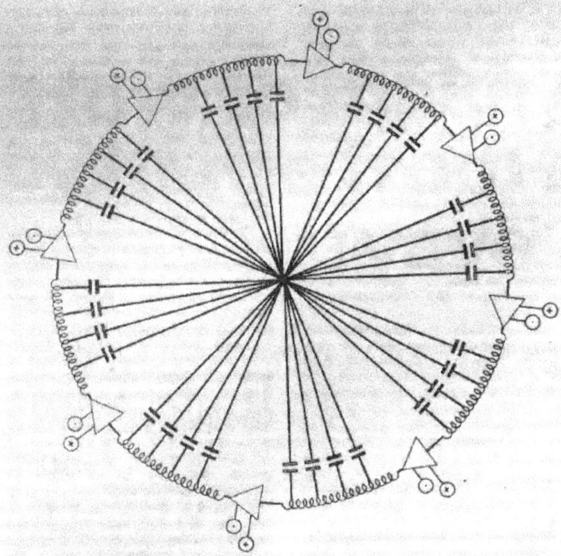

*Fig. 3. Artificial delay line whose construction is described in the text in which a wave runs in the clockwise direction. Rotation of the system in an anticlockwise direction at the same angular frequency as that of the wave produces a static field. Red and green light-emitting diodes, connected at the points marked + and -, indicate the stationary wave, as on the cover photograph.*

Figure 4: Reproduced from Electronics World http://www.electronicsworld.co.uk with permission[26].

One must always remember to play with toy models and try to develop one's physical intuition with them rather than start from the mathematics and be bogged down by equations. There was a famous account of Leon Foucault, who carried a model of the earth with a pendulum on it, daily playing with it while he took his walk in the park in Paris in 1851, ignoring the objections of all the famous mathematicians including Cauchy, Poisson, Binet etc. Foucault with no training in mathematics, came up with his sine law for the Foucault pendulum, ($T = \frac{24\text{hours}}{sin\theta}$), while the mathematicians were all still scratching their heads [30]. We would encourage students to build his table top model and try to refine it further. Other toy models are also available, Mark Davidson at this conference has also worked on similar models but based on line charges which we shall refer you to his work [32]. In an old unpublished preprint, one of us Choy[33], have also highlighted how useful it would be for physics education if students are brought to realise that the D field in Maxwell's equation and the electron charge $e$ are in fact the same, after all the acute student would have noticed that in SI units, they are measured in Coulombs per square metre and in Coulombs respectively. In addition Maxwell's wisdom in insisting on writing out all his equations in full component forms [9] and in fact adopting the quaternion formalism clearly showed he had deep intuition about the 'equivalence' of EM fields and charges [33].

## VI. A POOR CIVILIZATION'S QGD EXPERIMENTS - TESTING QUANTUM GRAVITY

As technology progresses, we are rapidly reaching a climax in experimental physics in which quantum gravity can be tested. So let's see how we can do that using a poor civilizations's graviton detector, modelling on section IV eqn(6). For gravity the analog of the E field is the strain field $\mathcal{H}$ or strain per unit length. To save you a lot of work we have written it out below, for derivation see for example Landau and Lifshitz[34]:

$$\mathcal{H} \leq \frac{8\pi}{c^4}\sqrt{G\hbar c}\;\nu^2 \approx 4.6 \times 10^{-39} F^2 \mathrm{km}^{-1}, \tag{7}$$

where $G$ is the universal gravitational constant, $\nu$ is the frequency in Hz, while $F$ as before is the (dimensionless) frequency in GHz i.e. $\nu = F \times 10^9$ Hz. The complicated factors here come from the fact that gravity wave sources are quadrupoles unlike electromagnetic wave sources which are dipoles. We can compare this to the latest Laser Interferometer Gravitational Wave Observatory (LIGO) specifications. LIGO currently has an aperture of about 4km and therefore a sensitivity to strain fields of $\mathcal{H} \sim 10^{-22}$/km and is designed for the 100-200Hz sweet spot to capture signals from black hole or neutron star mergers. Its proposed upgrade is LISA for Laser Interferometer Space Antenna to be placed in earth orbit. While its proposed long baseline of 2500 km is welcome, unfortunately current technology allows it to only operate in the milli-Hertz region, the wrong direction for detecting gravitons. Interestingly and perhaps not surprisingly, GHz LIGO observatories will require new technology in quantum metrology, super-entangled state detectors, superconducting high field magnetic levitating detectors etc to make them work at the GHz region, potentially detecting the quantum graviton signals from sources such as primordial black holes, cosmic strings and other exotic astrophysical phenomena we are yet to find out about, some possibly coming from even 'before' the beginning of the Big Bang? Young members at this conference who will live longer than most of us will be the ones privilaged to witness the discovery of the confirmation of quantum gravity as well as new astrophysical phenomena yet unknown.

## VII. CONCLUSION

In conclusion I would like to bring you back to 1972 just over 50 years ago at the Trieste symposium in celebration of Paul Dirac's 70th birthday and also close to the 50th anniversary of Heisenberg's quantum mechanics which we are all celebrating today. I refer to none other than our great man Heisenberg himself who gave an excellent paper entitled 'Development of concepts in the history of quantum theory[35]'. Heisenberg proposed that the concept of the elementary particle has to be greatly revised, he proposed the S matrix description at about this time. We have found here that the concept of the electro-tonic state discovered by Michael Faraday, a remarkable man with little formal education, but who had a profound idea that has evolved into the U(1) gauge field of quantum electrodynamics, and more. Now we think the gauge group of quantum gravitodynamics is the Diffeomorphism Group but this is still a matter under current research and not at all confirmed. Let us not forget that the history of physics is occasionally dotted with men like Michael Faraday, Leon Foucault, Heinrich Hertz and women like Agnes Pockels who were not household names or big guns in academia. So we leave you with some wise words from our three great men who were founders of quantum field theory 100 years ago, words spoken during the questions and answers session after Heisenberg's talk in which there was in fact quite some heated arguments:

Werner Heisenberg: There remains the question: What then has to replace the concept of a fundamental particle? I think we have to replace this concept by the concept of a fundamental symmetry.

Paul Dirac: I would like to ask Heisenberg a question. I am in general agreement with this point of view about elementary particles that a concept really doesn't exist, but there is one reservation. I wonder whether the electron should not be considered as an elementary particle. It may be that I am prejudiced because I have had success with the electron and no success with other particles. I would like to hear Heisenberg's view on that.

Werner Heisenberg: I cannot see that one could consider the electron as an elementary particle in the old sense, because an electron can produce light quanta. Light quanta can produce baryons. So actually the electron is connected with this world of baryons and hadrons and so on. So I don't see that you can separate it out. As soon as an electron has these interactions, then, of course, it is also surrounded by a cloud consisting of all these other things. Would you not agree?

Eugnene Wigner: The electron state, I believe, is surely orthogonal to the only other state with spin one-half, namely the proton, and it is surely also orthogonal to the neutron. The neutron is surely not, with a certain probability, just a single electron.

Werner Heisenberg: Is the electron orthogonal to helium ion which consists of a helium nucleus plus one electron? The helium ion then transforms like the electron, and I think these two things are not orthogonal to each other.

Eugene Wigner: Excuse me, I hate to argue with you, but the helium atom has four protons in it and therefore it has a baryon number four.

Werner Heisenberg: Oh yes, my example was wrong. Now let me see...

So we start again with our quote from Albert Einstein at the very beginning of this paper, is the Photon an elementary particle? Indeed may I also ask if the Graviton is an elementary particle?

## VIII. POSTSCRIPT

After the completion of this paper, we are delighted to receive correspondence from Andrey Akhmeteli regarding section II and for drawing our attention to earlier efforts towards Maxwell-Dirac unification of matter-electromagnetic fields [36],[37], as well as his own ongoing research [38] on this fascinating subject [39]. As of 2026, Dirac's fifty four years old comment concerning the electron above remains valid. The highest available energy accelerator scattering measurements and Penning trap anomalous g factor measurements suggest the electron remains an elementary particle down to some $10^{-19}$m or $10^{-4}$fm. Dr Andrew Smith G4OEP has also informed the authors that his recent low power 1 milliwatt radio transmissions on 10 MHz and 14 MHz wspr mode from Bristol UK were successfully picked up in Melbourne, Australia by Ross Wilkinson VK3ARW consistently with s/n between -13 to -29 dB. The ratio from his data shows 10MHz has 6:1 times more success than 14 MHz. No success was logged for 0.5 mW transmissions.

**Acknowledgments** The author is indebted to Basil Mahon for his assistance in writing paragraph 3 in the introduction. This paper is dedicated to the memory of Nancy Forbes (deceased), a nuclear physicist who co-authored with Basil Mahon the top seller 'Faraday, Maxwell, and the Electromagnetic field' (How two men revolutionized physics), Promethus Books, Connecticut USA (2019). For details of Nancy's unusual career path, see https://engage.aps.org/fgsa/resources/careers/non-academic-careers/nancy-forbes. Special thanks goes to Stella Josifovska, editor of Electronics World, UK for her permission to reuse in Figs 3 and 4 the original figures in Wireless World 1983 publication of Roger Jennison[26], gratis.

---

[1] Basil Mahon, "From Faraday and Maxwell to Quantum Physics. The early story of the electromagnetic Vector Potential." Preceding paper for the Heisenberg Centenary session IQSA 2025, Tropea, Italy 2025.

[2] Einstein A. in "The Born-Einstein Letters" Max Born, translated by Irene Born, Macmillan (1971).

[3] Heinrich Hertz, " Electric Waves" translated by D.E. Jones, Dover Publication, New York (1962) a reprint of the original by Macmillan and Co. 1893.

[4] Gordon Bussey in " Marconi's Atlantic Leap" page 72, (2000) , Marconi Communications, Coventry, UK.

[5] Max Born and Pascual Jordan, "Zur Quantenmerchanik" Z. Phys.**34** 858-888 (1925) reprinted in B.L. van der Waerden (editor), "Sources of Quantum Mechanics", vol V of classics of modern science series 277-306, Dover, New York 1968. Unfortunately the relevant section Chapter IV on electrodynamics has been left out in the translation of this reference. We are please to inform the reader that David Delphenich one of the speakers in this conference has the paper translated with this section included, see https://www.neo-classical-physics.info/electromagnetism.html. We thank Hajo Leschke and Helmut Fink for bringing this to our attention.

[6] D.H. Delphenich, "Solvay 1927: Quantum at the crossroads", invited talk for the Heisenberg Centenary session IQSA 2025, Tropea, Italy 2025 and accompanying paper in this volume.

[7] Maxwell, J.C., "On Faraday's Lines of Force", Transactions of the Cambridge Philosophical

Society, Vol. x, Part i. 20-22 (1855), also available in "The Scientific Papers of James Clerk Maxwell", Vol 1, 155-229, Dover, London (1965).

[8] Maxwell, J.C., "A Dynamical Theory of Electricity", Edited by Thomas F. Torrance, with an appreciation by Albert Einstein, Wipf and Stock Publishers, Oregon U.S.A. (1996)

[9] Basil Mahon, "The MAN WHO CHANGED EVERYTHING - The Life of James Clerk Maxwell" John Wiley and Sons, London (2003).

[10] C.N. Yang, "The conceptual origins of Maxwell's equations and gauge theory", Physics Today **67**(11), 45 (2014).

[11] The Planck Collaboration was the European Space Agency spacecraft launched in 2009 to map the cosmic microwave background. It completed it s mission in 2013.

[12] Weisskopf, Victor F., " Search for Simplicity: Atoms with several electrons", American Journal of Physics, **53**(4), 304-305 (1985).

[13] Choy, T.C., "A centennial reappraisal of Heisenberg's Quantum Mechanics with a perspective on Einstein's Quantum Riddle.", invited talk for the Heisenberg Centenary session IQSA 2025, Tropea, Italy 2025.

[14] Choy, T.C., "Effective Medium Theory - Principles and Applications" (second edition) Oxford University Press, Oxford, UK (2016).

[15] L. Landau and R. Peierls, "Extension of the Uncertainty Principle to Relativistic Quantum Theory", Zeitschrift für Physik **69**, 56-69 (1930).

[16] P.A.M. Dirac, "Principles of Quantum Mechanics", The International series of monographs in Physics, Oxford University Press, fourth edition (revised) (1958).

[17] J.W.S. Rayleigh, "The Theory of Sound", Vols 1 and 2, with a historical introduction by Robert Bruce Lindsay, Dover Publications, New York (1945).

[18] P.A.M. Dirac, "Lectures on Quantum Mechanics", 25-43, Dover Publications, New York, (2001).

[19] Ruei K.H., "Alternative formulation of electron-photon interaction in relativistic quantum mechanics", American Journal of Physics, **44**, 153-157, (1976). Erratum **44**, 1137, (1976).

[20] H.C. Pocklington, "Electrical Oscillations in Wires", Proc. Cambridge Phil. Soc. 324-332, (1897).

[21] Barrett, Terence W., "Topological Foundations of Electromagnetism", World Scientific Publications, Singapore (2008).

[22] Max Boirn and Leopold Infeld, "Foundations of the New Field Theory" Proc. Roy. Soc. London A, **144**(852), 425-451 (1934).

[23] MacDonald K.T., "On the history of the radiation reaction" Princeton archive https://web.archive.org/web/20221017154015/http://kirkmcd.princeton.edu/examples/selfforce.pdf

[24] Fritz Rohrlich, "Classical Charged Particles", World Scientific Publications, Singapore (2007)

[25] R.C. Jennison, "What is an electron?", Wireless World, 42-47, **85** (1522), June (1979).

[26] R.C. Jennison, "How to make electric charge from a radio wave?", Wireless World, 36-38, **89** (1571), August (1983).

[27] Paul Dirac, "An extensible model of the electron", Annals of Physics (New York), **268** 57-67, (1962).

[28] Barestetskii V.B., Lifshitz E.M. and Pitaevskii L.P., Quantum Electrodynamics, Landau and Lifshitz Course of Theoretical Physics Vol. 4 (2nd Edition) pg 15, Pergamon Press, Oxford (1979), translated from Russian by J.B.Sykes and J.S. Bell.

[29] Kim Huisung et al, " Smartphone-based low light detection for bioluminescnce application", Nature Scientific Reports Sci. Rep. **7**, 40203 (2017).

[30] see Amir D. Aczel, in "Pendulum, Léon Foucault and the Triumph of Science" Washington Square Press, 99-110, New York (2003)

[31] Paul Dirac, "Quantised singularities in the Electromagnetic Field", Proc. Roy. Soc. A **133**, 66-72, (1931).

[32] Mark Davidson "Classical Charged Particle Models Derived from Complex Shift Methods." Int. J. of Theoretical Physics,**62**(7), 154 (2023).

[33] Choy, T.C., "Comment on Electrons as field quanta: A better way to teach quantum physics to introductory general physics courses," by Art Hobson Am. J. Phys. 73 (7), 630-634, (2005), unpublished preprint now available on Research Gate, https://www.researchgate.net/profile/Tuck-Choy/publication

[34] Landau L.D. and Lifshitz E.M., The Classical Theory of Fields, Landau and Lifshitz Course of Theoretical Physics Vol. 2 (2nd Edition) pg 347, Pergamon Press, Oxford (1975), translated from Russian by M. Hamermesh.

[35] Heisenberg W., "Development of concepts in the history of quantum theory" in the proceedings of the symposium on the development of The Physicist's Conception of Nature in the Twentieth Century, dedicated to the seventieth birthday of Paul Dirac, 18-25 Sep (1972), D.

# Chapter 15

## Exploring Testable Quantumness of Massive Objects: Motivations, State of Play, and the Proposed Schrödinger's Pendulum Experiment

Dipankar Home
*Bose Institute, Kolkata, 700 009, India*

Sougata Bhattacharyya
*Indian Institute of Technology, Indore, 453 552, India*

**Abstract**

The motivations underlying investigation of the quantum nature of massive objects stem from some of the outstanding quantum foundational issues, which include the nature of the quantum-classical boundary, and the question of using table-top experiments to test the quantumness of gravity. It is crucial to emprically verify the validity of the quantum superposition principle at the macroscopic level to probe such foundational questions. The present article begins with a brief overview of these motivations, and then proceeds to discuss the usefulness of the notion of maocrorealism for probing the fundamental quantumness of macroscopic objects, and testing its incompatibility with quantum mechanics. The article then explains the two key ingredients in this enterprise, the Leggett-Garg Inequality (LGI) and the No Disturbance Condition (NDC), followed by a snapshot of the experimental tests done so far to test the quantum mechanically predicted violations of these conditions. However, the state-of-the-art demonstrations of quantum features have reached only up to objects of masses of about $10^4$ - $10^5$ amu. Hence, breakthrough ideas, which can be implemented experimentally in the near future, are needed to scale up the tests of macroscopic quantumness to ever more massive objects. It is precisely this challenge which has recently been addressed by D. Das, D. Home, H. Ulbricht, S. Bose [1] through the formulation of a novel procedure for demonstrating the quantum mechanically predicted violations of these LGI/NDC conditions for an oscillating object having any large mass. We will review the salient features of this important proposed scheme which has been called the "Schrödinger Pendulum Experiment" [2] and will assess the testability of this scheme.

# Contents

# 1 Introduction

While recovering from a serious bout of hay fever on the North Sea island of Helgoland, Werner Heisenberg developed the core ideas of the first coherent formulation of Quantum Theory. On July 11, 1925 Heisenberg handed Max Born the finished version of his groundbreaking paper. A few days later, in a letter to Einstein dated July 15, Born described it as "rather mystifying," yet "certainly true and profound." This marked the dawn of Quantum Mechanics (QM), shrouded in mystery. Intriguingly, this feeling of mystery has stuck around for a hundred years.

The "Second Quantum Revolution" [3,4] began with the discovery of Bell's Theorem in 1964 [5]. This was a direct response to the 1935 Einstein-Podolsky-Rosen (EPR) paper [6], which argued that the description of the physical reality provided by QM is incomplete. Bell provided a testable criterion, showing that any theory based on "local realism" must satisfy certain inequalities. QM, he showed, predicted violations of this inequality. This ushered in the "Second Quantum Revolution" marking the advent of modern research on the foundations of QM and leading to New Age Quantum Technologies like Quantum Computing, Quantum Communication and Quantum Sensing.

In this context, from the historical perspective, it is worth recalling that one of the fundamental questions Bohr had struggled with is the question of extrapolating the QM description to the macrolevel for describing any measurement process, as is evidenced by his various statements, such as the following one: "However far quantum effects transcend the scope of classical physical analysis, the account of the experimental arrangement and the record of the observations must always be expressed in common language supplemented with the terminology of classical physics" [7]. On the other hand, one of the founders of Quantum Mechanics, Schrödinger had argued to the contrary, contending that it is difficult to accept the *ad hoc dictum* that quantum theory ceases to be applicable in the case of a macro-object functioning as a measuring apparatus. For an instructive account of the interesting correspondence between Schrödinger and Bohr debating this question, see Moore [8]. In fact, Schrödinger's famous 'Cat Example' was formulated to highlight the intriguing questions raised by the extrapolation of the QM superposition principle to the macrolevel [9].

It is thus fitting that in this volume commemorating 100 years of the birth of QM, the present article provides an overview of the current theoretical and empirical enterprise critically investigating the applicability of QM for macrosystems, in particular the validity of the QM superposition principle for massive systems.

We begin by enumerating the pivotal foundational questions of QM which hinge on the vexed question of to what extent QM principles can be extrapolated from the microlevel to the macroscopic level.

# 2 Outstanding Quantum Foundational Issues

Despite a century of spectacular success, QM leaves us with deep unresolved and much debated questions that continue to challenge our understanding of the nature of physical reality. These include:

- **Quantum/Classical Boundary:**
  To what extent are the principles of QM valid in the macroscopic domain of our classical world?

- **How can one empirically probe the Quantum Measurement Problem?**
  The QM description of any measurement process involves the interaction of an observed microsystem with a macroscopic measuring apparatus, entailing extrapolation of the QM superposition of states to the macro-measuring system. Does the standard QM treatment ensure the occurrence of a definite outcome in any measurement? To ensure this, does one need to modify the standard QM framework?

- **Is Gravity really Quantum Mechanical in nature?**
  Can one test the Quantumness of Gravity by Table Top Experiments using QM superposition of states for massive systems?

A direct test of funadamental principles of quantum theory in the macroscopic realm is necessary to address these problems in an empirically relevant way for which it is crucial to test the validity of QM superposition of *macroscopically distinct states* of a *macro-object*. To this end, the Leggett-Garg inequality (LGI) provides a key tool [10], which is a temporal analogue of Bell's theorem, and its variants such as the No Disturbance Condition (NDC) [11]. The LGI and NDC offer experimentally feasible frameworks to explore the boundaries of quantum physics and to look for a potential breakdown of the superposition principle at the macroscopic scale. For a useful review, see, for example, C. Emary et. al. [12].

## 3 Hallmarks of the Classical Worldview

We begin by recapitulating the hallmarks of the classical world view:

- **Realism:** In any classical theory, it is assumed that all the measurable properties of an object have definite values, even when not measured.

- **Locality:** No action at a distance. For a pair of spatially separated correlated objects, the measurement of properties of any one of them cannot affect the properties of the other.

- **Absence of Retrocausality:** Future cannot affect the present or the past. Causality propagates only forward in time.

- **Non-Invasive Measurability (NIM):** The act of performing any measurement does not influence the outcome of a subsequent measurement. This is also known as the No-Disturbance Condition (NDC).

These are the conditions which characterize any classical worldview irrespective of the details of the framework of any classical theory.

## 4 Tests of Realism vis-à-vis Quantum Mechanics (QM)

Stephen Hawking commented that "Reality is not a quality you can test with litmus paper" [13]. But studies on the Bell Inequality (BI) and its temporal analogue, viz. the Leggett-Garg Inequality (LGI) show that the notion of Reality can indeed be tested, provided it is combined with some other suitable condition (Locality Condition for BI, Non-Invasive Measurability Condition for LGI).

### 4.1 Tests of Local Realism versus QM using BI

This line of studies was initiated by the celebrated Einstein-Podolsky-Rosen paper in 1935 [6], inspiring John Bell's seminal breakthrough in 1964 [5]. Bell combined Realism with the Locality Condition for spatially separated objects to derive his famous Bell Inequality (BI). This provided the first rigorous framework for a test of local realism. However, the empirically observed violation of this inequality, predicted by QM, forces us to abandon the joint assumption of both Realism and Locality. This violation serves as a direct testament to the holistic feature of Quantum Entanglement, where the correlations between measured results embodied in entangled states of spatially separated particles defy any classical explanation.

In a nutshell, the QM violation of Bell-type inequalities stems from Quantum Entanglement of states of the spatially separated particles, thereby signifying conflict between QM and Local Realism.

### 4.2 Tests of Macrorealism versus QM using LGI

Another equally profound line of studies emerged with the development of the LGI [10], which serves as a temporal counterpart to the Bell inequality. In this context, Realism is intertwined with the Non-Invasive Measurability (NIM) Condition for a single, time-evolving object. This setup paves the way for a test of Macrorealism. The subsequent empirical violation of the LGI, which is also predicted by QM, serves as a testament to the failure of Macrorealism.

Quantum violation of Leggett-Garg-type inequalities stems from the Quantum Superposition Principle applied to the state of a single particle evolving in time. It signifies conflict between QM and Macrorealism.

## 5 QM in the Macroscopic Limit

The underpinning question is: Why test QM in the macro-regime?

There are three primary motivations for this pursuit:

1. The testing of Macrorealism vis-à-vis QM.
2. It provides a stepping stone towards testing the Quantum Nature of Gravity.
3. Experimental oriented study of the QM Measurement Problem.

In addition, such studies can enable future developments of ultrasensitive Quantum Sensing Technology.

Here it is relevant to note that for addressing the QM Measurement Problem (for an overview, see for example, D. Home [14]), approaches and models have been proposed for modifying QM for massive objects, while keeping unaffected all the standard quantum predictions at the microlevel. For example, dynamical models of wave function collapse involving non-linear modifications of Schrödinger's equation [15–17] and models based on gravitationally induced wave function collapse have been proposed.

In recent years, experimental tests of quantum superposition of states for massive objects have been carried out [18–25]). Such experiments will be crucial to provide empirical constraints on these alternatives or to reveal possible limitations of the standard QM orthodox framework. For example, the underground test of gravity related wave function collapse by Sandro Donadi et al. [26].

### 5.1 Use of Macro-Superposition for Testing Quantum Gravity

One of the most daunting challenges of modern science is to reconcile General Relativity (GR) with QM. While various models of quantum gravity exist, there is no unambiguous well-formulated testable method to experimentally discriminate between them. On the other hand, GR leads to Newtonian gravity in the small-curvature, low-energy limit, which can even be described via graviton exchange in the linearised description of quantum gravity [27].

The unresolved challenging questions are:

- Is Gravity truly Quantum in nature?
- Can Cosmological evidences and Gravitational wave detection experiments yield any convincing signature of the Quantum Nature of Gravity?
- Can the Quantum Nature of Gravity be manifested at the laboratory scale?

Some recent proposals to address this last question are based on the ideas of detecting either gravitationally induced entanglement [28,29], or the quantum disturbance arising from the measurement of the gravitational field [30,31]. A key requirement in all these schemes is that *superposition of macroscopically distinct spatial states of a massive object* is critically involved, which underscores the necessity of *testing* the validity of the quantum idea of the superposition of macroscopically distinct spatial states in the macroscopic domain.

## 6 The Notion of Macrorealism

The notion of Macrorealism, in the context of a single, time-evolving system subjected to sequential measurements, is defined by two key assumptions:

- **Realism:** At any instant, even when not measured, a system is definitely in one of the available states and its observable properties have definite values.
- **Non-Invasive Measurability (NIM):** It is possible, at least in principle, to determine the system's state without disturbing it or affecting its future evolution.

The idea of NIM is thought to be ideally implemented by using the concept of Negative Result Measurement (NRM), for which NIM is argued to be a "natural corollary of Realism" [32].

## 7 Negative Result Measurement (NRM)

The concept of NRM is now examined. An example of such a measurement setup consists of a particle that oscillates between two regions, Region 1 and Region 2 (see Fig 1). An observable Q is defined as: Q = +1 if the particle is found in Region 1 and Q = -1 if the particle is found in Region 2.

How NRM works is that a probe is set to interact *only if* the particle is *in*, say, Region 1. If *no interaction* occurs, the particle is *inferred* to be in Region 2. Thus, the *absence of interaction* conveys the *measurement information*, while ideally leaving the system *classically undisturbed*. This procedure constitutes an effective implementation of the NIM condition.

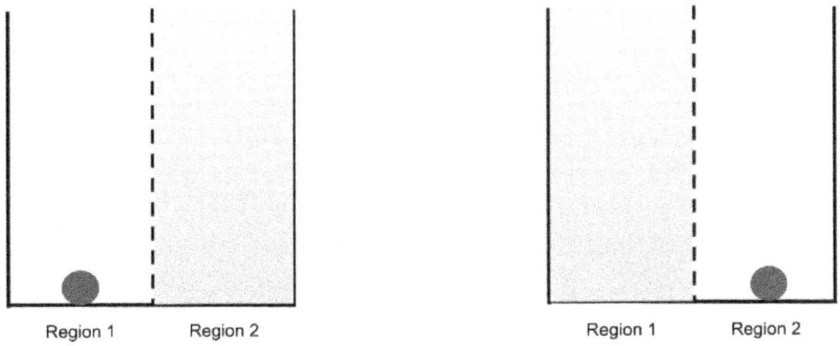


Figure 1: Schematic of Negative Result Measurement (NRM) setup showing the harmonic oscillator potential with two regions. A probe beam illuminates only Region 1. The absence of scattered light indicates the particle is in Region 2, implementing non-invasive measurement.

# 8 Macrorealism (MR) Testing using Leggett-Garg Inequality (LGI) and the No Disturbance Condition (NDC)

Having defined the concept of macrorealism, it is now possible to identify specific, testable conditions for its empirical validation. Several such conditions have been proposed, primarily based on analyzing temporal correlations measured in sequential experiments. These are:

- **The Leggett-Garg Inequality (LGI):**
  This was derived as a testable algebraic consequence of the deterministic form of Macrorealism [10].
- **The Wigner's Form of LGI (WLGI):**
  This was derived as a testable algebraic consequence of the probabilistic form of Macrorealism [33].
- **The No Disturbance Condition (NDC) or No-Signalling In Time (NSIT):**
  This condition is formulated as a statistical version of Non-Invasive Measurability (NIM) [11]. It is worth noting that NDC (or NSIT) can be derived from LGI [20].

Next, we provide a review of the significant experiments performed to date concerning LGI and NDC (NSIT) towards probing the validity of QM superposition of macroscopically distinct states of a macroscopic system involved in testing the notion of MR.

Table 1: Key elements of the Significant Experimental Tests of Macrorealism

| **System** | **Macroscopicity Parameter** | **Relevant Work** |
|---|---|---|
| Nitrogen vacancy center in diamond hosting a three level quantum system | Atomic mass | *R. E. George, L. M. Robledo, O. J. E. Maroney, M. S. Blok, H. Bernien, M. L. Markham, D. J. Twitchen, J. J. L. Morton, G. A. D. Briggs, and R. Hanson*, "Opening up three quantum boxes causes classically undetectable wavefunction collapse," *Proc. Natl. Acad. Sci. U.S.A., Vol. 110, No. 10, pp. 3777–3781, (2013).* [18] |
| A single cesium atom performing 'quantum walk' on a discrete lattice | Atomic mass | *C. Robens, W. Alt, D. Meschede, C. Emary, and A. Alberti*, Ideal negative measurements in quantum walks disprove theories based on classical trajectories, *Phys. Rev. X 5, 011003, (2015).* [19] |
| Spin-bearing phosphorus impurity in silicon | Atomic mass | *G. C. Knee, S. Simmons, E. M. Gauger, J. J. Morton, H. Riemann, N. V. Abrosimov, P. Becker, H.-J. Pohl, K. M. Itoh, M. L. Thewalt, G. A. D. Briggs, and S. C. Benjamin*, Violation of a Leggett-Garg inequality with ideal non- invasive measurements, *Nat. Commun. 3, 606, (2012).* [20] |
| Neutrino flavour oscillation | Length scale of the neutrino oscillation | *J. A. Formaggio, D. I. Kaiser, M. M. Murskyj, and T. E. Weiss*, Violation of the Leggett-Garg inequality in neutrino oscillations, *Phys. Rev. Lett. 117, 050402, (2016).* [21] |
| Three-level single photons | Spatial separation between the paths | *K. Wang, C. Emary, M. Xu, X. Zhan, Z. Bian, L. Xiao, and P. Xue*, "Violations of a Leggett-Garg inequality without signaling for a photonic qutrit probed with ambiguous measurements," *Phys. Rev. A, 97, 020101, (2018).* [22] |
| Superconducting flux qubit | Difference in magnetic moments of oppositely circulating superposed current states | *G. C. Knee, K. Kakuyanagi, M.-C. Yeh, Y. Matsuzaki, H. Toida, H. Yamaguchi, S. Saito, A. J. Leggett, and W. J. Munro*, A strict experimental test of macroscopic realism in a superconducting flux qubit, *Nat. Commun. 7, 13253, (2016).* [23] |
| Two and four qubit 'cat-states' studied in cloud-based quantum computing device 'IBM Quantum experience' | Number of constituent qubits | *H. Y. Ku, N. Lambert, F. J. Chan, C. Emary, Y. N. Chen, and F. Nori*, "Experimental test of non-macrorealistic cat states in the cloud," *NPJ Quantum Inf., 6, 98, (2020).* [24] |
| Single photons in interferometers | Spatial separation between two superposing single photon states | *K. Joarder, D. Saha, D. Home, and U. Sinha*, Loophole-free interferometric test of macrorealism using heralded single photons, *PRX Quantum 3, 010307, (2022).* [25] |

## 9 Mass as a Macroscopicity Parameter

Mass is one of the most natural parameters for characterizing macroscopicity in quantum systems. There were earlier tests of macrorealism (LGI/NDC violations) which involved single atoms, photons, etc. However, no test of macrorealism has been performed using systems that have significantly large mass. Currently, quantum interferometric tests of the quantum superposition principle using macromolecules have reached only up to the level of $10^4$ amu [34, 35]. Moreover, these experiments do not test macrorealism. This motivated the formulatation of an empirically realizable scheme for testing macrorealism vis-à-vis quantumness of a system having any arbitrary large mass [1].

## 10 The Proposed Schrödinger's Pendulum Experiment

The proposed scheme by D. Das, D. Home, H Ulbricht, S. Bose [1] has been called *Schrödinger's Pendulum Experiment* in an article in Scientific American [2]. In this scheme, a linear harmonic oscillator is considered initially to be in a Schrödinger coherent state, i.e., a minimum uncertainty product, with a non-spreading Gaussian wave packet whose peak follows classical motion. The formulated measurement scheme consists of coarse-grained spatial measurements to determine which half of the harmonic well the particle is in. Such determination is envisaged to be done using *Negative Result Measurements (NRM)* in order to satisfy the condition of *Non-Invasive Measurability (NIM)*.

The proposed scheme uses two-time LGI and two-time NDC with a key novelty. While the standard LGI/NDC schemes use the *same measurement of an observable* at successive instants, here in order to show Quantumness in a way that is independent of mass, angular frequency, and initial peak momentum, the scheme proposed by D. Das, D. Home, H. Ulbricht, S. Bose invokes *two different measurements of observables* at successive instants [1].

### 10.1 Experimental Setup and NRM Implementation

In this experiment, NRM is applied to a spatially dichotomized harmonic oscillator potential to determine whether the particle lies in one half of the well without disturbing its dynamics. For the NRM implementation, a high-intensity light-beam can be used to focus on one region of the well. Measurement interaction occurs only if the particle is present in that region. An absence of interaction implies the particle is in the other region, inferred without classical disturbance. This ensures that classical Non-Invasive Measurability (NIM) is achieved, a fundamental requirement for tests of Macrorealism.

In the proposed setup, the system evolves unitarily from time $t_1$ to $t_2$, with $t_1 < t_2$. Two distinct dichotomic observables are defined, $Q$ and $R$, which are measured at $t_1$ and $t_2$, respectively, with measurement outcomes: $\pm 1$.

The observable $Q$, measured at time $t_1$, is defined corresponding to the particle's position relative to a boundary at $x = \beta_1$:

$$Q = \begin{cases} +1 & \text{if the particle is found in the region } x > \beta_1 \\ -1 & \text{if the particle is found in the region } x < \beta_1 \end{cases} \tag{1}$$

This observable can be represented by the operator:

$$\hat{Q} = \int_{\beta_1}^{\infty} |x\rangle\langle x|dx - \int_{-\infty}^{\beta_1} |x\rangle\langle x|dx \tag{2}$$

Similarly, the observable $R$, measured at the later time $t_2$, is defined relative to a *different boundary* at $x = \beta_2$:

$$R = \begin{cases} +1 & \text{if the particle is found in the region } x > \beta_2 \\ -1 & \text{if the particle is found in the region } x < \beta_2 \end{cases} \tag{3}$$

This observable is represented by the operator:

$$\hat{R} = \int_{\beta_2}^{\infty} |x\rangle\langle x|dx - \int_{-\infty}^{\beta_2} |x\rangle\langle x|dx \tag{4}$$

Here we have taken the initial wave packet to be centered at $x = 0 \Rightarrow \beta_1 = 0$ at time $t_1 = 0$.

The coherent state is defined as

$$\psi(x,0) = \sqrt{\frac{1}{\sqrt{2\pi}\sigma_0}} \exp\left(-\frac{x^2}{4\sigma_0^2} + \frac{ip_0 x}{\hbar}\right) \tag{5}$$

where the initial wave packet with

$$\sigma_0 = \sqrt{\frac{\hbar}{2m\omega}}, \quad p_0 = \text{initial momentum}, \quad \omega = \text{frequency}.$$

## 10.2 Testing Macrorealism using LGI and NDC

The Two-time Non-Disturbing Condition (NDC) [11] is given by

$$N_{\pm} = p(R_{t_2}\pm) - [p(Q_{t_1}+, R_{t_2}\pm) + p(Q_{t_1}-, R_{t_2}\pm)] = 0 \tag{6}$$

where $p(Q_{t_1}+, R_{t_2}-)$ is the joint probability of getting $Q = +1$ at $t_1$ and $R = -1$ at $t_2$, while $p(R_{t_2}+)$ is the probability of getting $R = +1$ at $t_2$ without prior measurement at $t_1$. Therefore, its quantum violation would imply a non-zero value of $|N_{\pm}|$.

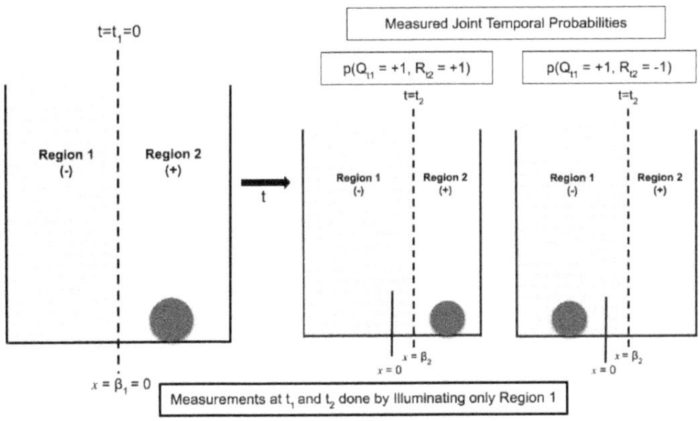


Figure 2: Measurement protocol for determining the joint probability $p(Q_{t_1} = +1, R_{t_2} = +1)$. At $t_1 = 0$, only Region 1 is illuminated (boundary at $x = \beta_1 = 0$). The absence of scattering indicates $Q = +1$. At $t_2$, only Region 1 is illuminated again (boundary at $x = \beta_2$), and absence of scattering yields $R = +1$.

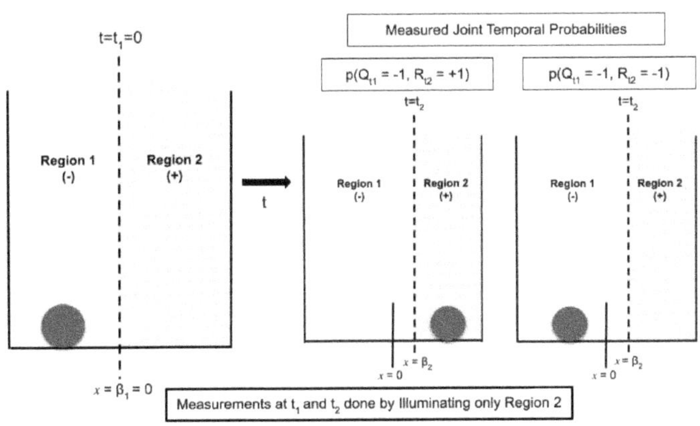


Figure 3: Measurement protocol for determining joint probabilities $p(Q_{t_1} = -1, R_{t_2} = \pm 1)$. At $t_1 = 0$, only Region 2 is illuminated. The absence of scattering indicates $Q = -1$. At $t_2$, illuminating only Region 2 yields the probabilities for both $R = +1$ and $R = -1$ outcomes.

Now, to satisfy a two-time Leggett-Garg Inequality (LGI) [10, 36]:

$$\text{Ł}_{s_1,s_2} = 1 + s_1\langle Q_{t_1}\rangle + s_2\langle R_{t_2}\rangle + s_1 s_2\langle Q_{t_1} R_{t_2}\rangle \geq 0, \quad s_1, s_2 \in \{\pm 1\} \tag{7}$$

where

$$\langle Q_{t_1} R_{t_2}\rangle = p(Q_{t_1}+, R_{t_2}+) - p(Q_{t_1}+, R_{t_2}-) - p(Q_{t_1}-, R_{t_2}+) + p(Q_{t_1}-, R_{t_2}-), \tag{8}$$

$$\langle Q_{t_1}\rangle = p(Q_{t_1}+) - p(Q_{t_1}-), \quad \langle R_{t_2}\rangle = p(R_{t_2}+) - p(R_{t_2}-). \tag{9}$$

Its quantum violation occurs if $\max_{s_1,s_2}(-L_{s_1,s_2}) > 0$.

Here it is worth stressing that in order to determine a typical temporal correlation function occurring in a test of MR, such as $\langle Q_{t_1}R_{t_2}\rangle$, a Negative Result Measurement (NRM) of $Q$ at the instant $t = t_1$ is performed. There is no need for NRM in the case of the second measurement of $R$ at $t_2$. It can be noted that the derivations of NDC and LGI do not require that the same observable be measured at different times.

## 11 Key Results and Mass-Independent Violations of LGI and NDC

The key results of this study demonstrate that quantum violations of both the two-time NDC and the two-time LGI can be made **independent of mass**, as well as of the initial momentum $p_0$ and angular frequency $\omega$ [1]. This is crucially achieved by a suitable choice of the boundary $\beta_2$ for the second measurement, specifically set at $\beta_2 = x_0^{(t_2)} \pm c\Delta^{(t_2)}$, where $x_0^{(t_2)}$ is the peak position at $t_2$ and $\Delta^{(t_2)}$ is the standard deviation at $t_2$.

For example, with $c = \sqrt{2}$, significant violations of NDC ($|N_\pm| \sim 0.12$–$0.17$) and LGI (up to 0.08) are observed for various fractions of the oscillation period $T$, as summarized in the table below:

| $m$ | $p_0$ | $\omega$ | $t_2$ | $\lvert N_\pm\rvert$ | Max LGI Violation |
|---|---|---|---|---|---|
| Any | Any | Any | $T/14$ | 0.12 | 0.08 |
| Any | Any | Any | $T/8$ | 0.15 | 0.04 |
| Any | Any | Any | $T/4$ | 0.17 | No Violation |
| Any | Any | Any | $T/3$ | 0.16 | No Violation |
| Any | Any | Any | $3T/8$ | 0.15 | 0.04 |
| Any | Any | Any | $2T/5$ | 0.14 | 0.07 |
| Any | Any | Any | $3T/4$ | 0.17 | No Violation |
| Any | Any | Any | $1/\omega$ | 0.16 | No Violation |
| Any | Any | Any | $1/(2\omega)$ | 0.13 | 0.08 |

Table 2: The predicted QM violations of the two-time NDC condition and the two-time LGI for different measurement times $t_2$, where $T = 2\pi/\omega$ is the time period of oscillation. These predicted QM violations are independent of mass $m$, initial momentum $p_0$, and angular frequency $\omega$ [1].

## 12 Proposed Experimental Realization and Feasibility

Feasible experimental systems include:

- Nano-objects of mass $\sim 10^{-14}$ kg using levitated nano-diamond crystals in low-noise traps [37].
- Macroscopic masses $\sim 10$ kg mirrors in Gravitational Wave Detectors (e.g., LIGO) [38].

The flexibility of the scheme proposed by D. Das, D. Home, H. Ulbricht, S. Bose [1] allows for the optimization of trap settings and minimization of decoherence and noise, as the quantum violation of MR is independent of mass $m$, oscillator frequency $\omega$, and initial momentum $p_0$ .

The following features are relevant in order to enhance detection efficiency and spatial control:

- One can use high numerical aperture collection optics such as parabolic mirrors [39].
- One can implement NIM by spatial selectivity: illuminate only one side of the boundary $x = \beta_i$ at time $t_i$ with a beam having a sharp intensity drop at $x = \beta_i$.
- One can fix $x = \beta_i$ with Angstrom precision, as demonstrated in single-molecule imaging [40] and optomechanical interferometry [41, 42].

# 13 Enhancing Detection Efficiency and Spatial Control

The successful experimental realization of the proposed Schrödinger's pendulum scheme hinges on two critical technical aspects: highly efficient detection and precise spatial control over the measurement regions. Significant advancements in experimental techniques make these requirements achievable.

## 13.1 Improving Detection Efficiency

To determine the position of the oscillating object via scattered light, a near-unity efficiency of photon collection is desirable. For an assumed detection of $n$ scattered photons with a wavelength $\lambda \sim \mu$m, the spatial resolution achieved is approximately $\lambda/\sqrt{n}$. For instance, reaching the spatial resolution of the ground-state spread $\sigma_0$ for a particle of mass $m \sim 10^{-14}$ kg in a trap with frequency $\omega \sim 1$ Hz requires collecting $n \sim 10^8$ photons. If the detected scattered power is 1 nW, this information is acquired in about $10^{-2}$ seconds. This necessitates that all undesirable heating rates, arising from undetected scattered photons, blackbody radiation, and gas collisions, are well below certain threshold which is determined by the number of gas molecules and undetected photons scattered per second remaining below $10^2$.

A promising approach to increase the critical detection efficiency is the use of collection optics with a high numerical aperture, such as parabolic mirrors [39]. This method enhances the fraction of scattered light collected, thereby improving the signal-to-noise ratio and the overall precision of the negative result measurement.

## 13.2 Ensuring Non-Invasive Measurability (NIM) by Spatial Selectivity

The core of the Non-Invasive Measurability (NIM) assumption in the scheme proposed by D. Das, D. Home, H. Ulbricht, S. Bose [1] can be implemented via Negative Result Measurement (NRM). To realize this in practice:

- Only one of the two spatial regions (e.g., either $x > \beta_i$ or $x < \beta_i$) is illuminated at the measurement instant $t_i$.
- A laser beam with an extremely sharp intensity drop at the boundary $x = \beta_i$ is used. The absence of any scattered light from this illuminated region is then used to infer the presence of the particle in the opposite, unilluminated region, thereby registering an outcome without classical interaction.

### 13.3 Precision in Boundary Fixation

The location of the boundary $x = \beta_i$ must be fixed with high precision, ideally at the Angstrom scale, to ensure the validity of the measurement. Such exquisite spatial resolution has already been demonstrated in other cutting-edge experiments, like the following ones:

- In single-molecule imaging, techniques like interferometric scattering microscopy have achieved Angstrom-level localization [40].
- In optomechanical interferometry, such precise position measurements of levitated particles have been realized [41, 42].

These existing technological capabilities provide a strong foundation for implementing the spatially selective, non-invasive measurements required for a definitive test of macrorealism using massive objects by using a scheme such as that suggested by D. Das, D. Home, H. Ulbricht, S. Bose [1].

## 14 Epilogue

Finally, in this context it is worth noting the very recent remarkable demonstration of quantum coherence in a 16-microgram mechanical oscillator using Schrödinger cat states [43]. However, this experiment does not test macrorealism, but nevertheless it is a major step towards the observation of quantum behavior at large mass scales. To go further by showing decisive violation of macrorealism for arbitrarily large masses, it is necessary to have a direct test of macrorealism such as the Schrödinger's pendulum experiment that has been proposed by D. Das, D. Home, H. Ulbricht, S. Bose [1].

We conclude by quoting John Bell:

> "It seems to me possible that the continuing anxiety about what quantum mechanics means will lead to still more and more tricky experiments which will eventually find some soft spot, some point where quantum mechanics is actually wrong."
> Interview in *'The Ghost in the Atom'* [44]

We hope that a research programme progressing in this direction will lead to a greater variety of experiments critically probing the validity of QM in the macroscopic world and will throw fresh light on the deeper questions about the nature of quantum reality .

## 15 Acknowledgement

This article is dedicated to the memory of late Peter Landsberg with whom DH had the pleasure of having many stimulating and instructive interactions on quantum foundational and other basic physics issues. Warm thanks are due to Mike Burt and Aris Alexoupoulos for their useful observations about the earlier version of this paper. We are deeply grateful to Tuck Choy for his valuable comments and suggestions.

The authors thank Andrew Smith for the magnificent job of proof reading and Anindya De for crucial help in finalizing the manuscript.

DH acknowledges support from the NASI Senior Scientist Fellowship. SB acknowledges support from the Indian Institute of Technology, Indore.

## 16 References

## References

[1] D. Das, D. Home, H. Ulbricht, and S. Bose. Mass-independent scheme to test the quantumness of a massive object. *Phys. Rev. Lett.*, 132:030202, (2024).

[2] T. Folger and C. Moskowitz. Schrödinger's pendulum experiment will search for the quantum limit. *Scientific American, (27 February 2024).*

[3] J. P. Dowling and G. J. Milburn. Quantum technology: the second quantum revolution. *Philos. Trans. A Math Phys Eng Sci*, 361(1809):1655–1674, (2003).

[4] Quantum physics, a novice guide: An interview with Alain Aspect, *Polytechnique Insights: A Review by Institut Polytechnique De Paris*, (5), (October, 2023).

[5] J. S. Bell. On the Einstein Podolsky Rosen paradox. *Physics Physique Fizika*, 1(3):195–200, (1964).

[6] A. Einstein, B. Podolsky, and N. Rosen. Can quantum-mechanical description of physical reality be considered complete? *Phys. Rev.*, 47:777–780, (1935).

[7] N. Bohr. On the notions of causality and complementarity. *Dialectica*, 2(3-4):312–319, (1948).

[8] W. J. Moore. *Schrödinger: Life and Thought.* Cambridge University Press, (1992).

[9] E. Schrödinger. The present status of quantum mechanics : English translation of 'Die Gegenwärtige Situation in der Quantenmechanik'. *Die Naturwissenschaften*, 23(48):807–812, (1935).

[10] A. J. Leggett and Anupam Garg. Quantum mechanics versus macroscopic realism: Is the flux there when nobody looks? *Phys. Rev. Lett.*, 54:857–860, (1985).

[11] J. Kofler and Č. Brukner. Condition for macroscopic realism beyond the Leggett-Garg inequalities. *Phys. Rev. A*, 87:052115, (2013).

[12] C. Emary, N. Lambert, and F Nori. Leggett–garg inequalities. *Rep. Prog. Phys.*, 77(1):016001, (2014).

[13] S. Hawking and R. Penrose. *The Nature of Space and Time.* Princeton University Press, Princeton, p. 121, (2019).

[14] D Home. *Conceptual Foundations of Quantum Physics, Chapter 2.* Plenum Press, New York, (1997).

[15] G. C. Ghirardi, A. Rimini, and T. Weber. Unified dynamics for microscopic and macroscopic systems. *Phys. Rev. D*, 34:470, (1986).

[16] P. Pearle. Combining stochastic dynamical state-vector reduction with spontaneous localization. *Phys. Rev. A*, 39:2277, (1989).

[17] A. Bassi, K. Lochan, S. Satin, T. P. Singh, and H. Ulbricht. Models of wave-function collapse, underlying theories, and experimental tests. *Rev. Mod. Phys.*, 85:471, (2013).

[18] R. E. George, L. M. Robledo, O. J. E. Maroney, M. S. Blok, H. Bernien, M. L. Markham, D. J. Twitchen, J. J. L. Morton, G. A. D. Briggs, and R. Hanson. Opening up three quantum boxes causes classically undetectable wavefunction collapse. *Proc. Natl. Acad. Sci. U.S.A.*, 110(10):3777–3781, (2013).

[19] C. Robens, W. Alt, D. Meschede, C. Emary, and A. Alberti. Ideal negative measurements in quantum walks disprove theories based on classical trajectories. *Phys. Rev. X*, 5:011003, (2015).

[20] G. C. Knee, S. Simmons, E. M. Gauger, J. J. Morton, H. Riemann, N. V. Abrosimov, P. Becker, H.-J. Pohl, K. M. Itoh, M. L. Thewalt, G. A. D. Briggs, and S. C. Benjamin. Violation of a Leggett-Garg inequality with ideal non-invasive measurements. *Nat. Commun.*, 3:606, (2012).

[21] J. A. Formaggio, D. I. Kaiser, M. M. Murskyj, and T. E. Weiss. Violation of the leggett-Garg inequality in neutrino oscillations. *Phys. Rev. Lett.*, 117:050402, (2016).

[22] K. Wang, C. Emary, M. Xu, X. Zhan, Z. Bian, L. Xiao, and P. Xue. Violations of a leggett-Garg inequality without signaling for a photonic qutrit probed with ambiguous measurements. *Phys. Rev. A*, 97:020101, (2018).

[23] G. C. Knee, K. Kakuyanagi, M. C. Yeh, Y. Matsuzaki, H. Toida, H. Yamaguchi, S. Saito, A. J. Leggett, and W. J. Munro. A strict experimental test of macroscopic realism in a superconducting flux qubit. *Nat. Commun.*, 7:13253, (2016).

[24] H. Y. Ku, N. Lambert, F. J. Chan, C. Emary, Y. N. Chen, and F. Nori. Experimental test of non-macrorealistic cat states in the cloud. *npj Quantum Inf.*, 6:98, (2020).

[25] K. Joarder, D. Saha, D. Home, and U. Sinha. Loophole-free interferometric test of macrorealism using heralded single photons. *PRX Quantum*, 3:010307, (2022).

[26] S. Donadi, K. Piscicchia, C. Curceanu, L. Diosi, M. Laubenstein, and A. Bassi. Underground test of gravity-related wave function collapse. *Nat. Phys.*, 17:74–78, (2021).

[27] R. J. Marshman, A. Mazumdar, and S. Bose. Locality and entanglement in table-top testing of the quantum nature of linearized gravity. *Phys. Rev. A*, 101:052110, (2020).

[28] S. Bose, A. Mazumdar, G. W. Morley, H. Ulbricht, M. Toroš, M. Paternostro, A. A. Geraci, P. F. Barker, M. S. Kim, and G. Milburn. Spin entanglement witness for quantum gravity. *Phys. Rev. Lett.*, 119:240401, (2017).

[29] C. Marletto and V. Vedral. Gravitationally induced entanglement between two massive particles is sufficient evidence of quantum effects in gravity. *Phys. Rev. Lett.*, 119:240402, (2017).

[30] M. Christodoulou, A. D. Biagio, M. Aspelmeyer, C. Brukner, C. Rovelli, and R. Howl. Locally mediated entanglement through gravity from first principles. *Phys. Rev. Lett.*, 130:100202, (2023).

[31] F. Hanif, D. Das, J. Halliwell, D. Home, A. Mazumdar, H. Ulbricht, and S. Bose. Testing whether gravity acts as a quantum entity when measured. *Phys. Rev. Lett.*, 133:180201, (2024).

[32] A. J. Leggett. Realism and the physical world. *Rep. Prog. Phys.*, 71:022001, (2008).

[33] D. Saha, S. Mal, P. K. Panigrahi, and D. Home. Wigner's form of the Leggett-Garg inequality, the no-signaling-in-time condition, and unsharp measurements. *Phys. Rev. A*, 91:032117, (2015).

[34] S. Gerlich, S. Eibenberger, M. Tomandl, S. Nimmrichter, K. Hornberger, P. Miguez, M. Mayor, and M. Arndt. Quantum interference of large organic molecules. *Nat. Commun.*, 2:263, (2011).

[35] Y. Y. Fein, P. Geyer, P. Zwick, F. Kialka, S. Pedalino, M. Mayor, S. Gerlich, and M. Arndt. Quantum superposition of molecules beyond 25 kda. *Nat. Phys.*, 15:1242, (2019).

[36] J. J. Halliwell. Leggett-Garg inequalities and no-signaling in time: A quasiprobability approach. *Phys. Rev. A*, 93:022123, (2016).

[37] C. Gonzalez-Ballestero, M. Aspelmeyer, L. Novotny, R. Quidant, and O. Romero-Isart. Levitodynamics: Levitation and control of microscopic objects in vacuum. *Science*, 374(6564), (2021).

[38] C. Whittle, E. D. Hall, S. Dwyer, N. Mavalvala, V. Sudhir, R. Abbott, A. Ananyeva, C. Austin, L. Barsotti, J. Betzwieser, et al. Approaching the motional ground state of a 10-kg object. *Science*, 372:1333, (2021).

[39] J. Vovrosh, M. Rashid, D. Hempston, J. Bateman, M. Paternostro, and H. Ulbricht. Parametric feedback cooling of levitated optomechanics in a parabolic mirror trap. *J. Opt. Soc. Am. B*, 34:1421, (2017).

[40] R. W. Taylor and V. Sandoghdar. Interferometric scattering microscopy: Seeing single nanoparticles and molecules via rayleigh scattering. *Nano Lett.*, 19:4827, (2019).

[41] A. Vinante, A. Pontin, M. Rashid, M. Toroš, P. F. Barker, and H. Ulbricht. Testing collapse models with levitated nanoparticles: Detection challenge. *Phys. Rev. A*, 100:012119, (2019).

[42] G. Cerchiari, L. Dania, D. S. Bykov, R. Blatt, and T. E. Northup. Position measurement of a dipolar scatterer via self-homodyne detection. *Phys. Rev. A*, 104:053523, (2021).

[43] M. Bild, M. Fadel, Y. Yang, U. von Lüpke, P. Martin, A. Bruno, and Y. Chu. Schrödinger cat states of a 16-microgram mechanical oscillator. *Science*, 380(6642):274–278, (2023).

[44] P. C. W. Davies and J. R. Brown. *The Ghost in the Atom.* Canto Edition, Cambridge University Press, p. 52, (1993).

# Chapter 16

# Minimalist versions of the Maxwell-Dirac electrodynamics and other quantum gauge theories

Andrey Akhmeteli
LTASolid Inc.
akhmeteli@ltasolid.com

**Abstract**

Heisenberg wrote in his groundbreaking work that the position and period of revolution of electrons are "apparently unobservable in principle", unlike the frequencies and amplitudes of their radiation. This raises the question of whether the electromagnetic field may, in fact, contain the complete physical information about an electron. This is indeed the case for some realistic quantum gauge theories, including the Maxwell-Klein-Gordon electrodynamics (scalar electrodynamics) and Maxwell-Dirac electrodynamics (spinor electrodynamics). A central result of this work used to prove this is that the Dirac equation in electromagnetic field (or in the Yang-Mills field) is generally equivalent to an equation involving a single real function describing the matter field.

## 1 Introduction

Heisenberg's work [1] laid the foundation of modern quantum mechanics. As a starting point, Heisenberg states that the position and period of revolution of electrons are "apparently unobservable in principle", in contrast to the frequencies and amplitudes of its radiation. Much later, van der Waerden ( [2], p. 33) commented: "This was an error, because according to the fully developed quantum mechanics the three coordinates $x, y, z$ of an electron are observable. However, this error was extremely fruitful, because it stimulated Heisenberg to look out for other, directly observable quantities."

Nevertheless, one may ask whether an alternative reading of Heisenberg's statement is possible, namely, that the electromagnetic field may, in fact, contain the complete physical information about the electron. The present work reviews several results that tentatively support this point of view. Most of them arise in the context of classical and quantum electrodynamics and use the choice of the gauge and modifications of the Maxwell equations to describe both the electromagnetic field and the electron field *as a single entity*. While approaches such as the Wheeler-Feynman absorber theory [3] or Barut's self-field quantum electrodynamics [4] eliminate the electromagnetic field, it is also possible to pursue the opposite direction and eliminate the matter field from the equations of electrodynamics (see, e.g., [5], where integrals involving the Green's function are used to eliminate the electron field, whereas this work uses algebraic and differential-algebraic elimination. Unification of the electromagnetic and electron fields is also discussed by Tuck Choy and Miguel Ortuño in Chapter 14 of this volume [6]).

Section 2 of the present paper begins with a discussion of Dirac's "new electrodynamics" [7] whose modified Maxwell equations describe both the independent evolution of the electromagnetic field and the motion of charged particles in agreement with the Lorentz equation. Schrödinger derived a related quantum theory by making the matter field of the Maxwell-Klein-Gordon electrodynamics real by a gauge transformation. The resulting theory also describes the independent evolution of the electromagnetic field.

Section 3 discusses the key result that enables an extension of Schrödinger's approach to the Maxwell-Dirac electrodynamics: the Dirac equation in electromagnetic field is generally equivalent to a fourth-order partial differential equation for a single component, and the latter can be made real by a gauge transformation.

In Section 4, arguments are presented suggesting the possibility of differential-algebraic elimination of the matter field from the Maxwell-Dirac electrodynamics. At present, however, such elimination is achieved in a less satisfactory manner, requiring the introduction of a complex electromagnetic four-potential. Again, the resulting equations describe the independent evolution of the electromagnetic field.

Section 5 describes an extension of Schrödinger's approach to the Dirac equation in a Yang-Mills field.

Section 6 describes a plasma-like model of quantum particles. This model is consistent with those described in the previous sections but captures discrete aspects of quantum mechanics naturally. In particular, it offers a natural interpretation of negativity of the Wigner distribution function and avoids the problem of wave function spreading.

Section 7 outlines tentative extensions of the models described here to multi-particle systems.

## 2 Classical electrodynamics and the Maxwell-Klein-Gordon electrodynamics

Dirac [7] derived the following equations of motion from a Lagrangian describing a free electromagnetic field subject to the constraint (nonlinear gauge) $A_\mu A^\mu = m^2/e^2$:

$$\Box A_\mu - A^\nu_{,\nu\mu} = \lambda A_\mu, \tag{1}$$

where the d'Alembertian $\Box = \partial^\mu \partial_\mu$ and $\lambda$ is the Lagrange multiplier ($c = 1, \hbar = 1$).

These equations describe the self-consistent dynamics of the electrodynamic field and of particles of mass $m$, charge $e$, and momentum $p^\mu = -eA^\mu$, creating the conserved current appearing on the right hand side of (1). The particles move in accordance with the Lorentz equation

$$\frac{dp^\mu}{d\tau} = \frac{e}{m} F^{\mu\nu} p_\nu. \tag{2}$$

The electrodynamic field evolves independently; that is, one can pose a well-defined Cauchy problem for $A^\mu$ (see [8] for details).

In a comment on Dirac's work, Schrödinger [9] considered the equations of the Maxwell-Klein-Gordon electrodynamics (scalar electrodynamics)

$$(\partial^\mu + ieA^\mu)(\partial_\mu + ieA_\mu)\psi + m^2\psi = 0, \tag{3}$$

$$\Box A_\mu - A^\nu_{,\nu\mu} = j_\mu, \tag{4}$$

$$j_\mu = ie(\psi^* \psi_{,\mu} - \psi^*_{,\mu}\psi) - 2e^2 A_\mu \psi^* \psi \tag{5}$$

and observed that the scalar field $\psi$ can be made real (at least locally) by a gauge transformation $\varphi = \psi \exp(ie\omega)$, $B_\mu = A_\mu - \omega_{,\mu}$. The equations will then have the following form:

$$\Box\varphi - (e^2 B^\mu B_\mu - m^2)\varphi = 0, \tag{6}$$

$$\Box B_\mu - B^\nu_{,\nu\mu} = j_\mu, \tag{7}$$

$$j_\mu = -2e^2 B_\mu \varphi^2. \tag{8}$$

Schrödinger made a remarkable comment: "That the wave function ... can be made real by a change of gauge is but a truism, though it contradicts the widespread belief about 'charged' fields requiring complex representation."

Notably, Eqs. (6,7,8) share a key property with the equations of Dirac's work [7]: the matter field $\varphi$ can be algebraically eliminated, yielding the independent evolution of the electromagnetic field $B^\mu$. Specifically, $\varphi$ can be eliminated from equation (7) for index $\mu = 0$, which contains no second-order time derivatives. Substituting the resulting expression in (7) for $\mu = 1, 2, 3$, one obtains the second-order time derivatives of $B^1$, $B^2$, $B^3$. The expression for the second-order time derivative of $B^0$ then follows from current conservation [8].

It is worth emphasizing a subtle but important aspect of Schrödinger's work [9]. The original Klein-Gordon equation (3) is equivalent to two real equations for the real and imaginary parts of $\psi$, whereas (6) is a single equation for the real field $\varphi$. This reduction occurs because a linear combination of the original equations corresponds to current conservation, which remains present after the gauge transformation and can be derived independently from the Maxwell equation (7).

# 3 The Dirac equation as an equation for one component

Schrödinger expressed an intention to extend his results [9] to the Dirac equation, but apparently this program was never completed. A direct application of Schrödinger's method is indeed impossible: the Dirac equation involves the four-component Dirac spinor, and in general it is not possible to make all four components real using a single gauge transformation.

Nevertheless, Schrödinger's ideas can be extended to the Dirac equation in a different and rather unexpected way. It turns out that three of the four complex components of the Dirac spinor can generally be eliminated algebraically, and that the remaining component can then be made real by a gauge transformation [10].

Let us start with the Dirac equation in electromagnetic field $A_\mu$:

$$(i\partial\!\!\!/ - A\!\!\!/)\psi = \psi, \tag{9}$$

where the Feynman slash notation is used ($\not{A} = A_\mu \gamma^\mu$), the units are chosen in such a way that $\hbar = c = m = 1$, and the electric charge $e$ is included in $A_\mu$. The chiral representation of the $\gamma$-matrices [11] is chosen:

$$\gamma^0 = \begin{pmatrix} 0 & -I \\ -I & 0 \end{pmatrix}, \gamma^i = \begin{pmatrix} 0 & \sigma^i \\ -\sigma^i & 0 \end{pmatrix}, \tag{10}$$

where index $i$ runs from 1 to 3, $I$ is the 2-dimensional identity matrix, and $\sigma^i$ are the Pauli matrices.

In terms of the four components of the Dirac spinor, the Dirac equation is as follows:

$$(A^0 + A^3)\psi_3 + (A^1 - iA^2)\psi_4 + i(\psi_{3,3} - i\psi_{4,2} + \psi_{4,1} - \psi_{3,0}) = \psi_1, \tag{11}$$

$$(A^1 + iA^2)\psi_3 + (A^0 - A^3)\psi_4 - i(\psi_{4,3} - i\psi_{3,2} - \psi_{3,1} + \psi_{4,0}) = \psi_2, \tag{12}$$

$$(A^0 - A^3)\psi_1 - (A^1 - iA^2)\psi_2 - i(\psi_{1,3} - i\psi_{2,2} + \psi_{2,1} + \psi_{1,0}) = \psi_3, \tag{13}$$

$$-(A^1 + iA^2)\psi_1 + (A^0 + A^3)\psi_2 + i\psi_{2,3} + \psi_{1,2} - i(\psi_{1,1} + \psi_{2,0}) = \psi_4. \tag{14}$$

After the obvious algebraic elimination of components $\psi_3$ and $\psi_4$, the equations for $\psi_1$ and $\psi_2$ can be written as follows:

$$\begin{aligned} &-\psi_{1,\mu}^{,\mu} + \psi_2(-iA^1_{,3} - A^2_{,3} + A^0_{,2} + A^3_{,2} + i(A^0_{,1} + A^3_{,1} + A^1_{,0}) + A^2_{,0}) + \\ &+\psi_1(-1 + A^\mu A_\mu - iA^\mu_{,\mu} + iA^0_{,3} - A^1_{,2} + A^2_{,1} + iA^3_{,0}) - 2iA^\mu \psi_{1,\mu} = 0, \end{aligned} \tag{15}$$

$$\begin{aligned} &-\psi_{2,\mu}^{,\mu} + i\psi_1(A^1_{,3} + iA^2_{,3} + iA^0_{,2} - iA^3_{,2} + A^0_{,1} - A^3_{,1} + A^1_{,0} + iA^2_{,0}) + \\ &+\psi_2(-1 + A^\mu A_\mu - i(A^\mu_{,\mu} + A^0_{,3} + iA^1_{,2} - iA^2_{,1} + A^3_{,0})) - 2iA^\mu \psi_{2,\mu} = 0. \end{aligned} \tag{16}$$

Thus, the Dirac equation is equivalent to a system of partial differential equations (PDE) of the second order for just two components of the Dirac spinor. This was known long ago [12, 13]. In particular, Feynman and Gell-Mann commented as follows:

"Why must the wave function have four components? It is usually explained by pointing out that to describe the electron spin we must have two, and we must also represent the negative-energy states or positrons, requiring two more. Yet this argument is unsatisfactory. For a particle of spin zero we use a wave function of only one component. The sign of the energy is determined by how the wave function varies in space and time. The Klein-Gordon equation is second order and we need both the function and its time derivative to predict the future. So instead of two components for spin zero we use one, but it satisfies a second order equation. Initial states require specification of that one and its time derivative. Thus for the case of spin 1/2, we would expect to be able to use a simple two-component spinor for the wave function, but have it satisfy a second order differential equation."

However, one notices that $\psi_2$ can be algebraically eliminated from (15), and the substitution in (16) yields a fourth-order PDE for just one component $\psi_1$ [10]:

$$\left((\Box' - iF^3)\left(iF^1 + F^2\right)^{-1}(\Box' + iF^3) - iF^1 + F^2\right)\psi_1 = 0, \tag{17}$$

where

$$F^i = E^i + iH^i, \tag{18}$$

the electric and magnetic fields $E^i$ and $H^i$ are defined by the formulae

$$F^{\mu\nu} = A^{\nu,\mu} - A^{\mu,\nu} = \begin{pmatrix} 0 & -E^1 & -E^2 & -E^3 \\ E^1 & 0 & -H^3 & H^2 \\ E^2 & H^3 & 0 & -H^1 \\ E^3 & -H^2 & H^1 & 0 \end{pmatrix}, \tag{19}$$

whereas the modified d'Alembertian $\Box'$ is defined as follows:

$$\Box' = \partial^\mu\partial_\mu + 2iA^\mu\partial_\mu + iA^\mu_{,\mu} - A^\mu A_\mu + 1. \tag{20}$$

Equation (17) is generally equivalent to the Dirac equation (9).

The remaining component $\psi_1$ can be made real by a gauge transformation, so the results of Schrödinger's work [9] can be extended to the Dirac equation.

A relativistically invariant form of (17) was derived in [14, 15]. One can choose arbitrary constant spinors $\xi$ and $\eta$ such that $\gamma^5\xi = \pm\xi, \gamma^5\eta = \pm\eta$ (so both $\xi$ and $\eta$ are right-handed or left-handed) and $\bar{\xi}\eta^c = 1$, where $\eta^c = C\bar{\eta}^T$ and $C$ is the charge conjugation matrix. Then the Dirac equation for one component $\bar{\xi}\psi$ of the Dirac spinor has the following form:

$$((\Box' - f_{\xi\eta})f_{\xi\xi}^{-1}(\Box' + f_{\xi\eta}) + f_{\eta\eta})(\bar{\xi}\psi) = 0, \tag{21}$$

where the relevant component of electromagnetic field for Dirac spinors $\alpha$ and $\beta$ is defined as

$$f_{\alpha\beta} = \bar{\alpha} F \beta^c \tag{22}$$

and

$$F = \frac{1}{2} F_{\nu\mu} \sigma^{\nu\mu}, \tag{23}$$

$$\sigma^{\mu\nu} = \frac{i}{2}[\gamma^\mu, \gamma^\nu]. \tag{24}$$

Furthermore, this result can be used to rewrite the Dirac equation as a tensor equation for one component [16]. Some tensor equivalents of the Dirac equation were proposed in the past, but they either were nonlinear [17] or included several components ( [18, 19, 20, 21]). A Dirac spinor $\psi$ can be represented as a sum of two chiral spinors $\psi_+$ and $\psi_-$:

$$\psi = \psi_+ + \psi_-, \gamma^5 \psi_+ = \psi_+, \gamma^5 \psi_- = -\psi_-. \tag{25}$$

These chiral spinors can be represented by antisymmetric second-rank tensors ( [17], [22], Section 154):

$$(\psi_\pm^{\mu\nu}) = (\overline{\psi_\pm^c} \sigma^{\mu\nu} \psi_\pm) = (\psi_\pm^T C \sigma^{\mu\nu} \psi_\pm). \tag{26}$$

It is possible to find constant antisymmetric tensors $u^{\mu\nu}, v^{\mu\nu}, w^{\mu\nu}$ satisfying the following conditions [16] (these tensors play a role similar to that of $\gamma$-matrices):

$$(u^{\mu\nu}) = (\mp i \star u^{\mu\nu}), u^{\mu\nu} u_{\mu\nu} = 0, \tag{27}$$

$$(v^{\mu\nu}) = (\mp i \star v^{\mu\nu}), v^{\mu\nu} u_{\mu\nu} = 0, u^\mu{}_\sigma v^{\sigma\nu} = -i u^{\mu\nu}, \tag{28}$$

$$(w^{\mu\nu}) = (\mp i \star w^{\mu\nu}), v^{\mu\nu} w_{\mu\nu} = 0, w^{\mu\nu} w_{\mu\nu} = 0, u^{\mu\nu} w_{\mu\nu} = -8, \tag{29}$$

where the Hodge dual $\star$ of a second-rank antisymmetric tensor is defined as [23]

$$\star F^{\alpha\beta} = \frac{1}{2} \epsilon^{\alpha\beta\gamma\delta} F_{\gamma\delta}. \tag{30}$$

Then the linear tensor equivalent to the Dirac equation for just one scalar function $\varphi_u$ has the following form:

$$((2\Box' - F_{\mu\nu} v^{\mu\nu})(F_{\mu\nu} u^{\mu\nu})^{-1}(2\Box' + F_{\mu\nu} v^{\mu\nu}) + F_{\mu\nu} w^{\mu\nu})\varphi_u = 0. \tag{31}$$

There is the following relationship between the scalar function and the Dirac spinor $\psi$:

$$\varphi_u = \left(-\frac{\psi_\mp^{\mu\nu} u_{\mu\nu}}{8}\right)^{\frac{1}{2}}. \tag{32}$$

# 4 The Maxwell-Dirac electrodynamics

We now consider the equations of the Maxwell-Dirac electrodynamics (spinor electrodynamics):

$$(i\partial\!\!\!/ - A\!\!\!/ - 1)\psi = 0, \tag{33}$$

$$\Box A_\mu - A^\nu_{,\nu\mu} = e^2 \bar{\psi}\gamma_\mu \psi. \tag{34}$$

As discussed in the previous section, three of the four components of the Dirac spinor can be eliminated algebraically. The elimination of the remaining component, and thus of the matter field as a whole, is not obvious. Nevertheless, there are indications that such an elimination may be possible.

After algebraically eliminating the components $\psi_3$ and $\psi_4$ as described above and imposing the gauge condition $\mathrm{Im}(\psi_1) = 0$, one obtains a system of eight PDEs for seven real dependent variables $A^0, A^1, A^2, A^3, \psi_{1r} = \mathrm{Re}(\psi_1), \psi_{2r} = \mathrm{Re}(\psi_2), \psi_{2i} = \mathrm{Im}(\psi_2)$. To analyze this system, we used the Maple *DifferentialThomas* package [24] to compute an equivalent so-called simple and integrable system of equations (the short code is deposited at [25]). The dependent variables were ordered as $[\psi_{2i}, \psi_{2r}, A^0, A^3, A^2, A^1, \psi_{1r}]$, the independent variables were $[x^0, x^1, x^2, x^3]$, and the ranking was chosen to be *DegRevLex* (the ranking *EliminateFunction* was found to require prohibitively large computational resources). The resulting system had the following leaders (the highest appearing variables in the polynomial with respect to the ranking) in each equation: $\psi_{2i,2000}$, $\psi_{2r,2000}$ ,$A^0_{,1000}$, $A^3_{,2000}$, $A^2_{,2000}$, $A^1_{,2000}$, $A^1_{,1100}$, $\psi_{1r,2000}$, where, for example,

$$A^1_{,1100} = \frac{\partial^2 A^1}{\partial x^0 \partial x^1}. \tag{35}$$

For the equation with leader $A^1_{,1100}$, the independent variables admissible for prolongation (those with respect to which one can repeatedly differentiate an equation) are $x^1, x^2, x^3$, whereas for all other equations prolongation with respect to $x^0, x^1, x^2, x^3$ is admissible. The total number of equations and their admissible prolongations containing derivatives of total order $k \leq 7$ is 1022, while the total number of derivatives of $\psi_{1r}, \psi_{2r}, \psi_{2i}$ of total order $k \leq 7$ is 990. This suggests that the matter field can be eliminated from the prolonged system of equations.

Solving such a large polynomial system explicitly does not look realistic. However, one can linearize the prolonged system about a particular solution and evaluate the rank of the resulting linear system. If this rank is equal to the number of derivatives of $\psi_{1r}, \psi_{2r}, \psi_{2i}$, it would imply that the matter field can be differential-algebraically eliminated from the equations of the Maxwell-Dirac electrodynamics. Such evaluation has not been performed yet.

Elimination of the matter field from the equations of the Maxwell-Dirac electrodynamics was performed previously in an imperfect form [26], at the cost of introducing a complex electromagnetic four-potential. This construction employed the following "generalized gauge transformation":

$$\psi = \exp(i\alpha)\varphi, \tag{36}$$

$$A_\mu = B_\mu - \alpha_{,\mu}, \tag{37}$$

where $\alpha = \alpha(x^\mu) = \beta + i\delta$ is a complex function, $\beta$ and $\delta$ are real functions. The resulting electromagnetic potential $B^\mu$ is complex, and its imaginary part is a gradient of a function.

In general, the generalized gauge transformation can be chosen so that the first component of the transformed Dirac spinor satisfies $\varphi_1 = 1$. The resulting equations then involve only the remaining spinor components and the function $\delta$, all of which can be eliminated. Moreover, the resulting equations describe independent evolution of the complex electromagnetic field $B^\mu$ in the following sense: if $B^\mu$ and its first and second time derivatives $\dot{B}^\mu$ and $\ddot{B}^\mu$ are specified on a spacelike hypersurface $x^0 = \text{const}$, then the third time derivatives of the third order $\dddot{B}^\mu$ can be determined on the same hypersurface. Consequently, a Cauchy problem for $B^\mu$ can be posed [26].

# 5 The Dirac equation in the Yang-Mills field

We have seen that Schrödinger's approach of making a scalar field real by gauge transformation can be adapted to replace a Dirac spinor function by a single real function. Since gauge theories are central to modern physics, it is natural to ask whether a similar construction is possible for the Dirac equation in a non-Abelian gauge field. For the Yang-Mills field, the answer is affirmative.

The Dirac equation in the Yang-Mills field reads ( [27], p.493):

$$(i\partial\!\!\!/ - A\!\!\!/)\psi = \psi, \tag{38}$$

where $A_\mu = A^a_\mu T^a$, $T^a$ are generators of the group SU($n$). The components $\psi_{i\alpha}$ of the field $\psi$ carry both an SU($n$) fundamental representation index ("color" index) $i$ and a spinor index $\alpha$.

Using the constant Dirac spinors $\xi$ and $\eta$ introduced in Section 3, satisfying $\gamma^5\xi = \pm\xi, \gamma^5\eta = \pm\eta$, and $\bar{\xi}\eta^c = 1$, one can derive [8] an equation for a single "spinor component" $\bar{\xi}\psi$ of the field $\psi$. This object carries $n$ color components $\xi^*_\alpha\gamma^0_{\alpha\beta}\psi_{i\beta}$ and satisfies the following equation:

$$((\Box' - \bar{\xi}F\eta^c)(\bar{\xi}F\xi^c)^{-1}(\Box' + \bar{\xi}F\eta^c) + \bar{\eta}F\eta^c)(\bar{\xi}\psi) = 0, \tag{39}$$

where

$$\Box' = \partial^\mu\partial_\mu + 2iA^\mu\partial_\mu + iA^\mu_{,\mu} - A^\mu A_\mu + 1 = -(i\partial_\mu - A_\mu)(i\partial^\mu - A^\mu) + 1, \tag{40}$$

$$F_{\mu\nu} = i[i\partial_\mu - A_\mu, i\partial_\nu - A_\nu] = A_{\nu,\mu} - A_{\mu,\nu} + i[A_\mu, A_\nu], \tag{41}$$

$$F = \frac{1}{2}F_{\nu\mu}\sigma^{\nu\mu}. \tag{42}$$

This equation is generally equivalent to the Dirac equation in a Yang-Mills field (38). The only "spinor component" in equation (39) has $n$ "color" components; however, by a gauge transformation, all of them but one can be made zero and the remaining component can be made real, as the action of the SU($n$) group on the unit sphere in $\mathbb{C}^n$ is transitive for $n \geq 2$ [28].

# 6 Plasma-like model of quantum particles

In previous sections, we discussed algebraic and differential-algebraic elimination of the matter field from the equations of scalar and spinor electrodynamics. These results can be used for wave-only interpretations of quantum theory. However, one may feel that an interpretation should also reflect discrete aspects of quantum phenomena, and the results described in this work can be compatible with such a requirement within a plasma-like model [26, 29, 8].

The analogy between quantum particles and plasmas has long been noted. For example, the dispersion relation for the Klein-Gordon equation

$$\omega^2 = m^2 + k^2 \tag{43}$$

is tantalizingly similar to the dispersion relation for a cold collisionless plasma:

$$\omega^2 = \omega_p^2 + k^2. \tag{44}$$

The results of the previous sections provide another justification for recognizing such an analogy, as real fields imply the existence of both positive and negative frequencies and, therefore, of positively and negatively charged particles. This justifies a model in which an electron is represented by a collection of $N+1$ electrons and $N$ positrons, with $N$ large. Such a model may be consistent with the notions of vacuum polarization and Zitterbewegung, however, in this model, the Zitterbewegung frequency is a "natural frequency", rather than a frequency of an "internal clock" [30].

The key issue is how a collection of point-like negative and positive charges can emulate a smooth charge density built from a wave function. The general idea is that a smoothed charge density of a collection of point-like charges should approximate a smooth charge density. Initially, the equality of truncated Fourier series was chosen as the criterion of the proximity of the smooth charge density and the charge density of a collection of discrete charges. It was proven [8] that, in the one-dimensional case, for any smooth charge density with an integer total charge and any $k$ one can find a collection of $\pm 1$ charges that has the same first $k$ terms of the Fourier series. Later, it was proven [31] for any number of dimensions.

The wonderful proof of [31] is important as it shows that plasma-like collections of [29, 8] can emulate any experimental effects described by, say, the Dirac equation. However, the proof is essentially not constructive, relying, for example, on Brouwer's fixed point theorem. A constructive approach is desirable as it would enable building plasma-like collections approximating arbitrary wave functions. Such collections can be used to emulate experiments described by quantum theory, such as the double-slit experiment.

To provide a constructive approach inspired by [31] (and also by [32]), a modified criterion of the proximity of a smooth probability density and a collection of $\delta$-functions with integer coefficients is used in this work. This criterion is mathematically different from the previous one but seems also adequate for applications to quantum physics.

The previous criterion was as follows.

Let us consider a smooth real function $f(\boldsymbol{x})$ in a $d$-dimensional cube $C: -\pi \leq x^i \leq \pi$, where $\boldsymbol{x} = (x^1, \ldots, x^d)$, $\int\limits_C f(\boldsymbol{x}) \boldsymbol{dx} = 1$, and $f(\boldsymbol{x})$ vanishes on the boundary of the cube.

Let us also consider distributions of the form $g(\boldsymbol{x}) = \delta(\boldsymbol{x}-\boldsymbol{x}_0)+\sum_{i=1}(\delta(\boldsymbol{x}-\boldsymbol{y}_i) - \delta(\boldsymbol{x} - \boldsymbol{z}_i))$, where $\delta(\boldsymbol{x})$ is the $d$-dimensional Dirac's $\delta$-function, and $\boldsymbol{x}_0, \boldsymbol{y}_i, \boldsymbol{z}_i$ belong to the interior of $C$.

Functions $f(\boldsymbol{x})$ and $g(\boldsymbol{x})$ in the cube can be expanded in Fourier series of the form

$$f(\boldsymbol{x}) = \sum_{\boldsymbol{k}\in\mathbb{Z}^d} f_{\boldsymbol{k}} \mathrm{e}^{i\boldsymbol{k}\boldsymbol{x}}, \tag{45}$$

where $d$-dimensional vectors $\boldsymbol{k} = (k^1, \ldots, k^d)$ have integer coordinates, $f_{\boldsymbol{k}} = f^*_{-\boldsymbol{k}}$ for real functions, and

$$f_{\boldsymbol{k}} = (2\pi)^{-d} \int_C f(\boldsymbol{x})\mathrm{e}^{-i\boldsymbol{k}\boldsymbol{x}} \boldsymbol{dx}. \tag{46}$$

For some natural $m$, the series can be truncated by keeping only terms with such $\boldsymbol{k}$ that $|k^i| \le m$:

$$f_{tr}(\boldsymbol{x}) = \sum_{\substack{\boldsymbol{k}\in\mathbb{Z}^d \\ |k^i|\le m}} f_{\boldsymbol{k}} \mathrm{e}^{i\boldsymbol{k}\boldsymbol{x}}. \tag{47}$$

Coincidence of the truncated series for $f(\boldsymbol{x})$ and $g(\boldsymbol{x})$ for some $m$ was used as the criterion of the proximity. The following statement was proven in [8] (for $d = 1$) and in [31] (for an arbitrary $d$):

$\forall m\ \exists n, \boldsymbol{x}_0, \boldsymbol{y}_i, \boldsymbol{z}_i\ (1 \le i \le n)$ such that the truncated Fourier series for $f(\boldsymbol{x})$ and $g(\boldsymbol{x})$ coincide.

The new criterion requires that $f(\boldsymbol{x})$ and $g(\boldsymbol{x})$ become close after Gaussian smoothing

$$g_\delta(\boldsymbol{x}) = \int g(\boldsymbol{\xi})G_\delta(\boldsymbol{x} - \boldsymbol{\xi})\boldsymbol{d\xi}, \tag{48}$$

a convolution with the Gaussian

$$G_\delta(\boldsymbol{x}) = (\sqrt{2\pi}\delta)^{-d} e^{-\frac{\boldsymbol{x}^2}{2\delta^2}}. \tag{49}$$

We do not require anymore that the smooth function $f(\boldsymbol{x})$ be only defined in the $d$-dimensional cube $C$. Then the modified statement (for the function $f(\boldsymbol{x})$ satisfying the conditions above) is as follows.

$\forall \delta > 0, \varepsilon > 0\ \exists n, \boldsymbol{x}_0, \boldsymbol{y}_i, \boldsymbol{z}_i\ (1 \le i \le n)$ such that $|f_\delta(\boldsymbol{x}) - g_\delta(\boldsymbol{x})| < \varepsilon$.

A sketch of the proof may be as follows. Let us define the following function:

$$\rho(\boldsymbol{x}) = G_{\delta_n}(\boldsymbol{x} - \boldsymbol{x}_0) - f(\boldsymbol{x}), \tag{50}$$

where $x_0$ is chosen arbitrarily and $\delta_n \ll \delta$. One can see that

$$\int \boldsymbol{dx}\rho(\boldsymbol{x}) = 0. \tag{51}$$

Let $\varphi(\boldsymbol{x})$ be a solution of the Poisson equation

$$\triangle\varphi(\boldsymbol{x}) = -\rho(\boldsymbol{x}), \tag{52}$$

such that $\varphi(\boldsymbol{x}) = 0$ at infinity. Using Green's first identity with a vanishing "surface" integral, we obtain

$$\int \boldsymbol{\nabla}_{\boldsymbol{x}'} G_\delta(\boldsymbol{x}' - \boldsymbol{x})\boldsymbol{\nabla}_{\boldsymbol{x}'}\varphi(\boldsymbol{x}')\boldsymbol{dx}' = -\int G_\delta(\boldsymbol{x}' - \boldsymbol{x}) \triangle_{\boldsymbol{x}'} \varphi(\boldsymbol{x}')\boldsymbol{dx}' = -\int G_\delta(\boldsymbol{x}' - \boldsymbol{x})\rho(\boldsymbol{x}')\boldsymbol{dx}'. \tag{53}$$

On the other hand, for $a \ll \delta_n$ (we also have $a \ll \delta$ in this case), $\boldsymbol{l} = (l_1, \ldots, l_j, \ldots, l_d)$, where $l_j$ are integer and $-\infty \leq l_j \leq \infty, \boldsymbol{x_l} = a\boldsymbol{l}$, we obtain, using a simple cubature formula and a slightly modified definition of the directional derivative as a limit:

$$\int \boldsymbol{\nabla}_{\boldsymbol{x}'} G_\delta(\boldsymbol{x}' - \boldsymbol{x})\boldsymbol{\nabla}_{\boldsymbol{x}'}\varphi(\boldsymbol{x}')\boldsymbol{dx}' \approx a^d \sum_{\substack{l^j=-\infty \\ 1\leq j\leq d}}^{\infty} \boldsymbol{\nabla}_{\boldsymbol{x_l}} G_\delta(\boldsymbol{x_l} - \boldsymbol{x})\boldsymbol{\nabla}_{\boldsymbol{x_l}}\varphi(\boldsymbol{x_l}) \approx$$

$$a^d \sum_{\substack{l^j=-\infty \\ 1\leq j\leq d}}^{\infty} \frac{G_\delta\left(\boldsymbol{x_l} + \frac{1}{2}a^d\boldsymbol{\nabla}_{\boldsymbol{x_l}}\varphi(\boldsymbol{x_l}) - \boldsymbol{x}\right) - G_\delta\left(\boldsymbol{x_l} - \frac{1}{2}a^d\boldsymbol{\nabla}_{\boldsymbol{x_l}}\varphi(\boldsymbol{x_l}) - \boldsymbol{x}\right)}{a^d} =$$

$$\sum_{\substack{l^j=-\infty \\ 1\leq j\leq d}}^{\infty} \left(G_\delta\left(\boldsymbol{x_l} + \frac{1}{2}a^d\boldsymbol{\nabla}_{\boldsymbol{x_l}}\varphi(\boldsymbol{x_l}) - \boldsymbol{x}\right) - G_\delta\left(\boldsymbol{x_l} - \frac{1}{2}a^d\boldsymbol{\nabla}_{\boldsymbol{x_l}}\varphi(\boldsymbol{x_l}) - \boldsymbol{x}\right)\right). \tag{54}$$

We replaced the derivatives with difference quotients. This approximation gets better with smaller $a$.

One can see from (50), (53), and (54) that

$$f_\delta(\boldsymbol{x}) \approx G_\delta(\boldsymbol{x} - \boldsymbol{x_0}) + \sum_{\substack{l^j=-\infty \\ 1\leq j\leq d}}^{\infty} \left(G_\delta\left(\boldsymbol{x_l} + \frac{1}{2}a^d\boldsymbol{\nabla}_{\boldsymbol{x_l}}\varphi(\boldsymbol{x_l}) - \boldsymbol{x}\right) - G_\delta\left(\boldsymbol{x_l} - \frac{1}{2}a^d\boldsymbol{\nabla}_{\boldsymbol{x_l}}\varphi(\boldsymbol{x_l}) - \boldsymbol{x}\right)\right) \tag{55}$$

(we used the condition $\delta_n \ll \delta$ to replace the result of Gaussian smoothing of function $G_{\delta_n}(\boldsymbol{x}-\boldsymbol{x}_0)$ with $G_\delta(\boldsymbol{x}-\boldsymbol{x}_0)$). Note that a convolution of Gaussian functions with standard deviations $\delta_1$ and $\delta_2$ is another Gaussian function with standard deviation $\sqrt{\delta_1^2+\delta_2^2}$ [33]. One can have an arbitrarily good approximation leaving only a finite number of "dipoles" in the sum in the right-hand side of (55) due to (51) and (52).

Thus, probability density functions built from one-particle wave functions of quantum theory can be approximated arbitrarily well by a set of $\delta$-functions with integer coefficients. A constructive approach to building such a set has been developed. This allows the simulation of quantum effects using plasma-like collections. In particular, quantum phase space distribution functions [34], such as the Wigner distribution function, find a natural interpretation in this approach, although these functions typically have negative values in some areas. Notably, the negativity of the Wigner function is regarded as a resource for quantum computing [35].

One can compare the plasma-like approach with the initial Schrödinger's interpretation of the wave function as a source of charge density (see, e.g., [36, 37] and references therein). Objections to Schrödinger's interpretation include the lack of explanation of discreteness of electric charge ([38], p. 23) and the well-known problem of spreading of the wave function. It seems that the plasma-like approach is immune to these problems.

A tentative mathematical model providing equations of motion for collections of charged particles and antiparticles was developed in [29]. For a sufficiently large cut-off constant, the model can emulate the Maxwell-Klein-Gordon electrodynamics arbitrarily well, so it can describe typical quantum phenomena, such as the double-slit experiment.

To fully define the mathematical model, one has to define the cutoff constant, but the model shares this problem of very high time and spatial frequencies with the standard quantum field theory [39].

It was important to resolve the following problem of the plasma-like model. Composite particles, such as nucleons or large molecules, also manifest quantum properties [40, 41], but the idea of molecule–antimolecule pairs being significant for diffraction of large molecules is not plausible. On the other hand, composite particles are subject to some kind of interactions (such as electromagnetic or strong interactions), so the modified plasma-like model [29] can involve a composite particle surrounded by a collection of (elementary) fermions and antifermions (for example, electrons and positrons for the electromagnetic interaction). Such picture is close to that of the de Broglie–

Bohm interpretation. The composite particle can be detected at all locations where the wave function does not vanish, but at most locations it is actually the fermions of the collection that interact with the instrument.

# 7 Tentative approaches to multi-particle systems

While models of this work can describe important single-particle quantum phenomena, there are currently only some tentative approaches to describing multi-particle systems. In particular, entanglement is one of the most important phenomena in multi-particle systems.

Bosons and fermions should be treated differently. The approach to bosons is described in [8], Section 2.9. It was proposed by nightlight [42] and uses Kowalski's generalization of Carleman linearization [43, 44]. It was shown that solutions of nonlinear PDEs describing independent evolution of electromagnetic field can be embedded into quantum field theories.

How can the Pauli principle emerge in models of this work? One can speculate that collections of discrete charges related to the same wave function need to have identical or similar coordinates of the charges, so configurations including two or more such collections would have very high energy, for example, due to Coulomb interaction.

The above reasoning implies that a theory similar to quantum field theory (QFT) can be built based on non-second-quantized scalar or spinor electrodynamics on the set of solutions of the latter. However, there is now a consensus opinion that violations of the Bell inequalities have been demonstrated experimentally, so no local realistic hidden variable theories are viable. This issue was discussed in [45, 46] using other people's arguments. Rather than repeat all the arguments, the author would like just to comment on more recent experiments widely considered to be loophole-free demonstrations of violations of the Bell inequalities [47, 48, 49, 50]. In all the experiments, the authors assume that the measurement outcomes are final at some specific moment. However, strictly speaking, these assumptions are not compatible with unitary evolution. For example, the authors of [47] state:

"Our observation of a statistically significant loophole-free Bell inequality violation thus indicates rejection of all local-realist theories that accept that the number generators produce a free random bit in a timely manner and that the outputs are final once recorded in the electronics.

Strictly speaking, no Bell experiment can exclude all conceivable local-realist theories, because it is fundamentally impossible to prove when and where free random input bits and output values came into existence [51]."

Furthermore the proof of violations of the Bell inequalities in quantum mechanics requires making assumptions about both the unitary evolution and the wave function collapse (projection postulate) [45, 46]. These assumptions are, strictly speaking, mutually contradictory. The models of this work emulate unitary evolution, but not the collapse, and the experimental status of the latter is controversial. For example, based on an analysis of experiments, Schlosshauer states [52]: "(i) the universal validity of unitary dynamics and the superposition principle has been confirmed far into the mesoscopic and macroscopic realm in all experiments conducted thus far; (ii) all observed "restrictions" can be correctly and completely accounted for by taking into account environmental decoherence effects; (iii) no positive experimental evidence exists for physical state-vector collapse".

# 8 Conclusion

The main results of this work can be summarized as follows:

- The Dirac equation in the electromagnetic field and in the Yang-Mills field is generally equivalent to an equation for a single real function. This extends Schrödinger's result [9] for a scalar field.

- In the Maxwell-Klein-Gordon electrodynamics (scalar electrodynamics) and, with certain caveats, in the Maxwell-Dirac electrodynamics (spinor electrodynamics), the matter field can be algebraically or differential-algebraically eliminated. The resulting systems of equations describe independent evolution of the gauge field.

- A plasma-like model can adequately emulate quantum theories, such as, the Maxwell-Klein-Gordon electrodynamics. In particular, it provides a natural interpretation of the negativity of the Wigner distribution function.

It should be noted that the above results are local in nature; global issues, such as those related to zeroes of denominators or global gauge obstructions, are not addressed.

The mathematical results discussed in this work yield classical models of quantum gauge theories, but hopefully they have their value irrespective of the reader's preferred interpretation of quantum theory.


## Acknowledgments

The author is grateful to Tuck Choy, Mark Davidson, and Andrew Smith for their interest in this work and valuable comments.


# References


[1] W. Heisenberg. Über quantentheoretische Umdeutung kinematischer und mechanischer Beziehungen. *Zeitschrift für Physik*, 33:879–893, 1925.

[2] B. L. van der Waerden (Ed.). *Sources of Quantum Mechanics* . Dover, New York, 1967.

[3] J. A. Wheeler and R. P. Feynman. Interaction with the Absorber as the Mechanism of Radiation. *Rev. Mod. Phys.*, 17:157–181, 1945.

[4] A. O. Barut. Foundations of Self-Field Quantumelectrodynamics. In A. O. Barut, editor, *New Frontiers in Quantum Electrodynamics and Quantum Optics*, page 345. Plenum Press, New York, 1990.

[5] K. H. Ruei. Alternate formulation of electron-photon interactions in relativistic quantum mechanics. *American Journal of Physics*, 44:153–157, 1976.

[6] Tuck Choy and Miguel Ortuño. From Faraday and Maxwell to Quantum Physics. The later story of the Vector Potential. In Tuck Choy, editor, *Quantum Mechanics: A Century Later. In Celebration of Heisenberg's Seminal 1925 Paper.* World Scientific, Singapore, 2026.

[7] P. A. M. Dirac. A new classical theory of electrons. *Proc. Roy. Soc. A*, 209:291–296, 1951.

[8] A. Akhmeteli. Some Classical Models of Particles and Quantum Gauge Theories. *Quantum Reports*, 4(4):486–508, 2022.

[9] E. Schrödinger. Dirac's New Electrodynamics. *Nature*, 169:538, 1952.

[10] A. Akhmeteli. One real function instead of the Dirac spinor function. *J. Math. Phys.*, 52:082303, 2011.

[11] C. Itzykson and J.-B. Zuber. *Quantum field theory.* McGraw-Hill, New York, 1980.

[12] O. Laporte and G. E. Uhlenbeck. Application of spinor analysis to the Maxwell and Dirac equations. *Phys. Rev.*, 37:1380, 1931.

[13] R. P. Feynman and M. Gell-Mann. Theory of the Fermi interaction. *Phys. Rev.*, 109:193, 1958.

[14] A. Akhmeteli. The Dirac equation as one fourth-order equation for one function – a general, manifestly covariant form, 2015. Preprint at `https://arxiv.org/abs/1502.02351`.

[15] A. Akhmeteli. The Dirac equation as one fourth-order equation for one function: A general, manifestly covariant form. In *Quantum Foundations, Probability and Information, A. Khrennikov, T. Bourama (Eds.)*, page 1, 2018.

[16] A. Akhmeteli. The Dirac equation as a linear tensor equation for one component. *Eur. Phys. J. C*, 84:488, 2024.

[17] E. T. Whittaker. On the Relations of the Tensor-calculus to the Spinor-calculus. *Proc. Roy. Soc. A*, 158:38–46, 1937.

[18] C. G. Darwin. The wave equations of the electron. *Proc. Roy. Soc. A*, 118:654–679, 1928.

[19] D. Iwanenko and L. Landau. Zur Theorie des magnetischen Electrons. I. *Zeitschrift für Physik*, 48:340–348, 1928.

[20] E. Kähler. Der innere Differentialkalkül. *Rendiconti di Matematica (3-4)*, 21:425, 1962.

[21] N. G. Marchuk. Dirac type tensor equations. *Nuovo Cim. B*, 116:1225–1248, 2001.

[22] E. Cartan. *The theory of spinors.* Dover Publications, Inc., New York, 1981.

[23] J. D. Jackson. *Classical Electrodynamics, Third Edition.* Wiley, New York, 1998.

[24] T. Bächler, V. Gerdt, M. Lange-Hegermann, and D. Robertz. Algorithmic Thomas decomposition of algebraic and differential systems. *J. Symbolic Computation*, 47:1233–1266, 2012.

[25] A. Akhmeteli. `https://github.com/akhmeteli1/MaxwellDirac`, 2025.

[26] A. Akhmeteli. No drama quantum electrodynamics? *Eur. Phys. J. C*, 73:2371, 2013.

[27] M. D. Schwartz. *Quantum Field Theory and the Standard Model*:. Cambridge University Press, 2014.

[28] A.W. Knapp. *Lie Groups Beyond an Introduction* . Birkhäuser, Boston e.a., 1996.

[29] A. Akhmeteli. Plasma-like description for elementary and composite quantum particles. *Entropy*, 24(2):261, 2022.

[30] P. Catillon, N. Cue, M.J. Gaillard, R. Genre, M. Gouanère, R.G. Kirsch, J.-C. Poizat, J. Remillieux, L. Roussel, and M. Spighel. A search for the de Broglie particle internal clock by means of electron channeling. *Found. Phys.*, 38:659, 2008.

[31] fedja. `https://mathoverflow.net/questions/476275/`.

[32] V. G. Koures. Gaussian basis expansion of the QED light cone Hamiltonian. *Phys. Lett. B*, 348:170–l 77, 1995.

[33] S. Vinga. `https://web.ist.utl.pt/susanavinga/renyi/convolution_normal.pdf`, 2004.

[34] Hai-Woong Lee. Theory and application of the quantum phase-space distribution functions. *Phys. Rep.*, 259:147, 1995.

[35] R. I. Booth, U. Chabaud, and P.-E. Emeriau. arxiv:2111.13218.

[36] C. T. Sebens. Electron Charge Density: A Clue from Quantum Chemistry for Quantum Foundations. *Foundations of Physics*, 51:75:1–39, 2021.

[37] A. O. Barut. Combining relativity and quantum mechanics: Schrödinger's interpretation of $\psi$. *Foundations of Physics*, 18:95–105, 1988.

[38] A. Khrennikov. *Beyond Quantum*. Pan Stanford Publishing, Singapore, 2014.

[39] H. Georgi. Effective field theory. *Annu. Rev. Nucl. Part. Sci.*, 43:209, 1993.

[40] A. Zeilinger, R. Gähler, C. G. Shull, W. Treimer, and W. Mampe. Single- and double-slit diffraction of neutrons. *Rev. Mod. Phys.*, 60:1067, 1988.

[41] Y. Y. Fein, P. Geyer, P. Zwick, F. Kiałka, S. Pedalino, M. Mayor, S. Gerlich, and M. Arndt. Quantum superposition of molecules beyond 25 kDa. *Nature Physics*, 15:1242, 2019.

[42] nightlight. http://www.physicsforums.com/ showthread.php?t=12311.

[43] K. Kowalski and W.-H. Steeb. *Nonlinear dynamical systems and Carleman linearization* :. World Scientific, Singapore, 1991.

[44] K. Kowalski. *Methods of Hilbert spaces in the theory of nonlinear dynamical systems* :. World Scientific, Singapore, 1994.

[45] A. Akhmeteli. quant-ph/1006.2578.

[46] A. Akhmeteli. Is no drama quantum theory possible? *Int. J. Quantum Inf.*, 9:Suppl. 17, 2011.

[47] B. Hensen, H. Bernien, A. E. Dréau, A. Reiserer, N. Kalb, M. S. Blok, J. Ruitenberg, R. F. L. Vermeulen, R. N. Schouten, C. Abellán, W. Amaya, V. Pruneri, M. W. Mitchell, M. Markham, D. J. Twitchen, D. Elkouss, S. Wehner, T. H. Taminiau, and R. Hanson. Loophole-free Bell inequality violation using electron spins separated by 1.3 kilometres. *Nature*, 526:682–686, 2015.

[48] L. K. Shalm, E. Meyer-Scott, B. G. Christensen, P. Bierhorst, M. A. Wayne, M. J. Stevens, T. Gerrits, S. Glancy, D. R. Hamel, M. S. Allman, K. J. Coakley, S. D. Dyer, C. Hodge, A. E. Lita, V. B. Verma, C. Lambrocco, E. Tortorici, A. L. Migdall, Yanbao Zhang, D. R. Kumor,

W. H. Farr, F. Marsili, M. D. Shaw, J. A. Stern, C. Abellán, W. Amaya, V. Pruneri, T. Jennewein, M. W. Mitchell, P. G. Kwiat, J. C. Bienfang, R. P. Mirin, E. Knill, and Sae Woo Nam. Strong Loophole-Free Test of Local Realism. *PRL.*, 115:250402, 2015.

[49] M. Giustina, M. A. M. Versteegh, S. Wengerowsky, J. Handsteiner, A. Hochrainer, K. Phelan, F. Steinlechner, Johannes Kofler, J.-Å. Larsson, Carlos Abellán, Waldimar Amaya, V. Pruneri, M. W. Mitchell, Jörn Beyer, T. Gerrits, A. E. Lita, L. K. Shalm, Sae Woo Nam, T. Scheidl, R. Ursin, B. Wittmann, and A. Zeilinger. Significant-loophole-free test of Bell's theorem with entangled photons. *Phys. Rev. Lett.*, 115:250401, 2015.

[50] W. Rosenfeld, D. Burchardt, R. Garthoff, K. Redeker, N. Ortegel, Markus Rau, and H. Weinfurter. Event-Ready Bell Test Using Entangled Atoms Simultaneously Closing Detection and Locality Loopholes. *PRL*, 119:010402, 2017.

[51] N. Brunner, D. Cavalcanti, S. Pironio, V. Scarani, and S. Wehner. Bell nonlocality. *Rev. Mod. Phys.*, 86:419, 2014.

[52] M. Schlosshauer. Experimental motivation and empirical consistency in minimal no-collapse quantum mechanics. *Annals of Physics*, 321:112, 2006.

Chapter 17

# QUANTUM MECHANICS, A HALF CENTURY LATER

PAPERS OF A COLLOQUIUM
ON FIFTY YEARS OF QUANTUM MECHANICS,
HELD AT THE UNIVERSITY LOUIS PASTEUR,
STRASBOURG, MAY 2–4, 1974

*Edited by*

JOSÉ LEITE LOPES
*Université Louis Pasteur, Strasbourg*

*and*

MICHEL PATY
*Centre de Recherches Nucleaires et Université Louis Pasteur, Strasbourg*

D. REIDEL PUBLISHING COMPANY
DORDRECHT-HOLLAND/BOSTON-U.S.A.

Library of Congress Cataloging in Publication Data

Colloquium on Fifty Years of Quantum Mechanics,
University Louis Pasteur, 1974.
Quantum mechanics.

(Epistème ; v. 5)
Bibliography: p.
Includes index.
1. Quantum theory—Congresses. I. Lopes, José Leite. II. Paty, Michel, 1938– III. Title.
QC173.96.C64 1974 530.1'2 77–8178
ISBN-13:978-94-010-1198-3 e-ISBN-13:978-94-010-1196-9
DOI: 10.1007/978-94-010-1196-9

---

Published by D. Reidel Publishing Company,
P.O. Box 17, Dordrecht, Holland

Sold and distributed in the U.S.A., Canada, and Mexico
by D. Reidel Publishing Company, Inc.
Lincoln Building, 160 Old Derby Street, Hingham,
Mass. 02043 U.S.A.

# TABLE OF CONTENTS

# FOREWORD

**The articles collected in this volume were written for a Colloquium on Fifty Years of Quantum Mechanics which was held at the University Louis Pasteur of Strasbourg on May 2–4, 1974, in commemoration of the original work by De Broglie in 1924.**

Full text can be read at: https://link.springer.com/book/10.1007/978-94-010-1196-9#accessibility-information

# PREFACE

**A few years ago, H. Barreau, G. Monsonego, M. Paty and myself organized the Seminar on the Foundations of Science at the University Louis Pasteur of Strasbourg.**

# Authors' Biographies

Andrew Briggs is Emeritus Professor of Nanomaterials at the University of Oxford and Chair of QuantrolOx. His research interests focus on nanomaterials for quantum technologies with machine learning for controlling the experiments, and the nanoscale thermodynamics of timekeeping and information. From 2002–2009, he directed the UK Interdisciplinary Research Collaboration in Quantum Information Processing (QIP IRC). From 2010–2020 he led grant making by Templeton World Charity Foundation. In 2021 he co-founded QuantrolOx to commercialise machine learning for tuning and characterizing quantum devices, with performance that exceeds what is feasible for humans and greatly accelerates progress in quantum computing.

He is a Fellow of St Anne's College, Oxford, Fellow of Wolfson College, Oxford, Fellow of the Cambridge Philosophical Society, Honorary Fellow of the Royal Microscopical Society, Fellow of the Institute of Physics, Member of Academia Europaea, Fellow of the International Society for Science and Religion and Honorary Lay Canon of Christ Church, Oxford, and he serves on many advisory boards. 40 former members of his laboratory hold faculty positions, and almost as many are employed in quantum companies around the world. From his laboratory and QIP IRC former members have between them founded 12 early-stage companies, of which 10 are in quantum computing.

He has over 650 publications, of which nearly 400 have been cited 10 times or more. His books for a general readership include "The Penultimate Curiosity: How Science Swims in the Slipstream of Ultimate Questions", for which there is a documentary film and a six-book series for children; "It Keeps Me Seeking: The Invitation from Science, Philosophy and Religion"; and "Human Flourishing: Scientific Insight and Spiritual Wisdom in Uncertain Times". His most recent book carries endorsements by the Archbishop of Canterbury and the Astronomer Royal, and has won an ISSR 2023 Book Prize.

Andrey Akhmeteli obtained his PhD in theoretical and mathematical physics from Moscow State University. He has held positions at Moscow State University, other research and educational institutions, and in industry. He is currently affiliated with LTASolid Inc., Houston. His research interests include the foundations of quantum theory, diffraction of electromagnetic waves, magnet science and technology, and ultralight structures.

David Delphenich was born in 1953 in San Antonio, Texas, but mostly spent his childhood in Midland, Michigan, where his father was a chemist for Dow Corning. The family then moved to Kettering, Ohio, where his father was a chemist for the Foreign Technology Division of the Air Force at Wright-Patterson Air Force Base and his mother taught music in a local elementary school. After high school, David studied physics and mathematics at Beloit College in Wisconsin and completed his Bachelor's diploma in 1975. He eventually got established as a member of the technical staff at the Aerospace Corporation in El Segundo, California in 1979, while pursuing a Master's degree in mathematics at Loyola Marymount University in Westchester, California at night, which he completed in 1981, while working on the first Space Shuttle at Rockwell's Space Division in Downey, California. After three years of graduate studies in mathematical physics at the University of California at Berkeley, he returned to aerospace work and eventually completed a master's degree in engineering mechanics at the University of Florida while working on a cruise missile project at Eglin Air Force Base, and later on a NASA spacecraft project for McDonnell Douglas in Huntsville, Alabama. That position ended with the collapse of the Soviet Union, which led David to return to graduate school in 1991 as a student of theoretical physics at Syracuse University, where he completed a Ph.D in 1998. After that, he taught physics, mathematics, engineering mechanics, and other subjects for nine years at various colleges and universities in the United States. He has published numerous research monographs and presented his research at various international conferences. He maintains a website with over one thousand of his own English translations of books and articles on topics in mathematics and physics that were originally published in French, German, and Italian (and a handful that were published in Russian). His primary research focus is on revisiting the dawn of quantum theory from a modern standpoint and expanding the scope of the optical-mechanical analogy that was at the foundations of wave mechanics, especially the analogy with the theory of electron optics and electron diffraction. He also has an interest in applying the geometrical and topological methods of mathematical physics to some of the problems in classical physics, such as the geometry of capillarity and the topological sources of stress fields.

Dipankar Home has been the earliest Indian Researcher who did his Doctoral work in India in the beginning of the 1980s on the then globally emerging line of research related to *experimentally oriented studies on Foundations of Quantum Mechanics*. Subsequently, after his post-doctoral studies in the UK with late Prof. David Bohm and Prof. Roger Penrose, he has been working in India, contributing to fostering this line of research in India and its evolution into the Frontier area of *Quantum Information based on Quantum Foundations*.

Apart from mentoring a considerable number of young Indian researchers, Home has authored more than *150 Research Publications* with his various collaborators in India and abroad, the most significant of these works including those involving theory-experiment collaboration works revealing novel Quantum Effects like *Duality in Entanglement*, *Quantum Cheshire Cat*, and proposals for probing fundamental quantum enigmas like the Schrödinger's Cat paradox in terms of *DNA molecules, facets of intraparticle entanglement between dynamical variables of a single particle* — all these have been cited in his website (https://dipankar home.com/).

Home's two notable books are:

1. **"Conceptual Foundations of Quantum Physics". Foreword** by **Anthony Leggett** (Plenum, 1997).
2. ***"Einstein's Struggles with Quantum Theory: A Reappraisal"*** with Andrew Whitaker. ***Foreword*** by **Roger Penrose** (Springer, 2007).

- Served in *Bose Institute*, Kolkata until superannuating as Senior Professor in 2018. Subsequently, he has been Senior Scientist, National Academy of Sciences India, attached to Bose Institute, Kolkata.
- Visiting Scientist at *Wilczek Quantum Centre*, Shanghai, China and *Institute of Fundamental and Frontier Sciences*, Chengdu, China.
- Dipankar Home's current research interests are as follows:
  - (A) Foundational issues of Quantum Mechanics and their experimentally oriented studies, towards unravelling novel features of Quantum Mechanics.
  - (B) Formulating laboratory based testable schemes for probing quantum aspects of gravity and the necessity of its quantization.
  - (C) Foundational aspects of the interplay between Quantum Foundational issues and Quantum Information, in particular the Quantum Entanglement enabled applications in information transfer/processing protocols.
  - (D) Devising efficient procedures for mitigating the effects of environment induced noise in Quantum Communication protocols like Quantum Teleportation, and in Quantum Foundations related experiments.
- The recent most significant works of Home and his collaborators are:
  - (a) Proposing laboratory-based test of Quantumness of Gravity using disturbance induced by Quantum Measurement of Gravity. *Physical Review Letters* 133 (18), 180201 (2024).
  - (b) Theory-Experimental collaboration work: Single System-based Generation of Certified Randomness using Leggett-Garg inequality. *Physical Review Letters* 133, 020802 (2024).

(c) Formulating Mass-independent scheme to test the Quantumness of spatial motion of an arbitrarily massive harmonic oscillator. *Physical Review Letters* 132 (3), 030202 (2024).
(d) Formulating scheme for using Massive Spatial Qubits towards testing Macroscopic Nonclassicality and Casimir interaction induced Entanglement. *Physical Review Research* 5, 033202 (2023).
(e) Theory-Experimental collaboration work achieving using weak measurement an unambiguous joint detection of spatially separated properties of a single photon in the two arms of an interferometer. *Communications Physics* (Nature Group) 6, 203 (2023).

Hajo Leschke (born 11 February 1945 in Wentorf bei Hamburg) is a German mathematical physicist and (semi-)retired professor of theoretical physics at the Friedrich-Alexander-Universität Erlangen-Nürnberg (FAU). He is mainly known for rigorous results on model systems in quantum (statistical) mechanics obtained through functional-analytic and probabilistic techniques, jointly with his (former) students and other coworkers. His research topics include: Peierls Transition, Functional Formulations of Quantum and Stochastic Dynamics, Pekar–Fröhlich Polaron, Quantum Spin Chains, Feynman–Kac Formulas, (Random)Schrödinger Operators, Landau-Level Broadening, Lifschitz Tails, Anderson Localization, Fermionic Entanglement Entropies, Quantum Spin Glasses.

Hans van Leeuwen lived from 1932 until 2024. He was a prominent internationally recognized scholar who has made major contributions to statistical physics over more than six decades. His PhD thesis in 1962 focussed on molecular correlation functions in fluids, resulted in the hypernetted chain approximation. In the same period his work with E.G.D. Cohen on mixtures of Fermi and Bose gases explained incomplete phase separation of $^{3}He$-$^{4}He$

mixtures and found application in helium-dilution refrigerators. Very influential was Van Leeuwen's contribution to the development of the renormalization theory for critical phenomena. Together with Th. Niemeijer he developed what became known as the Real Space Renormalization, a calculational scheme to implement scale transformations expressed in variables in position space, rather than their Fourier transforms, which was the accepted approach at the time. This work became the basis of numerous studies of universal and non-universal aspects of second order phase transitions. After his retirement he continued to work on a broad spectrum of problems, from polymer dynamics, the physics of ice scating, to radiation transport through the atmosphere.

Jean-Marc Lévy-Leblond is a (theoretical) physicist and (practical) epistemologist, Professor Emeritus at the University of Nice. Besides his main research work in mathematical physics (invariance principles and group theory, quantum theory, space-time structure), he has dealt with philosophy and history of science (foundational problems of quantum theory and relativity theory) and devoted many thoughts and writings to the social, political and cultural issues of modern science. Editor of several scientific series with the publisher Seuil (Paris), he is also founder and editor of the review *Alliage (Culture-Science-Technique)* since 1981.

Kurt Schönhammer was born in Munich in 1946. In school mathematics was his favorite subject from the very beginning. But after reading Fred Hoyle's book "Das grenzenlose All" (Frontiers of Astronomy) he decided to study physics. He obtained his diploma at the TU Munich in 1970 and received his PhD there in 1972 (adviser W. Brenig, topic: "Anderson localization"). He spent his post-doc time at the UC San Diego (with H. Suhl, 1975–76) and the University of Pennsylvania (with J.R. Schrieffer, 1976–77). After his Habilitation at the TU Munich he was staff member

(1978–79) at the IBM research lab, Rüschlikon (Switzerland). From 1979–1984 he had an associate professorship at the Univ. Hamburg and in 1984 became full professor at the Univ. Göttingen. In 1995 he was elected to be a member of the Göttingen Academy of Sciences and Humanities.

His work is in solid state theory with the main topic "electronic many body effects". Earlier work included e.g. correlation effects in surface physics, electronic friction and work on proper inclusion of many body effects in the theoretical description of electron spectroscopies, with application to mixed valent compounds and the non-Fermi-liquid behaviour in quasi-one-dimensional conductors ("Luttinger liquids"). The "functional renormalization group" was used as a tool to study e.g. transport through quantum wires and quantum dots. He also published various pedagogical papers in American Journal of Physics.

Mark Davidson is an independent scholar living in Silicon Valley, California, USA. He obtained his BS in physics from the City College of New York and his MS in physics at the University of Illinois, Champaign-Urbana. His PhD in theoretical physics was obtained from the University of Illinois, Champaign-Urbana. Subsequently he did his Postdoc in theoretical physics at the Lawrence Berkeley Laboratory and became a Staff Scientist in the same laboratory. His later academic career includes being Lecturer in physics, at the University of California, Berkeley and San Jose State University, California as well as working in Energy and environment related research at Lawrence Berkeley Laboratory. He has also worked for Teknekron, an energy related consulting firm and subsequently launched a software startup company called Sage Software where he wrote and marketed a Math program called DIFFEQ for the original Apple computer. He is a Co-founder of The Institute for Theoretical Studies, a non-profit corporation and has worked for KLA Instruments where he had designed and managed engineering products for wafer inspection, metrology, and process control for the semiconductor industry. KLA is the world leader in these fields. He then founded Spectel Research Corporation that did research in several fields including semiconductor metrology, manufacturing approaches, nanotechnology, some energy related work, light microscopy, scatterometry, numerical simulation of electromagnetic scattering, electron microscopy, novel lithography approaches. He has consulted with many semiconductor related companies and did extensive collaborations with NIST in nanotechnology. He has numerous patents and papers which are available on this link: https://scholar.google.com/citations?user=dfpjyKcAAAAJ&hl=en&oi=ao.

Miguel Ortuño was born in Yecla, in the south of Spain. He did his first degree in Physics at the Autonomous University of Madrid and a PhD at the Cavendish Laboratory at the University of Cambridge (1977–80) under the direction of John C. Inkson on local field effects in semiconductors. In the last year of his thesis, he collaborated assiduously with Sir Nevill Mott, who introduced him to the field of disordered materials. After a first postdoc at the University of Lund, on the subject of his PhD thesis, and a second postdoc at the University of California at Riverside under the supervision of Michael Pollak, he returned to the subject of disordered materials and, more specifically, to those in which interactions play a fundamental role. He worked for two years at the Autonomous University of Barcelona, and then moved to the University of Murcia in 1985, where he was awarded a chair in 1987 and where he had remained until the present day. Miguel has been a frequent visitor to the Universities of Cambridge, California at Riverside, Bar-Ilan, Oxford and Birmingham and the Rutherford Appleton Laboratory. In 2014, he was on sabbatical at the University of California in Santa Barbara. His main present line of research is the simulation of quantum system through integrals of motion. He has also worked on the superconductor to insulator transition, the study of relaxation in disordered systems and the theoretical interpretation of AFM experimental results. Over the years, his main interest moved on to electron glasses, systems with states localized by the disorder and Coulomb interactions between their charge carriers. In 2013, he wrote a book on the subject, together with Michael Pollak and Aviad Frydman, published by the Cambridge University Press. He has also worked on loop models, in which the description of the system is not through local entities, but through the loops, and allows an exact mapping of these systems to quantum systems that in this way can be simulated with unprecedented precision. He has published over 160 scientific papers. and translated into Spanish the book "Historia del tiempo" (*A Brief History of Time*) by Stephen Hawking.

Mike Burt studied at Hertford College, Oxford, UK, first to read Physics with the theory option (1965–8) and then to study for his DPhil. (Phd.) at The Department of Theoretical Physics under the supervision of Professor Sir Rudolf Peierls FRS (1968–71). His thesis topics were 'The Momentum of a Light Wave in a Refracting Medium' and 'The Surface Specific Heat of a Crystal Lattice'. From 1971 to 1973, he held a Fellowship in The Royal Society's European Programme at The Department for Theoretical Physics at The University of Utrecht, The Netherlands, working with Professors Nijboer and van Kampen. During that time he published on his DPhil thesis topics, and on the efficacy of cyclic boundary conditions. He also worked with the Utrecht group on the relation between the Lifshitz theory of van der Waals forces and more conventional approaches. From 1973 to 1979 he worked in Volker Heine's Theory of Condensed Matter Group at The Cavendish Laboratory, Cambridge, UK. His first project was the calculation, using pseudopotentials, of the escape probability of electrons from negative electron affinity semiconductor surfaces. He also worked with Volker Heine FRS to validate a simple model of the work-function of caesium and its plethora of oxides, the deposition of which on semiconductor surfaces produces negative electron affinity. His second project was the calculation of the photoionisation of deep levels in semiconductors using realistic bandstructures and wavefunctions. In 1979, he joined the Post Office Research Centre ( later BT Research Laboratories), at Martlesham Heath, Ipswich, UK, as its first theoretical physicist, applying the theory of electrons in semiconductors to the devices (lasers, detectors and modulators) used in optical fibre communications. He made his mark first by demonstrating that attempts, prevalent in the industry, to make antimonide based semiconductor lasers for optical fibre communications were doomed, in doing so dispelling the myth that static screening by free carriers was appropriate within auger recombination theory. He also identified how quantum well structures could be used to produce the holy grail of single longitudinal mode operation in semiconductors lasers. He also invented, successfully patented, and assisted subsequent prosecution of, the still widely used quantum well electroabsorption modulator using theory alone. He became an expert in all the optoelectronic and quantum electronic processes (optical absorption, gain and modulation, spontaneous emission, auger recombination, multiphonon recombination, impact ioinisation, nonlinear optics in semiconductors, both in the bulk and in nanostructures). He also developed an exact envelope function theory for electrons in semiconductor nanostructures. While such a theory was well established in the bulk, as effective mass or more generally k.p theory, it was widely assumed, that generalisation to structures containing heterostructures was impossible. This exact theory was endorsed by Volker Heine and Marshall Stoneham FRS who were both instrumental in enabling the publication of two substantial invited papers in the Journal of Condensed Matter Physics. The work also featured prominently in the Springer book 'The k.p Method' by Lew Yan Voon and Willatzen. In 1993 Mike was awarded the prestigious DSc (Doctor of Science ) degree by the University of Oxford as a mark of his authoritative status. In 1995, Mike was appointed visiting professor to the Physics Department of Durham University, UK an appointment he held for 20 years and has since converted to the less onerous position of academic visitor. In1995 he also initiated a very successful collaboration on HgTe nanocrystals between his group at BT and Professor Horst Weller's group at the Chemistry Department at Hamburg University, Germany. Mike took early retirement in 2002 to work part time as a consultant for a start up company to develop HgTe nanocrystal technology and also to develop other interests in physics. In 2012 he started work on understanding and clarifying the Peierls/ Heisenberg interpretation of quantum mechanics as described in this Heisenberg Centenary volume. He has also in the last decade demonstrated how to obtain a better understanding of, and a more rigorous approach to, the derivation of the so called Fermi Golden Rule highlighting its connection to adiabatic following. He has also developed a non-perturbative theory of the decay of quantum mechanical states.

After attending the City of London School from age 11, Richard Forbes studied at Trinity College, Cambridge. His first degree involved final-year specialisation in theoretical physics, taught by the Cavendish Laboratory. He is old enough to have attended lectures by Dirac and to have heard Heisenberg lecture. His PhD was in the then Cambridge Department of Metallurgy, initially under the supervision of Sir Alan Cottrell. His experimental/theoretical PhD project in the late 1960s helped us understand the theory of "how humanity first saw atoms", via the technique of field ion microscopy.

After a short non-academic gap, he moved to the Physics Department of the University of Aston in Birmingham, first as a postdoc. working on experimental energy analysis of field electron emission from semiconductors, and then as a temporary Lecturer. In 1982–83 he retrained into electrical engineering (including a three-month industrial placement relating to integrated-circuit design). He then joined and spent most of his career at the then Department of Electrical Engineering at Surrey, serving the Department and University in many different roles. In particular, he was responsible for leading the introduction of nanotechnology teaching.

Richard holds BA, MA and PhD degrees from Cambridge University and a DSc from the University of Aston. He is a professional engineer (UK and European), and a Fellow or Member of various learned societies. He is an Inaugural Fellow of the International Field Emission Society (IFES) and has twice been the elected IFES President. Inter alia, he has been Chairman of the UK Institute of Physics Midland Branch, and of the UK Engineering Council Surrey Region (when that existed), and has also been Chairman of the Board of Surrey SATRO (a local school support organization established as a UK limited company).

His research focus can be described generally as the "quantum science and technologies associated with high-magnitude electric fields", with particular theoretical emphasis on the basic theory of field-induced electron and ion emission and related electrostatics. Central aims of his research are "the theory of related technologies": he aims to explore and establish the theoretical details of how the many related techniques and technologies work, and of how their results may be interpreted. He has published nearly 200 items and has made about 600 other scientific presentations (see ResearchGate entry for details). In field electron emission, according to the ScholarGPS citation analysis software, and based on their weighted 2024 analysis of citations in the five years 2020 to 2024 inclusive, he emerges as the most highly ranked researcher in the world in this specialism.

Richard has co-authored a 2014 book entitled: "Atom Probe Tomography: The Local Electrode Atom Probe", and has contributed to and co-edited a 2020 handbook entitled: "Modern Developments in Vacuum Electron Sources".

Richard has been officially retired for many years but remains attached to the University of Surrey and remains research active as an Emeritus Associate Professor. Recently he has transferred from the University's Advanced Technology Institute into the School of Mathematics and Physics. Present interests include: advocating correction of a conceptual error in the definition of the concept of "measured voltage" in the International Standard for Electromagnetism (so that we can have a "fully quantum" theory of electricity that integrates the different approaches currently taken in physics, electrical engineering and electrochemistry); and identifying deep fundamental issues in the quantum mechanics of field electron and ion emission (the subject of the chapter here), including an issue that might also be relevant to quantum biology and quantum botany.

After attending the Royal Military Academy Sandhurst Basil Mahon took a degree in electrical engineering at the Royal Military College of Science and served for 17 years as an officer in the Royal Electrical and Mechanical Engineers. While on a posting in London he gained a master's degree in statistics by attending evening classes at Birkbeck College. Soon after that he joined the Government Statistical Service, where he ran the Census of Population in England and Wakes in 1991. After retiring from regular work, he followed up a lifetime interest in the physical sciences by writing. He is the author of The Man Who Changed Everything: the Life of

James Clerk Maxwell, and The Forgotten Genius of Oliver Heaviside, and is co-author, with Nancy Forbes, of Faraday, Maxwell, and the Electromagnetic Field.

Sougata Bhattacharyya completed his undergraduate studies at Presidency University, Kolkata, and is currently pursuing his M.Sc. in Astronomy at the Indian Institute of Technology (IIT), Indore. His research portfolio includes work in quantum foundations, quantum entanglement, and topology, with his primary research interest dedicated to cosmology. (Contact: msc2503121011@iiti.ac.in)

Born in Singapore's Chinatown in 1954, Tuck Choy is the grandson of immigrants from Canton, China. His grandparents emigrated soon after Sun Yat-Sen founded the modern Republic of China in 1921. As a child he had his early education at the Pearl's Hill English school for boys near Chinatown and then at Gan Eng Seng high school for boys at Anson Road. At 14, he developed interest in physics, mathematics, literature and history, due to the influence of his grandfather and teachers. His family was poor, so he was at first unable to go to university. He worked as a laboratory technician for 2 years after National Service and managed to pursue his studies via the London University External degrees program, for a Bachelor of Science (BSc) in Physics. He finished his final year BSc at Queen Mary College in London and was perhaps one of only two post-war External degree students who were awarded First-Class Honours in Physics. London University then offered him a postgraduate scholarship and in 1981 he graduated with a PhD in theoretical physics with a thesis on the Bethe Ansatz in exactly soluble models like the Hubbard chain. Then after, he was a postdoctoral fellow at Imperial College London (1982–1986), Visiting Assistant Professor in Rhode Island (1984–1986) USA, Research Fellow in Canberra, Australia (1986–1988), then as Harwell-Wolfson College Fellow in Oxford (1988–1990). He returned to Australia as lecturer at

Monash University in 1990. From 1998 to 2000 he was also visiting professor at the Centre for Theoretical Physics in Hsinchu, Taiwan. He returned to the UK in 2000 and joined Philips Semiconductors Inc. at Stockport and then in 2003 the Centre for Nano-Materials Research at the University College London in 2003. He has over 80 publications and the second edition of his book: Effective Medium Theory, Principles and Applications was published by Oxford University Press in 2015. His research interest is broadly in the areas of Condensed Matter Physics, Superconductivity, Electronic Properties of Semiconductors and Metals, Statistical Mechanics, Many-Body Theory, Electrodynamics, Continuum Mechanics, Polymers, Computational Materials Modelling, Foundations of Quantum Mechanics and Quantum Field Theory, Exactly Soluble Models, Spin Glasses, Density Functional Theory and van der Waals forces. In 1999 he was elected to be a Senior member of the IEEE, USA. Tuck Choy's other relaxational interests are in classical music, Opera and Shakespeare. He is also a licensed amateur radio operator since 1990 currently holding a call sign F4WCV in France and has also published several articles in amateur radio journals.

## I. EPILOGUE - THE END OF AN ERA

> The time will come when diligent research over long periods will bring to light things which now lie hidden. A single lifetime, even though entirely devoted to the sky , would not be enough for the investigation of so vast a subject ... And so this knowledge will be unfolded only through long successive ages. There will come a time when our descendants will be amazed that we did not know things that are so plain to them... Many discoveries are reserved for ages still to come, when memory of us will have been effaced. Our universe is a sorry little affair unless it has in it something for every age to investigate...Nature does not reveal her mysteries once and for all.
>
> Seneca, "The Younger" 4 BC - 65 AD in Natural Questions, Book 7, Chapter 25.4-5 ; Thomas H. Corcoran Loeb translation (1971)

I am humbled and indeed very honoured by this unique opportunity to be the editor for the 100th anniversary Heisenberg centenary volume and grateful to my predecessors, the late José Leite Lopes and Michel Paty [1] who laid their pioneering groundwork in 1974 which guided me throughout this mission. I only wish they had left some notes, documenting the history and human stories associated with such an enormous enterprise. It was a hard lesson in many ways, not just in physics but in the tremendous effort required on many fronts to make this volume a success. From the beginning, my experience in academic life has instructed me that if I did not do it, most probably Lopez and Paty would have had no followers. To the editors of any future volumes in 2075 or more importantly 2125, I wish the best of luck. Einstein famously stated that it will take 100 years for the world to catch up with him on his objections to quantum mechanics [2]. With the end of an old era, a new one has now just begun.

The Editor, April 2026

---

[1] See "Quantum Mechanics, a half century later", papers of a colloquim on fifty years of quantum mechanics, held at the University of Louis Pasteur, Strasbourg, May 2-4, 1974, edited by José Leite Lopes and Michel Paty, D. Riedel Publishing, Dordrecht-Holland/Boston-USA 1977.

[2] Abraham Pais, "Subtle is the Lord...", The Science and Life of Albert Einstein , paper back edition Oxford University Press, Oxford (1983), reissued 2005, pg 489.

www.ingramcontent.com/pod-product-compliance
Ingram Content Group UK Ltd.
Pitfield, Milton Keynes, MK11 3LW, UK
UKHW061933100926
8556IPUK00007B/897

9 789819 827565